Applied Mathematical Sciences

Volume 177

For further volumes:
http://www.springer.com/series/34

Preface

This book is devoted to an account of theories of heat conduction where the temperature may travel as a wave with a finite speed. This area of non-classical diffusion is very topical in the research literature. With the advent of micro-scale technology there is increasing evidence that thermal motion is via a wave mechanism as opposed to by diffusion. We survey many of the theories which have been proposed as candidates to describe thermal motion as a wave. These theories are linked to solid mechanics (elasticity) and also to fluid mechanics.

Wave motion in the form of acceleration waves and of shock waves is discussed. An exposition of numerical work in the area of thermal waves is also included. Analytical methods for establishing uniqueness, continuous dependence, growth, spatial decay and other results are described.

Two important chapters are the final two. These focus firstly on where nanofluids and heat transfer are relevant. Hyperbolic temperature equations have been linked to the recent and "hot" area of nanofluids. The final chapter investigates applications of "heat wave - like" ideas to other areas, particularly those in mathematical biology are also investigated.

I should like to thank a referee for several pointed remarks and suggestions for rewriting which have substantially helped with this book. My early work on heat waves was influenced greatly by discussions with the late Dario Graffi of the University of Bologna. I have benefitted over the years by many discussions on heat waves with several people and I would especially like to thank Stan Chirita, Christo Christov, Ivan Christov, Michele Ciarletta, Mauro Fabrizio, Franca Franchi, Pedro Jordan, Kenneth Lindsay, Angelo Morro, Larry Payne, Ramon Quintanilla and Jaime Muñoz Rivera.

In particular, I would like to thank Mauro Fabrizio and Ivan Christov for helping me locate copies of the paper by Cisotti and also some by Dario Graffi. It is a pleasure to thank Achi Dosanjh of Springer for her advice with editorial matters.

This research was in part supported by a grant from the Leverhulme Trust, “Tipping points: mathematics, metaphors and meanings”.

Durham Brian Straughan

Contents

1

Introduction

1.1 Heat waves in a rigid conductor

1.1.1 Second sound

First sound is the classical mechanism which allows us to hear, i.e. a disturbance of pressure (or density) which propagates through a continuous medium such as air or water. Second sound is a more recent phenomenon involving the propagation of heat as a temperature wave. The classical theory of heat propagation is via diffusion where a temperature field diffuses through a continuous body. However, experiments in the late 1960's and early 1970's showed that a thermal disturbance could travel as a wave and this has acted as an impulse to much subsequent theoretical work in this area. As (Caviglia et al., 1992) remark thermal pulse propagation has been experimentally observed under accurate conditions in solid helium (He^3 and He^4) by (Ackerman and Overton, 1969) (see also the references therein), in sodium fluoride, (Jackson et al., 1970), (McNelly et al., 1970), (Hardy and Jaswal, 1971), in bismuth, (Narayanamurti and Dynes, 1972), and in sodium iodide and in lithium fluoride, see (McNelly et al., 1970). This aspect of second sound is a low temperature phenomenon, the experiments just cited having been performed in the 1-20°K range.

In addition to a thermal wave, the experiments of (McNelly et al., 1970) and of (Jackson et al., 1970) showed that second sound was also important in thermoelasticity. They employed a very pure crystal of sodium fluoride and evaporated manganin heaters and lead detectors onto opposing faces of the crystal and were thereby able to transmit heat pulses through their sample. These careful experiments revealed the existence of three distinct waves. There was a longitudinal elastic wave which travels fastest, a transverse elastic wave, and also a thermal wave. When the temperature was below 8°K, three distinct waves were observed, the fastest being the longitudinal one, the transverse one next fastest, while the thermal wave was slowest.

Within fluid mechanics, the effect of temperature upon wave propagation of a disturbance has a long history. As (Lindsay and Straughan, 1978) remark, a critical review of the early literature in this field is provided by (Truesdell, 1953). (Stokes, 1851) investigated the behaviour of disturbances in a perfect fluid when the fluid is subject to radiation effects. (Kirchoff, 1868) and Langevin (see (Biquard, 1936)) studied the behaviour of disturbances in a fluid taking heat conduction and viscosity into account. They obtained a fourth order characteristic equation for the wavespeed, with that of (Kirchoff, 1868) being for an equation of state appropriate to a perfect gas whereas Langevin adopted an arbitrary equation of state. (Rayleigh, 1896), eq. (247.18), also obtained disturbance solutions for a heat conducting fluid without viscosity, his characteristic equation likewise being fourth order. For the physics literature on temperature waves in low temperature Helium II one might consult e.g. (Peshkov, 1944) or (Donnelly, 2009). Undoubtedly the article of (Truesdell, 1953) and the experiments described above inspired much theoretical work on the propagation of a thermal wave (heat wave). Much of this work is described in the reviews of (Chandrasekharaiah, 1986), (Chandrasekharaiah, 1998), (Dreyer and Struchtrup, 1993), (Hetnarski and Ignaczak, 1999), (Joseph and Preziosi, 1989; Joseph and Preziosi, 1990) and (Jou et al., 2010a). The paper of (Green and Naghdi, 1991) brought a new way of thinking to the area of heat wave propagation and their article has influenced many subsequent developments.

A lot of the recent interest in second sound is due to discoveries that it may have relevance in mundane areas other than low temperature physics. For example, (Mitra et al., 1995), (Vedavarz et al., 1992) suggest thermal relaxation effects may be important in biological tissues, (Lebon and Dauby, 1990) remark that second sound should be detectable in any material, in addition second sound may be used to dry sand, (Meyer, 2006), it may be important in nanofluids, (Vadasz et al., 2005), in cooling or heating in stars, (Herrera and Falcón, 1995), (Falcón, 2001), in cryovulcanology in one of Saturn's moons, (Bargmann et al., 2008b), in phase changes, (Miranville and Quintanilla, 2009), (Liu et al., 2009), in nuclear reactor technology, (Espinosa-Paredes and Espinosa-Martinez, 2009), in skin burns, (Dai et al., 2008), in the medical technique of radiofrequency heating, (López Molina et al., 2008), (Tung et al., 2009), and this technique is important as a surgical procedure in the elimination of cardiac arrhythmias, tumours, in heating of the cornea, or in the treatment of gastroesophageal reflux disease. Additionally, the mathematical theories derived to describe second sound, especially that of (Cattaneo, 1948), have been adapted to study biological problems such as chemotaxis, (Dolak and Hillen, 2003), (Hillen and Levine, 2003), (Wang and Hillen, 2008), the spread of the hantavirus, (Abramson et al., 2001), (Barbera et al., 2008), to traffic flow, (Jordan, 2005b), (Bellomo and Dogbé, 2008), and to the control of fish stocks, (Niwa, 1998).

This book looks at a variety of issues connected with heat waves and, in particular, we do include accounts of the contemporary issues just mentioned.

To understand the ideas we begin with some simple examples.

Let us consider the classical diffusion equation on $x \in \mathbb{R}, t > 0$, i.e.

$$\frac{\partial \theta}{\partial t} = D \frac{\partial^2 \theta}{\partial x^2}, \qquad x \in \mathbb{R},\ t > 0. \tag{1.1}$$

As initial data we can consider either a point source, $\theta = N$ at $x = 0$, or a finite distribution of θ at $t = 0$, i.e.

$$\theta(x, 0) = N\delta(x), \tag{1.2}$$

or

$$\theta(x, 0) = f(x), \quad x \in \mathbb{R}. \tag{1.3}$$

In (1.2) $\delta(x)$ is the Dirac delta function and we are thinking of f as having a finite support, i.e. f vanishes outside a finite region. One may solve equation (1.1) together with (1.2) or (1.3) by using a Fourier transform, see e.g. (Sneddon, 1995). Then for the initial data condition (1.2) we obtain

$$\theta(x, t) = \frac{N}{2\sqrt{\pi D t}} e^{-x^2/4Dt}, \quad t > 0, \tag{1.4}$$

whereas with (1.3) one may show that

$$\theta(x, t) = \frac{1}{2\sqrt{\pi D t}} \int_{-\infty}^{\infty} f(\xi) \exp\left[\frac{-(x-\xi)^2}{4Dt}\right] d\xi. \tag{1.5}$$

For both solutions (1.4) or (1.5) we see that as soon as $t > 0$, $\theta \neq 0$ everywhere. Thus, we can think of θ as having an infinite speed of propagation. This is thought of as being an undesirable effect and, therefore, we seek to find a method whereby θ will propagate with a finite speed of propagation. In sections 1.2 - 1.12 we present a variety of models which have been studied widely in the literature to attempt to overcome the problem of infinite speed of propagation.

At this juncture we simply present three simple ways to help understand the process. As a first step we might argue that the diffusion coefficient D in (1.1) should depend on temperature, as it does in real life. Hence, recollect equation (1.1) arises from the two equations

$$\frac{\partial \theta}{\partial t} = -\frac{\partial q}{\partial x} \qquad \text{and} \qquad q(x, t) = -D \frac{\partial \theta}{\partial x}(x, t). \tag{1.6}$$

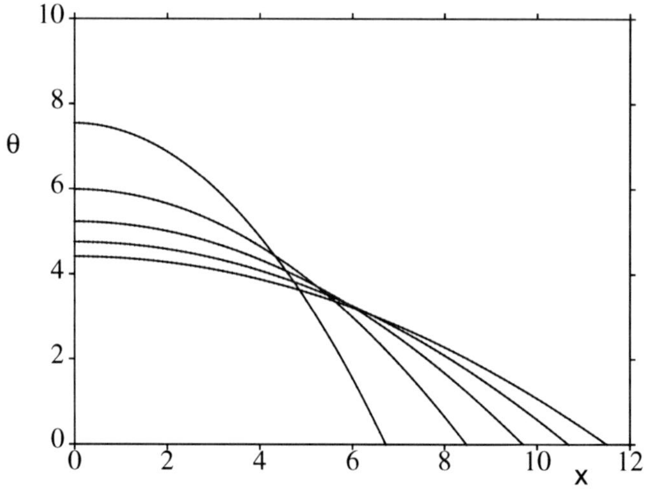

Figure 1.1. θ profile as t increases for solution (1.8). The curves are for $t = 1, 2, 3, 4, 5$, moving downward at $x = 0$. Only the right hand part of θ is shown, the left hand part being a mirror image in the $\theta-$axis. The solution starts as a point height 60 at $t = 0$, $x = 0$. The values of N and h are $N = 60, h = 1$.

The function q is the one-dimensional heat flux. We now suppose D is a linear function of θ, i.e. $D = h\theta$, for h a constant. Then, instead of equation (1.1) we find equations (1.6) lead to

$$\frac{\partial \theta}{\partial t} = h\frac{\partial}{\partial x}\left(\theta \frac{\partial \theta}{\partial x}\right). \tag{1.7}$$

One may show that if equation (1.7) is posed on the domain $\{x \in \mathbb{R}\} \times \{t > 0\}$ then the solution with the initial condition (1.2) is

$$\theta(x,t) = \frac{N^{2/3}3^{1/3}}{2h^{1/3}\pi^{1/3}}\frac{1}{t^{1/3}} - (6h)^{-1}\frac{x^2}{t}. \tag{1.8}$$

Thus, θ starts at $t = 0$ with $\theta = N$ at $x = 0$ and spreads out as t increases, keeping $\theta > 0$, and the edge where $\theta = 0$ is at time t,

$$x = t^{1/3}\,\frac{h^{1/3}N^{1/3}3^{2/3}}{\pi^{1/6}}.$$

Clearly, θ is moving with a finite wavespeed, as may be seen from figure 1.1. We see from figure 1.1 that the solution flattens out as t increases but the temperature is moving with a finite wavespeed.

Another procedure might be to say $(1.6)_2$ is too restrictive and to argue that q should not be proportional to θ_x at the same time, but there should be a slight time lag. Thus, we might replace $(1.6)_2$ by

$$q(x, t+\tau) = -D\frac{\partial \theta}{\partial x}(x,t). \tag{1.9}$$

Rather than use (1.9) we expand the left hand side using Taylor series and retain only the first two terms, to find

$$\tau\frac{\partial q}{\partial t}(x,t) + q(x,t) = -D\frac{\partial \theta}{\partial x}(x,t)\,. \tag{1.10}$$

If we now combine this equation with equation $(1.6)_1$ then instead of the classical diffusion equation (1.1) we find θ satisfies

$$\tau\frac{\partial^2\theta}{\partial t^2} + \frac{\partial \theta}{\partial t} = D\frac{\partial^2\theta}{\partial x^2}\,. \tag{1.11}$$

A solution to equation (1.11) travels with a finite speed of propagation as is shown below.

An alternative (third) way is to again argue $(1.6)_2$ does not adequately describe the situation, and argue that $q(x,t)$ is not only not proportional to $\partial\theta/\partial x$ at time t, but instead argue that it depends on the history of $\partial\theta/\partial x$ over some previous time interval, say $(0,t)$. Then we replace equation $(1.6)_2$ by, for example,

$$q(x,t) = -\frac{D}{\tau}\int_0^t e^{-(t-s)/\tau}\frac{\partial\theta}{\partial x}(x,s)\,ds. \tag{1.12}$$

This expression means that q depends more on the recent history of $\partial\theta/\partial x$ since the dependence decays exponentially as one goes further into the past. When one combines equation $(1.6)_1$ with equation (1.12), then we again arrive at equation (1.11).

To see why (1.11) removes the infinite speed of propagation issue, we consider equation (1.11) on the domain $\{x \in \mathbb{R}\} \times \{t > 0\}$ with the initial conditions

$$\theta(x,0) = f(x), \qquad \frac{\partial\theta}{\partial t}(x,0) = g(x), \tag{1.13}$$

where f and g are non-zero only on a finite interval (x_1, x_2), say. The solution of (1.11) together with the initial conditions (1.13) may be found by writing the equation as a hyperbolic system. Thus, we put $w = \theta_t, v = \theta_x$ and equation (1.11) is equivalent to

$$\begin{aligned} &w_t - \lambda^2 v_x + \frac{1}{\tau}\,w = 0,\\ &v_t - w_x = 0, \end{aligned} \tag{1.14}$$

with $\lambda^2 = D/\tau$. This is in the classical form of a hyperbolic system

$$\frac{\partial u_i}{\partial t} + a_{ij}\frac{\partial u_j}{\partial x} + b_i = 0$$

where

$$\mathbf{a} = \begin{pmatrix} 0 & -\lambda^2 \\ -1 & 0 \end{pmatrix}, \qquad \mathbf{b} = \begin{pmatrix} 1/\tau \\ 0 \end{pmatrix}.$$

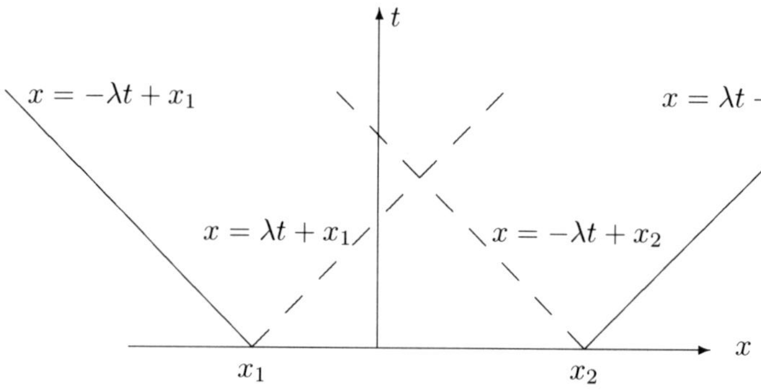

Figure 1.2. Characteristics for equation (1.15)

(Hyperbolic systems are studied in general in the books of (Dafermos, 2010) and of (Whitham, 1974).) The determinant equation $|\mathbf{a} - c\mathbf{I}| = 0$ yields $c = \pm\lambda = \pm\sqrt{D/\tau}$. This yields the characteristics $dx/dt = c$, i.e. $dx/dt = \pm\lambda$. We follow the classical procedure and multiply by the left eigenvectors of $\mathbf{a}$ to arrive at the characteristic equations

$$
\begin{aligned}
&\frac{d}{dt}(w - \lambda v) + \frac{1}{\tau}\,w = 0 \qquad \text{on} \quad \frac{dx}{dt} = \lambda\,,\\
&\frac{d}{dt}(w + \lambda v) + \frac{1}{\tau}\,w = 0 \qquad \text{on} \quad \frac{dx}{dt} = -\lambda\,.
\end{aligned}
\tag{1.15}
$$

The characteristic system (1.15) may now be integrated (numerically) to find θ and θ_t for increasing t. The solution moves with a finite wavespeed because it is contained between the limiting characteristics $x = \lambda t + x_2$, $x = -\lambda t + x_1$ as shown in figure 1.2.

One may, in fact, derive the exact solution to (1.11) by introducing the variables $T = t/2\tau$ and $y = x/2\sqrt{\tau D}$. Then equation (1.11) transforms to

$$\theta_{TT} + 2\theta_T = \theta_{yy}\,. \tag{1.16}$$

The initial conditions (1.13) must also be transformed and we denote these by

$$\theta(y, T = 0) = F(y), \qquad \frac{\partial \theta}{\partial T}(y, T = 0) = G(y), \tag{1.17}$$

where F, G denote the functions equivalent to f and g. Then, as (Cattaneo, 1948), p. 96, shows, the exact solution to (1.16), (1.17) is

$$\theta(y, T) = \frac{e^{-T}}{2}\left\{F(y + T) + F(y - T) + \int_{y-T}^{y+T} I(s, y, T)ds\right\}. \tag{1.18}$$

The function I is given by

$$I(s,y,T) = \{G(s) + F(s)\}\,\Psi\{(s-y)^2 - T^2\} - 2TF(s)\Psi'\{(s-y)^2 - T^2\}, \tag{1.19}$$

where Ψ is defined by

$$\Psi(X) = J_0(\sqrt{X}), \tag{1.20}$$

with J_0 being the Bessel function of zeroth order.

Solution (1.18) shows how the function θ is limited to the domain within the characteristics, but also displays dissipation due to the exponentially decaying in time term.

1.1.2 Notation, definitions

Standard indicial notation is used throughout this book together with the Einstein summation convention for repeated indices. Standard vector or tensor notation is also employed where appropriate. For example, we write

$$u_x \equiv \frac{\partial u}{\partial x} \equiv u_{,x} \qquad u_{i,t} \equiv \frac{\partial u_i}{\partial t} \qquad u_{i,i} \equiv \frac{\partial u_i}{\partial x_i} \equiv \sum_{i=1}^{3} \frac{\partial u_i}{\partial x_i}$$

$$u_j u_{i,j} \equiv u_j \frac{\partial u_i}{\partial x_j} \equiv \sum_{j=1}^{3} u_j \frac{\partial u_i}{\partial x_j}, \qquad i = 1, 2 \text{ or } 3.$$

In the case where a repeated index sums over a range different from 1 to 3 this will be pointed out in the text. Note that

$$u_j u_{i,j} \equiv (\mathbf{u}\cdot\nabla)\mathbf{u} \qquad \text{and} \qquad u_{i,i} \equiv \operatorname{div}\mathbf{u}.$$

As indicated above, a subscript t denotes partial differentiation with respect to time. When a superposed dot is used it either means partial differentiation with respect to time, or when dealing with nonlinear fluid theories the material derivative will often be used. The material derivative is given by,

$$\dot{f}_i \equiv \frac{\partial f_i}{\partial t} + u_j \frac{\partial f_i}{\partial x_j},$$

where u_i in the equation above is the velocity field. For linear theories we may use a superposed dot to denote $\partial/\partial t$. The exact use will be made clear in the text.

The letter Ω will denote a fixed, bounded region of 3-space with boundary, Γ, sufficiently smooth to allow applications of the divergence theorem. The symbols $\|\cdot\|$ and $(\cdot,\cdot)$ will denote, respectively, the L^2 norm on Ω, and the inner product on $L^2(\Omega)$, i.e.

$$\int_\Omega f^2 dV = \|f\|^2 \quad \text{and} \quad (f,g) = \int_\Omega fg\,dV.$$

We sometimes have recourse to use the norm on $L^p(\Omega)$, $1 < p < \infty$, and then we write

$$\|f\|_p = \left(\int_\Omega |f|^p dx\right)^{1/p}.$$

We introduce the ideas of stability and instability in the context of a nonlinear damped wave equation (which would be defined with suitable boundary conditions), which is placed into context as an equation for temperature wave propagation in section 1.2,

$$\frac{\partial^2 u}{\partial t^2} + \mu \frac{\partial u}{\partial t} - \nabla(\kappa(u)\nabla u) = 0, \tag{1.21}$$

where μ is a positive constant and κ is a known nonlinear function, where $\mathbf{x} \in \Omega \subset \mathbb{R}^3$, and where $\Delta = \partial^2/\partial x^2 + \partial^2/\partial y^2 + \partial^2/\partial z^2$ is the Laplace operator.

We introduce notation in the context of a steady solution to (1.21), namely a solution $\bar{u}$ satisfying

$$\nabla(\kappa(\bar{u})\nabla\bar{u}) = 0. \tag{1.22}$$

(We could equally deal with the stability of a time-dependent solution, but many of the problems encountered here are for stationary solutions and at this juncture it is as well to keep the ideas as simple as possible.) Let w be a perturbation to (1.22), i.e. put $u = \bar{u} + w(\mathbf{x}, t)$. Then, it is seen from (1.21) and (1.22) that w satisfies the system

$$\frac{\partial^2 w}{\partial t^2} + \mu \frac{\partial w}{\partial t} - \left\{\nabla\left[\kappa(\bar{u} + w)\nabla(\bar{u} + w)\right] - \nabla\left[\kappa(\bar{u})\nabla\bar{u}\right]\right\} = 0. \tag{1.23}$$

To discuss linearized instability we linearize (1.23) which means we keep only the terms which are linear in w. From a Taylor series expansion of κ we have

$$\kappa(\bar{u} + w) = \kappa(\bar{u}) + w\kappa'(\bar{u}) + O(w^2). \tag{1.24}$$

Then, using (1.22), (1.24) in (1.23) we derive the linearized equation satisfied by w, namely

$$\frac{\partial^2 w}{\partial t^2} + \mu \frac{\partial w}{\partial t} - \nabla\left[w\kappa'(\bar{u})\nabla\bar{u} + \kappa(\bar{u})\nabla w\right] = 0. \tag{1.25}$$

Since (1.25) is a linear equation we may introduce an exponential time dependence in w so that $w = e^{\sigma t}s(\mathbf{x})$. Then (1.25) yields

$$\sigma^2 s + \mu\sigma s - \nabla\left[s\kappa'(\bar{u})\nabla\bar{u} + \kappa(\bar{u})\nabla s\right] = 0. \tag{1.26}$$

We say that the steady solution $\bar{u}$ to (1.22) is *linearly unstable* if

$$Re(\sigma) > 0,$$

where $Re(\sigma)$ denotes the real part of σ. Equation (1.26) (together with appropriate boundary conditions) is an eigenvalue problem for σ. For many of the problems discussed in this book the eigenvalues may be ordered so that

$$Re(\sigma_1) > Re(\sigma_2) > \dots$$

For linear instability we then need only ensure $Re(\sigma_1) > 0$.

Let $w_0(\mathbf{x}) = w(\mathbf{x}, 0)$ be the initial data function associated to the solution w of equation (1.23). The steady solution $\bar{u}$ to (1.22) is *nonlinearly stable* if and only if for each $\epsilon > 0$ there is a $\delta = \delta(\epsilon)$ such that

$$\|w_0\| < \delta \Rightarrow \|w(t)\| < \epsilon \tag{1.27}$$

and there exists γ with $0 < \gamma \leq \infty$ such that

$$\|w_0\| < \gamma \Rightarrow \lim_{t\to\infty} \|w(t)\| = 0. \tag{1.28}$$

If $\gamma = \infty$, we say the solution is *unconditionally* nonlinearly stable (or simply refer to it as being asymptotically stable), otherwise for $\gamma < \infty$ the solution is *conditionally* (nonlinearly) stable. For nonlinear stability problems it is an important goal to derive parameter regions for unconditional nonlinear stability, or at least conditional stability with a finite initial data threshold (i.e. finite, non-vanishing, radius of attraction). It is important to realise that the linearization as in (1.25) and (1.26) can only yield linear *instability.* It tells us nothing whatsoever about stability. There are many equations for which nonlinear solutions will become unstable well before the linear instability analysis predicts this. Also, when an analysis is performed with $\gamma < \infty$ in (1.28) this yields conditional nonlinear stability, i.e. nonlinear stability for only a restricted class of initial data.

We have only defined stability with respect to the $L^2(\Omega)$ norm in (1.27) and (1.28). However, sometimes it is convenient to use an analogous definition with respect to some other norm or positive-definite solution measure. It will be clear in the text when this is the case. When we refer to continuous dependence on the initial data we mean a phenomenon like (1.27). Thus, a solution w to equation (1.23) depends continuously on the initial data if a chain of inequalities like (1.27) holds.

Throughout the book we make frequent use of inequalities. In particular, we often use the Cauchy-Schwarz inequality for two functions f and g, i.e.

$$\int_\Omega fg\,dx \leq \left(\int_\Omega f^2 dx\right)^{1/2} \left(\int_\Omega g^2 dx\right)^{1/2}, \tag{1.29}$$

or what is the same in L^2 norm and inner product notation,

$$(f, g) \leq \|f\|\,\|g\|. \tag{1.30}$$

The arithmetic-geometric mean inequality (with a constant weight $\alpha > 0$) is, for $a, b \in \mathbb{R}$,

$$ab \leq \frac{1}{2\alpha}\,a^2 + \frac{\alpha}{2}\,b^2, \tag{1.31}$$

and this is easily seen to hold since

$$\left(\frac{a}{\sqrt{\alpha}} - \sqrt{\alpha}b\right)^2 \geq 0.$$

Another inequality we frequently have recourse to is Young's inequality, which for $a, b \in \mathbb{R}$ we may write as

$$ab \leq \frac{|a|^p}{p} + \frac{|b|^q}{q}, \qquad \frac{1}{p} + \frac{1}{q} = 1, \quad p, q \geq 1. \tag{1.32}$$

1.1.3 Overview

In the remainder of the current chapter we present eleven ways in which heat may travel as a wave in a rigid heat conductor. The next chapter reviews models which couple some of the theories discussed in the rigid body case to the situation of a deformable elastic body. This thus presents theories appropriate to thermal wave propagation in thermoelasticity, both from a nonlinear and a linear viewpoint. Chapter 3 reviews where some of the ideas discussed in the rigid heat conductor case are coupled to fluid mechanics theories. The focus is on nonlinear theories, and both inviscid fluids (gases) and viscous fluids are considered. Chapter 4 analyses the propagation of an acceleration wave in a rigid heat conductor, in thermoelasticity, and in fluid mechanical theories, in each case employing a theory capable of allowing heat to travel as a wave. Some new results are included in this book, such as those in section 4.3 on acceleration waves in a rigid body with microtemperatures, those in sections 4.4 and 4.7 dealing with type II thermoelasticity or a type III fluid, respectively, or those for the nonlinear theories of fluid mechanics in section 6.8. The next chapter investigates thermal shock waves in a rigid heat conductor and also in a thermoelastic body, always employing a theory where a thermal wave may propagate. The development of an acceleration wave into a shock wave is also analysed and a brief review is given of some of the (considerable) numerical work which has been performed. Chapter 6 focusses on qualitative results for second sound theories for a rigid heat conductor, in thermoelasticity, and also in fluid mechanics. The following chapter reviews work on spatial decay in a rigid heat conductor and also in thermoelasticity. Again, results appropriate to second sound theories are emphasized. Special attention is given to recent work in thermoelasticity when the elastic coefficients are not positive - definite but merely satisfy conditions of strong ellipticity. The penultimate chapter, chapter 8, concentrates on heat transfer in nanofluids, thermal convection in nanofluids, and convection in fluid mechanical theories which allow heat to travel as a wave. Finally, in chapter 9, we report on recent work specifically using the hyperbolic - like theories which are discussed earlier in the book. In addition to specific areas in continuum mechanics, we review work on convection in stars, heat transfer in a moon of a planet, hyperbolic motion of traffic flow, and hyperbolic theories which have been employed in biology. These include population dynamics, the motion of a school of fish, spread of viruses, chemically driven movement of cells, the medical technique of radiofrequency heating of human tissue, and flash burns of human skin.

1.2 Maxwell-Cattaneo theory

The paper by (Cattaneo, 1948) is one which has had a major influence on virtually every paper on thermal waves. (Cattaneo, 1948) begins with the classical diffusion equation for heat, for example in one space dimension,

$$\frac{\partial \theta}{\partial t} = D \frac{\partial^2 \theta}{\partial x^2}, \qquad x \in \mathbb{R}, t > 0, \tag{1.33}$$

with the initial data

$$\theta(x, 0) = f(x), \qquad x \in \mathbb{R}. \tag{1.34}$$

(Cattaneo, 1948) observes that the well known solution to this equation is given by (1.5). He further observes that the solution (1.5) essentially has an infinite speed of propagation, for example, f might be 0 outside a finite set but (1.5) implies $\theta \neq 0 \; \forall x \in \mathbb{R}$ for $t > 0$.

If θ denotes the temperature of a rigid solid, ρ its density, c its specific heat, and k its thermal conductivity, then (Cattaneo, 1948) notes that equation (1.33) arises from the energy balance law

$$\rho c \frac{\partial \theta}{\partial t} = -\frac{\partial q}{\partial x}, \tag{1.35}$$

together with the Fourier law of heat conduction,

$$q = -k \frac{\partial \theta}{\partial x}. \tag{1.36}$$

In equations (1.35) and (1.36), q is the heat flux. In equation (1.33) we would take $D = k/\rho c$. In order to obtain a finite speed of propagation, (Cattaneo, 1948) employs a very interesting argument essentially based on statistical mechanics for a gas. He argues that q in (1.36) may be replaced by the relation

$$q = -k \frac{\partial \theta}{\partial x} + \sigma \frac{\partial^2 \theta}{\partial x \partial t}. \tag{1.37}$$

The coefficient σ is given in equation (17) of (Cattaneo, 1948). To derive a further relation (Cattaneo, 1948) differentiates (1.37) with respect to t to obtain

$$\frac{\partial q}{\partial t} = -k \frac{\partial^2 \theta}{\partial x \partial t} + \sigma \frac{\partial^3 \theta}{\partial x \partial t^2}$$

and then forms the relation

$$\sigma\frac{\partial q}{\partial t} + kq = -k^2\frac{\partial \theta}{\partial x} + \sigma^2\frac{\partial^3 \theta}{\partial x \partial t^2}\,. \tag{1.38}$$

To bring this into line with modern terminology we put $\tau = \sigma/k(>0)$ and write as

$$\tau\frac{\partial q}{\partial t} + q = -k\frac{\partial \theta}{\partial x} + \tau\sigma\frac{\partial^3 \theta}{\partial x \partial t^2}\,. \tag{1.39}$$

The coefficient τ may be calculated from equations (12), (14) and (17) of (Cattaneo, 1948) and is a ratio of statistical mechanical averages over molecular velocities and positions.

At this point (Cattaneo, 1948) argues that one may discard the last term in (1.39) due to its smallness by comparison to the other terms. However, if we retain it the resulting system of equations is (1.35) and (1.39) which may be combined to yield the equation for the temperature

$$\tau\frac{\partial^2 \theta}{\partial t^2} + \frac{\partial \theta}{\partial t} = \frac{k}{\rho c}\frac{\partial^2 \theta}{\partial x^2} - \frac{\sigma^2}{\rho c k}\frac{\partial^4 \theta}{\partial x^2 \partial t^2}\,. \tag{1.40}$$

This equation has some similarity to the equations derived for dual and triple phase lag theories in section 1.5.

To return to the mainstream argument of (Cattaneo, 1948), on page 93 he argues that the last term in equation (1.39) may be discarded. His famous system of equations is then derived from equations (1.35) and (1.39) as

$$\begin{aligned} \rho c\frac{\partial \theta}{\partial t} &= -\frac{\partial q}{\partial x}\,,\\ \tau\frac{\partial q}{\partial t} + q &= -k\frac{\partial \theta}{\partial x}\,. \end{aligned} \tag{1.41}$$

(Cattaneo, 1948) observes that eliminating q leads to the damped wave equation

$$\tau\frac{\partial^2 \theta}{\partial t^2} + \frac{\partial \theta}{\partial t} = \frac{k}{\rho c}\frac{\partial^2 \theta}{\partial x^2}\,. \tag{1.42}$$

In addition, (Cattaneo, 1948) derives a three-dimensional version of system (1.41) and studies acceleration waves in his system, cf. chapter 4, section 4.1.

A way to derive equation (1.41) which is often used in the current literature is to argue to replace the Fourier law (1.36) by a delay equation

$$q(x, t+\tau) = -k\frac{\partial \theta}{\partial x}(x,t)\,. \tag{1.43}$$

In other words, the heat flux does not depend instantaneously on the temperature gradient at a point; there is a short time lag before the effect is felt. One then expands the left hand side of equation (1.43) using a Taylor series so that

$$q(x, t + \tau) = q(x, t) + \tau \frac{\partial q}{\partial t}(x, t) + O(\tau^2). \tag{1.44}$$

The $O(\tau^2)$ terms are neglected and then employing equation (1.43) one may arrive at equation (1.41). However, care must be taken with this approach since (Jordan et al., 2008) show that if one combines the energy balance law $(1.41)_1$ together with equation (1.43) with no approximation then the resulting delay equation displays a lack of continuous dependence on the initial data, i.e. an instability. (For interest, we point out that (Quintanilla, 2008b) and (Quintanilla and Jordan, 2009) show that well posedness may be recovered if one combines the equation (1.43) with a two-temperature energy balance equation. The two-temperature theory is explained in section 1.7.)

The Cattaneo system (1.41) and its three-dimensional equivalent,

$$\begin{aligned} \rho c \frac{\partial \theta}{\partial t} &= -\frac{\partial q_i}{\partial x_i}, \\ \tau \frac{\partial q_i}{\partial t} + q_i &= -k \frac{\partial \theta}{\partial x_i}, \end{aligned} \tag{1.45}$$

have been the subject of immenense study in the literature. If one wishes to study a fully nonlinear equivalent of (1.41) or (1.45) then the coefficients will, in general, cease to be constants. This was shown by (Coleman et al., 1982) and subsequently analysed by (Franchi, 1985), (Coleman et al., 1986), (Morro and Ruggeri, 1987; Morro and Ruggeri, 1988), (Coleman and Newman, 1988). An appealing way to see why non constant coefficients may arise was given by Dario Graffi in (Graffi, 1984), see (Franchi and Straughan, 1994a). If $\mathbf{g}$ denotes temperature gradient then (Graffi, 1984) noted that for θ constant, thermodynamics requires

$$q_i g_i \leq 0. \tag{1.46}$$

In one space dimension if $g = G_0 \sin \omega t$ then equation $(1.41)_2$ becomes

$$\tau q_t + q = -k G_0 \sin \omega t$$

which has solution

$$q = A(\sin \omega t - \tau\omega \cos \omega t)$$

for

$$A = -\frac{k G_0}{\tau^2 \omega^2 + 1} < 0.$$

This leads to

$$qg = A G_0 \omega t (\sin \omega t - \tau\omega \cos \omega t). \tag{1.47}$$

Equation (1.47) is not compatible with inequality (1.46) since qg may be positive. This leads (Graffi, 1984) to suggest replacing inequality (1.46) by

$$q(\alpha\theta q_t + g) \leq 0$$

for a suitable α and then setting

$$q = -k(\alpha\theta q_t + g).$$

Note that the Cattaneo equation $(1.41)_2$ still holds, but τ must be a function of θ. The above arguments are deduced rigorously using internal variables in continuum thermodynamics by (Franchi, 1985).

(Morro and Ruggeri, 1988) derive a nonlinear temperature dependent system akin to (1.45) which has form

$$\begin{aligned} F(\theta)q_{i,t} + (1 + \Gamma(\theta)\theta_t)q_i &= -k\theta_{,i} \\ c_0(\theta)\theta_t &= -q_{i,i} \end{aligned} \tag{1.48}$$

in which the functions F, Γ and c_0 take the forms

$$\begin{aligned} F &= k(A\theta^{-3} + B\theta^{n-3}), \\ \Gamma &= -k(5A\theta^{-4} + (5-n)B\theta^{n-4}), \qquad c_0 = \epsilon_1\theta^3 \end{aligned}$$

for suitable constants A, B, n and ϵ_1. Notably F is a nonlinear function of θ which replaces the constant relaxation time τ in equation $(1.45)_2$. Another notable difference with equation $(1.45)_2$ is the presence of the $\Gamma(\theta)\theta_t$ term multiplying q_i in $(1.48)_1$. Acceleration waves and shock waves are considered in a nonlinear system not dissimilar to (1.48) by (Morro and Ruggeri, 1987).

It is worth observing that the thermodynamic development of (Coleman et al., 1982) leads to the conclusion that the internal energy, ϵ, and entropy, η, are not simply functions of temperature, θ. They must also depend on the heat flux, q_i, cf. also (Franchi, 1985), (Coleman et al., 1986), (Morro and Ruggeri, 1987; Morro and Ruggeri, 1988), (Coleman and Newman, 1988). (Ruggeri, 2001) addresses carefully the question of the thermal inertia, i.e. the $\tau\dot{q}_i$ term, such that τ depends on temperature. He provides a physical explanation for the meaning of the thermal inertia by a development based on a mixture of simple fluids, see also (Ruggeri, 2010), (Gouin and Ruggeri, 2008), (Ruggeri and Simić, 2005).

A recent interesting derivation of the Maxwell-Cattaneo equation is due to (Ostoja-Starzewski, 2009).

1.3 Guyer-Krumhansl theory

A generalization of equation $(1.45)_2$ which accounts for space correlation, being based on the Boltzmann equation, was derived by (Guyer and

Krumhansl, 1964; Guyer and Krumhansl, 1966a; Guyer and Krumhansl, 1966b). This equation may be written

$$\tau q_{i,t} + q_i = -k\theta_{,i} + \hat{\tau}\Delta q_i + 2\hat{\tau} q_{k,ki}. \tag{1.49}$$

Here $\hat{\tau} = \tau\tau_N c_s^2/5$ where τ_N is a relaxation time and c_s is the mean speed of the phonons. Equation (1.49) is derived by (Lebon and Dauby, 1990) by means of a variational argument in the context of extended thermodynamics. Another derivation based on hidden variables is presented by (Morro et al., 1990). A recent derivation using a generalized Coleman & Noll principle may be found in (Cimmelli et al., 2010b), cf. also (Triani et al., 2010). (Morro et al., 1990) allow the coefficients $\hat{\tau}$ and $2\hat{\tau}$ in equation (1.49) to be more general and they replace them by μ and ν. In general, the coefficients in (Morro et al., 1990), τ, k, μ and ν depend on temperature θ and are related to a Helmholtz free energy function ψ of form

$$\psi = \Psi(\theta) + \frac{1}{2} f(\theta)\mathbf{\Lambda}.\mathbf{\Lambda},$$

where $\mathbf{\Lambda}$ is a hidden variable which coincides with $\nabla\theta$ in stationary homogeneous conditions. (Morro et al., 1990) also refer to the generalization of equations (1.45) as a generalized Maxwell-Cattaneo system, which has form

$$\begin{aligned} \rho c\theta_t &= -q_{i,i} \\ \tau q_{i,t} &= -q_i - k\theta_{,i} + \mu\Delta q_i + \nu q_{k,ki}\,. \end{aligned} \tag{1.50}$$

In the general case, the coefficients in (1.50) are functions of temperature θ, although in their subsequent analysis, (Morro et al., 1990) study a linearized form for which c, τ, k, μ and ν are constants. From a mathematical point of view, the extra derivatives in $(1.50)_2$ usually lead to a greater degree of stabilization in a heat wave problem.

In the case where (1.50) is linearized about a constant thermodynamic state, constant temperature, the coefficients are constants and then we may take the divergence of equation $(1.50)_2$ and eliminate $q_{i,i}$ to derive a single equation for θ of form

$$\tau\theta_{tt} + \theta_t = \kappa\Delta\theta + (\mu + \nu)\Delta\theta_t\,, \tag{1.51}$$

where $\kappa = k/\rho c$. In one space dimension this is

$$\tau\frac{\partial^2\theta}{\partial t^2} + \frac{\partial\theta}{\partial t} - (\mu+\nu)\frac{\partial^3\theta}{\partial x^2\partial t} = \kappa\frac{\partial^2\theta}{\partial x^2}\,. \tag{1.52}$$

This equation should be contrasted with equation (1.42) which arises from Maxwell-Cattaneo theory.

1.4 High order relaxation dynamics

(Alvarez-Ramirez et al., 2006; Alvarez-Ramirez et al., 2008) are two interesting contributions which deal with extension of the Cattaneo equations (1.41) or (1.45). (Alvarez-Ramirez et al., 2006) observe that one may take the Laplace transform of equation $(1.41)_2$ and generalize the class of fluxes in transform space. They deal with diffusion in general rather than simply diffusion of temperature. But we here describe their work in terms of heat transport. If $F(s)$ denotes the Laplace transform of $f(t)$, cf. (Sneddon, 1995), then

$$F(s) = \int_0^\infty e^{-st} f(t)dt.$$

Thus, denote by $Q(s)$ and $\Theta(s)$ the Laplace transforms of q and θ, in t (with the x variable still present). Then transforming equation $(1.41)_2$, results in the equation for suitably normalized initial data,

$$(s\tau + 1)Q(s, x) = -k\Theta_x(s, x).$$

(Alvarez-Ramirez et al., 2006) observe that this equation may be rewritten as

$$Q(s, x) = -kF_1(s; \tau)\Theta_x(s, x), \tag{1.53}$$

where the function F_1 has form

$$F_1(s; \tau) = \frac{1}{\tau s + 1}\,. \tag{1.54}$$

(Alvarez-Ramirez et al., 2006) propose extending equation (1.53) to one with a more general class of functions $F(s)$, so they put

$$Q(s, x) = -kF(s)\Theta_x(s, x), \tag{1.55}$$

where, in particular, F is a rational function of form

$$F(s) = \frac{b_m s^m + b_{m-1} s^{m-1} + \ldots + b_0}{s^n + a_{n-1} s^{n-1} + \ldots + a_0}\,. \tag{1.56}$$

for coefficients $b_0, \ldots, b_m, a_0, \ldots, a_{n-1}$, with $m \leq n$.

(Alvarez-Ramirez et al., 2006) note that (1.56) leads to a system of equations in the time domain (rather than the Laplace domain) which contains higher derivatives than that of the classical diffusion equation. For example, their equivalent of the Cattaneo system (1.41) has form

$$\begin{aligned}
\frac{\partial \theta}{\partial t} &= \zeta \\
\frac{\partial \zeta}{\partial t} &= -\frac{1}{\tau}(\zeta - k\theta_{xx})\,.
\end{aligned}$$

The higher order systems which result have much in common with those which arise from approximation of phase-lag models, as discussed in section 1.5.

(Alvarez-Ramirez et al., 2006) discuss various applications of their theory, for example to electrical circuits, and interestingly investigate a fractal version of equation (1.54) of form

$$F(s;\tau) = \frac{1}{(\tau s)^\gamma + 1}, \qquad \text{for } 0 < \gamma < 1.$$

(Alvarez-Ramirez et al., 2008) develops a Lattice-Boltzmann scheme appropriate to Cattaneo's equation. They show that this approach has some distinct advantages over other schemes. For example, in numerical approximation of solutions the computer codes are inherently parallelizable. They extend naturally to higher dimensions provided one chooses a suitable lattice framework. They also discuss their Lattice-Boltzmann approach to the higher order model of (Alvarez-Ramirez et al., 2006) where $F(s)$ has a form like (1.56). Several numerical results are presented in (Alvarez-Ramirez et al., 2008).

1.5 Phase lag models

There has been much recent interest in developing theories of heat propagation which extend the phase lag heat flux law given in (1.43) and, in particular, which consider extensions of the Taylor series for the heat flux as given in equation (1.44). Much of this stems from the work of (Tzou, 1995b; Tzou, 1995a), and we cite in particular, (Han et al., 2006), (Jou and Criado-Sancho, 1998), (Quintanilla, 2002a), (Quintanilla and Racke, 2006a; Quintanilla and Racke, 2007; Quintanilla and Racke, 2008), (Serdyukov, 2001), (Serdyukov et al., 2003) and the references therein. The key would appear to be the assertion that equation (1.43) be replaced by an equation of form

$$q_i(\mathbf{x}, t+\tau_q) = -\kappa\theta_{,i}(\mathbf{x}, t+\tau), \tag{1.57}$$

where τ_q and τ will have (in general) different values. Various truncations of the Taylor series expansion are considered. For example, (1.57) is replaced by

$$q_i(\mathbf{x}, t) + \tau_q q_{i,t}(\mathbf{x}, t) = -\kappa\theta_{,i}(\mathbf{x}, t) - \kappa\tau\theta_{,it}(\mathbf{x}, t), \tag{1.58}$$

(Han et al., 2006), (Jou and Criado-Sancho, 1998), (Serdyukov, 2001), (Serdyukov et al., 2003), (Quintanilla and Racke, 2006a), section 4. Combined with the energy equation for a rigid heat conductor,

$$\rho\epsilon_\theta\theta_t = -q_{i,i}, \tag{1.59}$$

equation (1.58) yields (for $\rho\epsilon_\theta = c$, constant)

$$c\theta_t + c\tau_q\theta_{tt} = \kappa\Delta\theta + \kappa\tau\Delta\theta_t\,. \tag{1.60}$$

Let us observe that equation (1.60) is equivalent to equation (1.52) obtained in section 1.3 from the Guyer-Krumhansl equations. Thus, one may assert that in a precise linear sense the Guyer-Krumhansl model and one form of the approximate dual phase lag model lead to the same equations. (Quintanilla, 2002a), (Quintanilla and Racke, 2006a) and (Serdyukov et al., 2003) consider adding a further term in the expansion of $q_i(t+\tau_q)$ to the left of (1.58) so that

$$q_i(\mathbf{x},t) + \tau_q q_{i,t}(\mathbf{x},t) + \frac{\tau_q^2}{2} q_{i,tt}(\mathbf{x},t) = -\kappa\theta_{,i}(\mathbf{x},t) - \kappa\tau\theta_{,it}(\mathbf{x},t). \tag{1.61}$$

Together with (1.59) this leads to the hyperbolic equation

$$\frac{c\tau_q^2}{2}\theta_{ttt} + c\tau_q\theta_{tt} + c\theta_t = \kappa\Delta\theta + \kappa\tau\Delta\theta_t\,. \tag{1.62}$$

A very interesting derivation of equation (1.62) for gas flow through a package of heat conducting plates is given by (Serdyukov et al., 2003). These writers use a Cattaneo theory for the plates and a Newton cooling - like law for the gas, of form

$$\begin{aligned} \rho c(\tau\theta_{tt} + \theta_t) &= \kappa\Delta\theta - \beta_1(\theta - \theta_g), \\ \rho_g c_g \theta_t^g &= \beta_2(\theta - \theta_g), \end{aligned}$$

where θ and θ_g are the temperatures of the plates and gas, respectively.

(Quintanilla and Racke, 2006a; Quintanilla and Racke, 2007) consider a further extension to (1.61) of form

$$\begin{aligned} q_i(\mathbf{x},t) + \tau_q q_{i,t}(\mathbf{x},t) &+ \frac{\tau_q^2}{2} q_{i,tt}(\mathbf{x},t) \\ &= -\kappa\theta_{,i}(\mathbf{x},t) - \kappa\tau\theta_{,it}(\mathbf{x},t) - \frac{\kappa\tau^2}{2}\theta_{,itt}(\mathbf{x},t). \end{aligned} \tag{1.63}$$

They show that this together with equation (1.59) leads to the following equation for the temperature field θ,

$$\frac{c\tau_q^2}{2}\theta_{ttt} + c\tau_q\theta_{tt} + c\theta_t = \kappa\Delta\theta + \kappa\tau\Delta\theta_t + \frac{\kappa\tau^2}{2}\Delta\theta_{tt}\,. \tag{1.64}$$

(Quintanilla and Racke, 2006a) note that if one employs the approximation to (1.57) of form

$$q_i(\mathbf{x},t) + \tau_q q_{i,t}(\mathbf{x},t) = -\kappa\theta_{,i}(\mathbf{x},t) - \kappa\tau\theta_{,it}(\mathbf{x},t) - \frac{\kappa\tau^2}{2}\theta_{,itt}(\mathbf{x},t) \tag{1.65}$$

together with the energy equation (1.59) then one derives the temperature equation

$$\tau_q\theta_{tt} + \theta_t = \kappa\Delta\theta + \kappa\tau\Delta\theta_t + \frac{\kappa\tau^2}{2}\Delta\theta_{tt}\,. \tag{1.66}$$

In a pertinent article (Jordan et al., 2008) argue that the dual phase lag equation (1.57) is equivalent to a single lag model like that of (1.43). We recall that (Jordan et al., 2008) show that if one does not employ a Taylor expansion in the equation (1.43) then a non-well posed problem arises. They infer that the dual phase lag model equation (1.57) also has this behaviour. This does draw attention to the important point that there is a major distinction between a true phase lag model and the approximations which arise through the use of Taylor expansions.

(Quintanilla and Racke, 2008) observe that (Roy Choudhuri, 2007) proposes an extension to the phase lag equation (1.57) of the form

$$q_i(\mathbf{x}, t+\tau_q) = -\kappa\theta_{,i}(\mathbf{x}, t+\tau) - \kappa^*\alpha_{,i}(\mathbf{x}, t+\tau_v) \tag{1.67}$$

where τ_q, τ, τ_v are positive constants, and α is a thermal displacement variable defined by $\alpha_t = \theta$. They refer to this as a three phase lag theory. By using Taylor expansions in equation (1.67), (Quintanilla and Racke, 2008) show that coupled with the energy balance law (1.59) one may derive two further temperature equations of form

$$c\theta_{tt} + \tau_q c\theta_{ttt} = \kappa^*\Delta\theta + (\kappa^*\tau_v + \kappa)\Delta\theta_t + \kappa\tau\Delta\theta_{tt}\,, \tag{1.68}$$

and

$$c\theta_{tt} + \tau_q c\theta_{ttt} + \frac{\tau_q^2 c}{2}\theta_{tttt} = \kappa^*\Delta\theta + (\kappa^*\tau_v + \kappa)\Delta\theta_t + \kappa\tau\Delta\theta_{tt}\,. \tag{1.69}$$

(Quintanilla and Racke, 2008) show how equations (1.68) and (1.69) may be related to equations derived earlier in this section from the dual phase lag theory and also how they may be related to linearized versions of type II and type III equations of Green & Naghdi which are discussed in sections 1.10, 1.11.

(Quintanilla, 2009) studies a well posed problem for a three dual phase lag model for heat transfer.

1.6 Heat flux history models

The models for producing thermal waves which travel with a finite wavespeed have so far, in some sense, all been based on a time delay between the heat flux $\mathbf{q}$ and the gradient of temperature field $\nabla\theta$, or have involved Taylor expansions which lead to the introduction of a thermal relaxation time. Our aim in this section is to introduce the beautiful model of (Gurtin and Pipkin, 1968) in which they do not employ simply a delay, but allow the heat flux to depend on the past history of the temperature gradient. Before doing this we briefly discuss a simple example motivated by the work of (Abramson et al., 2001). (Abramson et al., 2001) consider a nonlinear version of the classical diffusion equation (1.33) but allow the diffusion term to be spread over the history from a fixed time to the current time. Thus, they consider the equation

$$\frac{\partial u}{\partial t} = D\int_0^t \phi(t-s)\frac{\partial^2 u}{\partial x^2}(x,s)ds + k\,f(u),$$

for a nonlinear function $f(u)$. We restrict attention to the case where $f \equiv 0$ so in terms of θ,

$$\frac{\partial \theta}{\partial t} = D\int_0^t \phi(t-s)\frac{\partial^2 \theta}{\partial x^2}(x,s)ds\,. \tag{1.70}$$

This, in general, leads to a finite speed of propagation. In particular, as (Abramson et al., 2001) observe, the choice $\phi(t) = \alpha e^{-\alpha t}$ reduces equation (1.70) to

$$\frac{\partial^2 \theta}{\partial t^2} + \alpha\frac{\partial \theta}{\partial t} = D\alpha\frac{\partial^2 \theta}{\partial x^2}\,. \tag{1.71}$$

This is the same as the equation for θ obtained using the Cattaneo theory in section 1.2, equation (1.42). Thus, having a heat flux $\mathbf{q}$ which depends on the history of the temperature gradient $\nabla\theta$ will, in general, lead to a finite speed of propagation of a thermal disturbance. (Gurtin and Pipkin, 1968) developed a nonlinear theory along similar lines which we now briefly describe.

1.6.1 Gurtin - Pipkin theory

To recount the theory of (Gurtin and Pipkin, 1968) we need a little of their notation. Let f be a real function, i.e. $f:\mathbb{R}\to\mathbb{R}$, then introduce the *history* of f, f^t, and the *summed history* of f up to time t, $\bar{f}^t$, by

$$f^t(s) = f(t-s) \qquad \text{and} \qquad \bar{f}^t(s) = \int_0^s f^t(a)da = \int_{t-s}^t f(a)da. \tag{1.72}$$

An influence function h is a continuous, monotone decreasing function with $s^2h(s)$ integrable in s on $[0,\infty)$ and then (Gurtin and Pipkin, 1968) define $\|f\|$ to be the norm of a scalar or vector valued function f on $[0,\infty)$, given by

$$\|f\|^2 = \int_0^\infty |f(s)|^2 h(s)\,ds.$$

They then let H be the set of all measureable real-valued functions f on $[0,\infty)$ with $\|f\| < \infty$ and let $\mathbf{H}$ be the equivalent set when $\mathbf{f}$ is a real vector-valued function. Further, H^+ is the cone in H of essentially strictly positive functions and H^{++} is the cone in H^+ of essentially strictly monotone increasing functions. (Gurtin and Pipkin, 1968) then introduce a smooth scalar valued functional on $\mathbb{R}^+\times H^{++}\times\mathbf{H}$ and define their Helmholtz free energy functional $\psi(t)$ by

$$\psi(t) = \Psi(\theta, \bar{\theta}^t, \bar{\mathbf{g}}^t)$$

where $\mathbf{g}$ is shown to amount to the temperature gradient of θ.

(Gurtin and Pipkin, 1968) postulate for a stationary rigid heat conductor, an energy balance law

$$\epsilon_t = -q_{i,i} + r \tag{1.73}$$

and an entropy inequality of form

$$\eta_t \geq -\left(\frac{q_i}{\theta}\right)_{,i} + \frac{r}{\theta}.$$

With the free energy (Helmholtz) satisfying $\psi = \epsilon - \theta\eta$ and $\mathbf{g} = \nabla\theta$ being the temperature gradient, (Gurtin and Pipkin, 1968) give constitutive equations of form

$$\begin{aligned}
\psi(t) &= \Psi(\theta, \bar{\theta}^t, \bar{\mathbf{g}}^t),\\
\eta &= N(\theta, \bar{\theta}^t, \bar{\mathbf{g}}^t),\\
q_i &= Q_i(\theta, \bar{\theta}^t, \bar{\mathbf{g}}^t),
\end{aligned}$$

where N and Q_i are functionals on $\mathbb{R}^+ \times H^{++} \times \mathbf{H}$.

By using thermodynamic arguments (Gurtin and Pipkin, 1968) determine precise forms for N and Q_i in terms of the functional Ψ. In fact, they show

$$\begin{aligned}
\eta &= -D_\theta \Psi(\theta, \bar{\theta}^t, \bar{\mathbf{g}}^t),\\
\mathbf{q} &= -\theta J_{\mathbf{g}} \Psi(\theta, \bar{\theta}^t, \bar{\mathbf{g}}^t),
\end{aligned}$$

where D_θ is the partial derivative of Ψ with respect to θ, and

$$J_{\mathbf{g}}\Psi \cdot \mathbf{v} = \delta_2 \Psi(\mathbf{v}^c)$$

is the partial Frechet derivative of Ψ with respect to $\bar{\mathbf{g}}^t$. (Gurtin and Pipkin, 1968) remark that this was the first ever theory where the heat flux was determined by the functional for the free energy. (Gurtin and Pipkin, 1968) show some important results regarding what is essentially a thermal conductivity tensor and investigate material symmetry properties. They also determine the wavespeeds of an acceleration wave, what they term a temperature-rate wave, and they show that such a wave has a finite speed. Complete determination of the amplitude of such a temperature-rate wave in the one-dimensional case is achieved by (Chen, 1969a), while (Chen, 1969b) determines the wavespeed and amplitude in detail for a curved wave.

(Gurtin and Pipkin, 1968) also develop a linearized version of their theory. They show that the internal energy ϵ in that case has form

$$\epsilon = b + c\theta - \int_0^\infty \beta'(s)\bar{\theta}^t(s)ds.$$

They also show that the heat flux and energy balance equations reduce to

$$q_{i,t} = -a(0)\theta_{,i} - \int_0^\infty a'(s)\theta_{,i}(t-s)ds \tag{1.74}$$

and

$$\epsilon_t = c\theta_t + \beta(0)\theta + \int_0^\infty \beta'(s)\theta(t-s)(s)ds. \tag{1.75}$$

The heat flux itself has equation

$$q_i(t) = -\int_0^\infty a(s)\theta_{,i}(t-s)ds. \tag{1.76}$$

(Gurtin and Pipkin, 1968) also note that the choice $a(s) = \kappa\sigma e^{-\sigma s}$ reduces equation (1.76) to the Maxwell-Cattaneo equation $(1.41)_2$.

Further interesting results for the class of rigid linear heat conductors with memory as outlined above are established by (Fabrizio et al., 1998) and by (Gentili and Giorgi, 1993). (Morro, 2006) derives general jump relations for discontinuous derivative solutions to the equations for heat conductors with memory. He derives further thermodynamic restrictions on the coefficients and analyses in depth the behaviour of singular surface temperature-rate waves.

1.6.2 *Graffi - Fabrizio theory*

We commence with a very brief description of work of (Graffi, 1936a) in a generalized theory of Maxwell's equations in electromagnetism which involves memory terms. That this might have relevance to temperature waves is explained by (Fabrizio, 2011), as we detail below.

(Graffi, 1936a) proposes a generalization of Maxwell's equations which involve memory terms. In fact, in the medium outside an antenna, (Graffi, 1936a) begins with Faraday's law

$$\frac{\partial \mathbf{B}}{\partial t} = -\text{curl}\,\mathbf{E}, \tag{1.77}$$

where $\mathbf{E}$ is the electric field and $\mathbf{B}$ is the magnetic induction. He also writes Ampère's law with the electric displacement correction of Maxwell, namely

$$\text{curl}\,\mathbf{H} = \frac{\partial \mathbf{D}}{\partial t} + \mathbf{j}, \tag{1.78}$$

where $\mathbf{H}$ is the magnetic field, $\mathbf{j}$ is the current density, and $\mathbf{D}$ is the electric displacement current.

For the current density (Graffi, 1936a) proposes

$$\mathbf{j} = \sigma\mathbf{E} + \int_0^t \beta(t-s)\mathbf{E}(s)ds, \tag{1.79}$$

where σ is the electrical conductivity and the β term represents the history dependence of the current on the electric field. (Graffi, 1936a) notes that certain materials display dielectric hysteresis and so proposes the dependence of $\mathbf{D}$ upon $\mathbf{E}$ as

$$\mathbf{D} = \epsilon\mathbf{E} + \int_0^t \gamma(t,s)\mathbf{E}(s)ds. \tag{1.80}$$

Here ϵ is the usual electric permittivity and the γ term represents the history dependence on the electric field. It is worth observing that a basis for equations like (1.80) and the analogous one involving $\mathbf{B}$ and $\mathbf{H}$ is discussed in detail in a thermodynamic context in the book of (Fabrizio and Morro, 2003).

(Graffi, 1936a) observes that defining the functions λ and α by

$$\lambda(t) = \gamma(t,t) + \sigma \quad \text{and} \quad \alpha = \gamma'(t,s) + \beta(t-s)$$

then equations (1.78) - (1.80) lead to the equation

$$\operatorname{curl}\mathbf{H} = \epsilon\frac{\partial \mathbf{E}}{\partial t} + \lambda\mathbf{E} + \int_0^t \alpha(t,s)\mathbf{E}(s)ds. \tag{1.81}$$

Furthermore, if $\mathbf{B} = \mu\mathbf{H}$ where μ is the magnetic permeability, then equation (1.77) becomes

$$\operatorname{curl}\mathbf{E} = -\mu\frac{\partial \mathbf{H}}{\partial t}. \tag{1.82}$$

Equations (1.81) and (1.82) are Graffi's system for the electric and magnetic fields $\mathbf{E}$ and $\mathbf{H}$.

(Graffi, 1936a) establishes a general uniqueness theorem for equations (1.81) and (1.82) on an unbounded domain. We observe that if $\alpha \equiv 0$, i.e. if $\gamma'(t,s) = -\beta(t-s)$, then (1.81) is very like the Cattaneo law, equation $(1.45)_2$. To see that $\alpha \equiv 0$ is not meaningless we may select the realistic and frequently made choices

$$\gamma(t,s) = \gamma_0\, e^{-\delta(t-s)}, \qquad \beta = \beta_0\, e^{-\delta(t-s)},$$

which lead to

$$\alpha = \gamma' + \beta = (\beta - \gamma_0\delta)\, e^{-\delta(t-s)},$$

and so when $\beta = \gamma_0\delta$ then $\alpha \equiv 0$. We further mention that (Graffi, 1928a; Graffi, 1928b; Graffi, 1999a) also employs equations like (1.80) in elasticity, and a nonlinear version of (1.80) is used by him, (Graffi, 1936b; Graffi, 1999c), in an inspiring paper where he explains the Luxemburg effect.

(Graffi, 1936a) explains that (Cisotti, 1911) derives an analogue of (1.81) and (1.82) but with the generalization of history dependence in the $\mathbf{B} = \mathbf{B}(\mathbf{H})$ relationship. However, we note that while (Cisotti, 1911) has a relation like (1.80) and its analogue involving $\mathbf{B}$ and $\mathbf{H}$, he does not derive a system like equations (1.81) and (1.82). (Cisotti, 1911) concentrates on deriving an expression for the energy density.

The relevance of this section on the topic of heat waves is observed by (Fabrizio, 2011). (Fabrizio, 2011) notes that one could replace **E** by **q** and curl **H** by $-k\nabla\theta$ in equation (1.81), with $\alpha \equiv 0$, to, in fact, obtain Cattaneo's equation. Furthermore, one could adapt the argument of (Fabrizio, 2011) to propose that instead of (1.81) one write for the heat flux

$$\tau q_{i,t} + q_i + \int_0^t \alpha(t,s) q_i(\mathbf{x},s)\,ds = -k\nabla\theta. \tag{1.83}$$

Then, with the energy balance equation $(1.45)_1$, one has

$$\rho c \frac{\partial \theta}{\partial t} = -\frac{\partial q_i}{\partial x_i}\,. \tag{1.84}$$

Of course, with $\alpha \equiv 0$ equations (1.83), (1.84) reduce to the Cattaneo system, equations (1.45). We shall refer to the system of equations (1.83) and (1.84) as the Graffi - Fabrizio system.

It is of interest to observe that eliminating q_i between (1.83) and (1.84) leads to the following equation for the temperature field $\theta(\mathbf{x},t)$,

$$\tau\theta_{tt} + \theta_t + \int_0^t \alpha(t,s)\theta_s(s) = \frac{k}{\rho c}\,\Delta\theta. \tag{1.85}$$

In the case where $\alpha = \alpha_0\, e^{-\omega(t-s)}$ one may then find θ satisfies the equation

$$\tau\theta_{ttt} + (1+\omega\tau)\theta_{tt} + (\alpha_0 + \omega)\theta_t = \frac{k}{\rho c}\,\Delta\theta_t + \frac{\omega k}{\rho c}\,\Delta\theta. \tag{1.86}$$

1.7 Two temperature model

(Chen and Gurtin, 1968) develop another interesting theory in a rigid body which allows transient behaviour of a heat wave. In this theory they introduce a conductive temperature, ϕ, and a thermodynamic temperature, θ. Their nonlinear development assumes that the internal energy, ϵ, entropy, η, heat flux, q_i, and thermodynamic temperature depend on the constitutive variables $\phi, \phi_{,m}, \phi_{,mn}$, i.e.

$$\begin{aligned} \epsilon &= \epsilon(\phi, \phi_{,m}, \phi_{,mn}) \qquad & \eta &= \eta(\phi, \phi_{,m}, \phi_{,mn}) \\ \theta &= \theta(\phi, \phi_{,m}, \phi_{,mn}) \qquad & q_i &= q_i(\phi, \phi_{,m}, \phi_{,mn}). \end{aligned}$$

They have the balance of energy equation

$$\epsilon_t = -q_{i,i} + r$$

and the entropy inequality

$$\eta_t \geq -\left(\frac{q_i}{\phi}\right)_{,i} + \frac{r}{\theta}\,.$$

By using the two above relations (Chen and Gurtin, 1968) deduce that

$$\psi = \epsilon - \eta\theta = \psi(\theta), \qquad \eta = \eta(\theta), \qquad \epsilon = \epsilon(\theta),$$

and

$$\eta(\theta) = -\frac{\partial\psi}{\partial\theta}(\theta).$$

They also deduce various relations involving q_i and derive a reduced entropy inequality. Material symmetry is exploited and a fully nonlinear theory is developed.

(Chen and Gurtin, 1968) also derive a linearized theory, linearized about a fixed reference temperature ϕ_0. With $c = \partial\epsilon/\partial\theta$ the basic equations then become

$$\begin{aligned} c\theta_t &= -q_{i,i}\,, \\ q_i &= -k\phi_{,i}\,, \\ \theta &= \phi - a\Delta\phi, \end{aligned} \tag{1.87}$$

for constants k, a. A uniqueness theorem is established and (Chen and Gurtin, 1968) further investigate wave motion for their theory.

We have already mentioned in section 1.2 the developments by (Quintanilla, 2008b) and by (Quintanilla and Jordan, 2009) involving equation (1.87) and time lags. It is instructive to recall them at this point. (Jordan et al., 2008) showed that the phase lag constitutive theory

$$q_i(\mathbf{x}, t + \tau) = -k\theta_{,i}(\mathbf{x}, t)$$

coupled with the classical balance of energy equation leads to an improperly posed problem. However, (Quintanilla, 2008b) and (Quintanilla and Jordan, 2009) study what amounts to (1.87) but with the equation $(1.87)_2$ replaced by

$$q_i(\mathbf{x}, t + \tau) = -k\phi_{,i}(\mathbf{x}, t).$$

The analyses of (Quintanilla, 2008b) and (Quintanilla and Jordan, 2009) show that this is now a well posed problem.

1.8 Green-Laws theory

The starting point of (Green and Laws, 1972) is to postulate a new entropy inequality. Suppose $\mathcal{B}$ is a rigid body and let $\mathcal{P} \subset \mathcal{B}$ be any sub-body which has boundary $\partial\mathcal{P}$. (Green and Laws, 1972) assert that

$$\frac{d}{dt}\int_{\mathcal{P}} \rho\eta\, dV - \int_{\mathcal{P}} \frac{\rho r}{\phi}\, dV + \oint_{\partial\mathcal{P}} \frac{q_i n_i}{\phi}\, dA \geq 0$$

where ρ, η, r, q_i are density, entropy, external heat supply, and heat flux. In addition, n_i is the unit outward normal to $\partial\mathcal{P}$. The quantity ϕ is a new function which is specified by a constitutive equation. (Green and Laws, 1972) require that $\phi = \theta$, the temperature, *in equilibrium*. The pointwise version of the above entropy inequality may be written as

$$\rho\phi\dot{\eta} - \rho r + q_{i,i} - \frac{q_i\phi_{,i}}{\phi} \geq 0, \tag{1.88}$$

where $\dot{\eta} = \partial\eta/\partial t$.

For a rigid body the balance of energy equation is

$$\rho\dot{\epsilon} + q_{i,i} = \rho r \tag{1.89}$$

where ϵ is the internal energy function. Using the Helmholtz free energy function $\psi = \epsilon - \eta\phi$ the inequality (1.88) may be transformed to

$$-\rho(\dot{\psi} + \eta\dot{\phi}) - \frac{q_i\phi_{,i}}{\phi} \geq 0. \tag{1.90}$$

As constitutive theory (Green and Laws, 1972) suppose that

$$\psi, \eta, \phi \quad \text{and} \quad q_i$$

are functions of the variables

$$\theta, \dot{\theta}, \quad \text{and} \quad \theta_{,i}\,.$$

Inequality (1.90) is now expanded using this constitutive theory to obtain

$$\begin{aligned} &- \rho(\psi_\theta + \eta\phi_\theta)\dot{\theta} - \rho(\psi_{\dot{\theta}} + \eta\phi_{\dot{\theta}})\ddot{\theta} \\ &- \rho(\psi_{\theta_{,i}} + \eta\phi_{\theta_{,i}})\dot{\theta}_{,i} - \frac{q_i}{\phi}\,(\phi_{\theta_{,i}} + \phi_{\dot{\theta}}\dot{\theta}_{,i} + \phi_{\theta_{,k}}\theta_{,ik}) \geq 0. \end{aligned} \tag{1.91}$$

(Green and Laws, 1972) now argue that the linearity of $\ddot{\theta}, \dot{\theta}_{,i}$ and $\theta_{,ik}$ in inequality (1.91) allow them to select these quantities arbitrarily, balancing the energy equation (1.89) by a suitable choice of r. Since they may be selected arbitrarily, keeping other quantities fixed, inequality (1.91) may be violated unless the coefficients of these terms vanish. Thus (Green and Laws, 1972) deduce that

$$\frac{\partial\psi}{\partial\dot{\theta}} + \eta\frac{\partial\phi}{\partial\dot{\theta}} = 0, \tag{1.92}$$

$$\rho\left(\frac{\partial\psi}{\partial\theta_{,i}} + \eta\frac{\partial\phi}{\partial\theta_{,i}}\right) + \frac{q_i}{\phi}\frac{\partial\phi}{\partial\dot{\theta}} = 0, \tag{1.93}$$

$$q_i\frac{\partial\phi}{\partial\theta_{,k}} + q_k\frac{\partial\phi}{\partial\theta_{,i}} = 0. \tag{1.94}$$

In addition, there remains the residual entropy inequality

$$-\rho(\psi_\theta + \eta\phi_\theta)\dot{\theta} - \frac{q_i}{\phi}\,\phi_\theta\theta_{,i} \geq 0. \tag{1.95}$$

From (1.94) (Green and Laws, 1972) deduce that for $q_i \neq 0$,

$$\frac{\partial \phi}{\partial \theta_{,i}} = 0$$

and so one must have

$$\phi = \phi(\theta, \dot{\theta}). \tag{1.96}$$

The function ϕ so defined is often referred to as a generalized temperature. (Green and Laws, 1972) then deduce that

$$q_i = -\frac{\rho \phi \psi_{\theta_{,i}}}{\phi_{\dot{\theta}}} \tag{1.97}$$

and then further,

$$\frac{\partial q_i}{\partial \theta_{,j}} = \frac{\partial q_j}{\partial \theta_{,i}}\,.$$

Thus, if q_i satisfies a linear relation in $\theta_{,r}$, say

$$q_i = -\kappa_{ij}(\theta, \dot{\theta})\theta_{,j}\,, \tag{1.98}$$

then necessarily the conduction tensor κ_{ij} is symmetric.

(Green and Laws, 1972) define (thermal) equilibrium in the rigid body to be when $\dot{\theta} = 0$ and $\theta_{,i} = 0$. They require

$$\phi|_E = \theta$$

where $|_E$ denotes thermal equilibrium and they then deduce that

$$\left.\frac{\partial \phi}{\partial \theta}\right|_E = 1.$$

(Green and Laws, 1972) derive further results in equilibrium, in particular

$$\eta|_E = -\frac{\partial \psi}{\partial \theta}\,, \qquad q_i|_E = 0, \qquad \left.\frac{\partial \psi}{\partial \theta_{,i}}\right|_E = 0$$

and

$$-\left(\left.\frac{\partial q_i}{\partial \theta_{,j}}\right|_E + \left.\frac{\partial q_j}{\partial \theta_{,i}}\right|_E\right)$$

is a positive semi-definite tensor.

(Green and Laws, 1972) show that their energy equation in a linearized theory becomes

$$\rho(\phi\eta_{\dot{\theta}})|_E\ddot{\theta} + \rho(\phi\eta_\theta)|_E\dot{\theta} + \left\{\left.\frac{\partial q_i}{\partial \dot{\theta}}\right|_E + \rho(\phi\eta_{\theta_{,i}})|_E\right\}\dot{\theta}_{,i} = \left.\frac{\partial q_i}{\partial \theta_{,k}}\right|_E \theta_{,ik} - \rho r. \tag{1.99}$$

This is effectively a damped linear wave equation, and so permits the passage of a thermal wave.

A fully nonlinear acceleration wave theory for the (Green and Laws, 1972) model was developed by (Lindsay and Straughan, 1976).

A further development of the ideas of allowing higher derivatives of θ to be present in the constitutive theory is due to (Batra, 1974; Batra, 1975) and also (Meixner, 1974). They based their work on the analysis of (Müller, 1971a; Müller, 1971b).

1.9 Temperature dependent conductivity

We recall that the classical diffusion equation (1.33)

$$\frac{\partial\theta}{\partial t} = D\frac{\partial^2\theta}{\partial x^2}, \qquad x \in \mathbb{R}, t > 0,$$

leads to what is essentially infinite speed of propagation. In terms of energy balance and heat flux we see that this equation is equivalent to equations (1.35) and (1.36), i.e.

$$\rho c\frac{\partial\theta}{\partial t} = -\frac{\partial q}{\partial x}, \qquad q = -k\frac{\partial\theta}{\partial x}, \tag{1.100}$$

with $D = k/\rho c$.

One effective way to achieve a finite speed of propagation for θ is to allow D to depend on the temperature θ itself. Since in reality the thermal conductivity k does depend on temperature this is realistic. Thus, with $D = D(\theta)$, equations (1.100) lead to

$$\frac{\partial\theta}{\partial t} = \frac{\partial}{\partial x}\left(D(\theta)\frac{\partial\theta}{\partial x}\right). \tag{1.101}$$

The solution to this equation when $D = D_0\theta^m/\theta_0^m$, for m, D_0, θ_0 positive numbers is conveniently located in (Murray, 2003a), pp. 402–405 (although Murray applies the equation to the phenomenon of insect dispersal). For this choice of D we might consider the initial value problem

$$\begin{aligned} &\frac{\partial\theta}{\partial t} = \frac{D_0}{\theta_0^m}\frac{\partial}{\partial x}\left(\theta^m\frac{\partial\theta}{\partial x}\right), \qquad x \in \mathbb{R}, t > 0, \\ &\theta(x,0) = N\delta(x), \end{aligned} \tag{1.102}$$

where N is the total initial temperature and $\delta(x)$ is the Dirac delta function. (Murray, 2003a) gives the solution to this as

$$\theta(x,t) = \begin{cases} \dfrac{\theta_0}{\lambda(t)}\left[1 - \left(\dfrac{x}{r_0\lambda(t)}\right)^2\right]^{1/m}, & |x| \le r_0\lambda(t), \\ 0, & |x| > r_0\lambda(t), \end{cases}$$

where $\lambda(t)$ and the constants r_0 and t_0 are given by

$$\lambda(t) = \left(\frac{t}{t_0}\right)^{1/(2+m)}, \quad r_0 = \frac{N\Gamma(m^{-1} + 3/2)}{\sqrt{\pi}\,\theta_0\Gamma(m^{-1}+1)}, \quad t_0 = \frac{r_0^2 m}{2D_0(m+2)},$$

with Γ being the gamma function, cf. (Sneddon, 1980), p. 21. Clearly, the temperature θ "spreads out" with a finite wavespeed, the edge of the temperature field being located at $r_0\lambda(t)$.

A common case is a linear thermal conductivity for which $m = 1$, and then

$$\theta = \frac{\hat{\theta}_0}{t^{1/3}}\left(1 - \frac{x^2}{c_0 t^{2/3}}\right)$$

where the constants $\hat{\theta}_0$ and c_0 may be calculated from the general solution.

1.10 Type II rigid body

(Green and Naghdi, 1991) do not have a generalized temperature ϕ which depends on θ and $\dot{\theta}$. Instead they state T is an empirical temperature and then define a positive scalar valued function θ by

$$\theta = \theta(T, \alpha), \qquad \theta > 0, \qquad \partial\theta/\partial T > 0,$$

where the variable α is called a thermal displacement and is defined by

$$\alpha(\mathbf{x}, t) = \int_{t_0}^{t} T(\mathbf{x}, s)ds + \alpha_0 \,. \tag{1.103}$$

We only deal with the case of a rigid body and so there is no need for the distinction between coordinates $\mathbf{X}$ in the reference configuration and $\mathbf{x}$ in the current configuration. (Green and Naghdi, 1991) define the variables β_i and γ_i by

$$\beta_i = \frac{\partial\alpha}{\partial x_i}\,, \qquad \gamma_i = \frac{\partial T}{\partial x_i}\,. \tag{1.104}$$

The premise of (Green and Naghdi, 1991) is to use a *balance of entropy equation*

$$\rho\dot{\eta} = \rho s + \rho\xi - p_{i,i} \tag{1.105}$$

where ρ, η, s are the density, entropy, and an external supply of entropy. The vector p_i is the entropy flux, $p_i = q_i/\theta$, where q_i is the heat flux. In addition ξ is an intrinsic supply of entropy which requires a constitutive equation and must be determined during a thermodynamic process.

(Green and Naghdi, 1991) introduce a Helmholtz free energy function ψ and employ a thermal cycle argument to derive a balance of energy equation from the equation (1.105). Their energy balance equation has form

$$\rho\dot{\psi} + \rho\eta\dot{\theta} + p_i\theta_{,i} + \rho\theta\xi = 0. \tag{1.106}$$

To define heat flow of type II in a rigid body, (Green and Naghdi, 1991) specify the constitutive theory that

$$\psi, \theta, \eta, p_i \qquad \text{and} \qquad \xi \tag{1.107}$$

depend on the independent variables

$$T, \alpha, \alpha_{,i} = \beta_i \, . \tag{1.108}$$

They then expand the energy balance law (1.106) as

$$\begin{aligned} &\rho(\psi_T + \eta\theta_T)\dot{T} + \rho(\psi_\alpha + \eta\theta_\alpha)T + \rho(\psi_{\beta_i} + \eta\theta_{\beta_i})\gamma_i \\ &\qquad + p_i(\gamma_i\theta_T + \beta_i\theta_\alpha + \theta_{\beta_j}\beta_{i,j}) + \rho\theta\xi = 0. \end{aligned} \tag{1.109}$$

By requiring this to hold for all heat flows and employing the independent externally supplied entropy s of equation (1.105) they are then able to deduce from equation (1.109), since $\dot{T}, \gamma_i, \beta_{i,j}$ are linear, the relations

$$\frac{\partial\theta}{\partial\beta_i} = 0, \qquad \psi_T + \eta\theta_T = 0, \qquad p_i\theta_T + \rho\psi_{\beta_i} = 0, \tag{1.110}$$

and the residual from (1.109) is

$$\rho(\psi_\alpha + \eta\theta_\alpha)T + p_i\beta_i\theta_\alpha + \rho\theta\xi = 0. \tag{1.111}$$

The first of (1.110) shows θ cannot depend on $\alpha_{,i}$ and so

$$\theta = \theta(T, \alpha). \tag{1.112}$$

From $(1.110)_{2,3}$ we derive expressions for the entropy η and the entropy flux p_i of form

$$\eta = -\frac{\psi_T}{\theta_T}, \qquad p_i = -\rho\frac{\psi_{\alpha,i}}{\theta_T} \, . \tag{1.113}$$

The equation (1.111) then yields the instrinsic entropy supply ξ as

$$\xi = -\frac{T}{\theta}(\psi_\alpha + \eta\theta_\alpha) - \frac{1}{\rho\theta}\, p_i\alpha_{,i}\theta_\alpha. \tag{1.114}$$

The fully nonlinear equation governing heat flow of type II in a rigid body is then derived by employing equations (1.112) - (1.114) in the entropy balance law (1.105). A complete nonlinear acceleration wave analysis for this was performed by (Jaisaardsuetrong and Straughan, 2007) and details are given in section 4.2.1 of this book.

(Green and Naghdi, 1991) do not employ their nonlinear theory. Instead, they investigate special cases leading to linear theories. They consider the special case

$$\psi = \psi(\theta) = c(\theta - \theta \ln\theta) + \frac{1}{2}k\alpha_{,i}\alpha_{,i} \, , \qquad \theta = a + bT, \tag{1.115}$$

for a, b, c positive constants. This leads to the linear equation for the variable α of form

$$cb\ddot{\alpha} = r + \frac{ka}{b}\,\Delta\alpha, \tag{1.116}$$

for a supply function r. This is clearly a wave equation and the thermal displacement then travels as a wave with finite speed with no dissipation.

This is known as a theory without heat dissipation. The second special case of (Green and Naghdi, 1991) is where

$$\begin{aligned} \theta &= a + bT + d_1\alpha \\ \psi &= \frac{1}{2}k\alpha_{,i}\alpha_{,i} - \frac{1}{2}d_2\alpha^2 - b_2\alpha T - \frac{1}{2}b_3T^2. \end{aligned} \tag{1.117}$$

They show that this leads to a wave equation with dissipation, of form

$$\frac{a}{b}(b_2\dot{\alpha} + b_3\ddot{\alpha}) = r + \frac{ka}{b}\,\Delta\alpha. \tag{1.118}$$

1.11 Type III rigid body

The difference between heat flow of type III and heat flow of type II materials is that the variable $\gamma_i = T_{,i} = \dot{\alpha}_{,i}$ is added to the list (1.108). Thus, (Green and Naghdi, 1991) define heat flow of type III in a rigid body by stating that

$$\psi, \theta, \eta, p_i \quad \text{and} \quad \xi \tag{1.119}$$

depend on the variables

$$T, \alpha, \beta_i \quad \text{and} \quad \gamma_i. \tag{1.120}$$

The list (1.120) can alternatively be thought of as

$$T, \alpha, \alpha_{,i}\, T_{,i} \qquad \text{or} \qquad T, \alpha, \alpha_{,i}\, \dot{\alpha}_{,i}\,.$$

The energy balance law (1.106) still holds and substitution of the list (1.120) and expanding the time derivatives leads to the equation

$$\begin{aligned} &\rho(\psi_T + \eta\theta_T)\dot{T} + \rho(\psi_\alpha + \eta\theta_\alpha)T + \rho(\psi_{\gamma_i} + \eta\theta_{\gamma_i})\dot{\gamma}_i + \rho(\psi_{\beta_i} \\ &+ \eta\theta_{\beta_i})\gamma_i + p_i(\gamma_i\theta_T + \beta_i\theta_\alpha + \theta_{\gamma_j}\gamma_{j,i} + \theta_{\beta_j}\beta_{j,i}) + \rho\theta\xi = 0. \end{aligned} \tag{1.121}$$

(Green and Naghdi, 1991) next employ the fact that $\dot{T}, \dot{\gamma}_i, \gamma_{i,j}\,, \beta_{i,j}$ appear linearly in (1.121) and may be selected arbitrarily using the external entropy supply s to balance equation (1.105). In this way they deduce the relations

$$\frac{\partial\theta}{\partial\gamma_i} = 0, \quad \frac{\partial\theta}{\partial\beta_i} = 0, \quad \frac{\partial\psi}{\partial\gamma_i} = 0, \quad \frac{\partial\psi}{\partial T} + \eta\frac{\partial\theta}{\partial T} = 0, \tag{1.122}$$

from which it follows that

$$\theta = \theta(T, \alpha), \qquad \psi = \psi(T, \alpha, \beta_i). \tag{1.123}$$

Then, from $(1.122)_4$,

$$\eta = -\frac{\psi_T}{\theta_T} = F(T, \alpha, \beta_i). \tag{1.124}$$

What remains from equation (1.121) is the relation

$$\rho(\psi_\alpha + \eta\theta_\alpha)T + \rho\psi_{\beta_i}\gamma_i + p_i(\gamma_i\theta_T + \beta_i\theta_\alpha) + \rho\theta\xi = 0. \tag{1.125}$$

The (in general genuinely nonlinear) equation of heat flow is then obtained using (1.123) - (1.125) in the entropy balance equation (1.105) and (Green and Naghdi, 1991) show this is

$$\rho\theta\dot{\eta} - \rho r + q_{i,i} - \rho(\psi_\alpha + \eta\theta_\alpha)T - \rho\psi_{\beta_i}\gamma_i = 0. \tag{1.126}$$

I am not aware of where this (fully nonlinear) equation has been studied in the literature. At this point (Green and Naghdi, 1991) develop a linearized version of their theory.

(Green and Naghdi, 1991) continue by considering the special case where

$$\begin{aligned} \theta &= a + bT + d_1\alpha, \qquad & \psi &= \frac{k}{2}\beta_i\beta_i - \frac{d_2}{2}\alpha^2 - b_2\alpha T - \frac{b_3}{2}T^2, \\ q_i &= -(a_1\beta_i + a_2\gamma_i), \qquad & \eta b &= b_2\alpha + b_3 T, \end{aligned}$$

the coefficients a, b, etc. being constants. Then, equation (1.126) may be shown to reduce to the linear equation

$$\frac{\rho a}{b}(b_2\dot{\alpha} + b_3\ddot{\alpha}) = \rho r + a_1\Delta\alpha + a_2\Delta\dot{\alpha}, \tag{1.127}$$

where $r = \theta s$. It is interesting to observe that if we differentiate this equation with respect to t, then an equation of the same form results for the temperature T. Thus, the temperature satisfies the same linear equation as was found from Guyer-Krumhansl theory in section 1.3, equation (1.51), or from dual phase lag theory in section 1.5, equation (1.60).

It would appear that there may be a lot of potential from equation (1.126). For example, if we assume

$$\psi = \frac{k}{2}\alpha_{,i}\alpha_{,i} + G(T,\alpha) \tag{1.128}$$

for an arbitrary nonlinear function G, then

$$\eta = -\frac{G_T(T,\alpha)}{\theta_T(T,\alpha)} = H(T,\alpha) \tag{1.129}$$

where H is defined as indicated. Suppose also

$$q_i = -A(T,\alpha)\alpha_{,i} - B(T,\alpha)\dot{\alpha}_{,i}\,. \tag{1.130}$$

Then equation (1.126) still remains nonlinear and leads to

$$\begin{aligned} (\rho\theta H_\alpha - \rho G_\alpha - \rho H\theta_\alpha)\dot{\alpha} + \rho\theta H_T\ddot{\alpha} - \frac{k\rho}{2}\frac{\partial}{\partial t}(\alpha_{,i}\alpha_{,i}) \\ = (A\alpha_{,i})_{,i} + (B\dot{\alpha}_{,i})_{,i} + \rho r. \end{aligned} \tag{1.131}$$

This is a damped nonlinear wave equation with the dissipative spatial damping term $(B\dot{\alpha}_{,i})_{,i}$.

1.12 Microtemperatures

In this section we describe a theory for temperature wave propagation in a rigid heat conductor which allows for variation of thermal properties at a microstructure level. As we discuss in chapter 8 nanofluids are prevalent in the heat transfer industry. These are typically very fine particles of a metallic oxide held in suspension in a carrier fluid, cf. (Vadasz et al., 2005). The possibility that the suspension might have a different thermal microstructure to the carrier fluid should therefore be investigated. However, nanostructures in solids are also important. Cryogenic liquids are heavily involved in space research and such liquids must be stored in stainless steel vessels known as run-tanks, see (Jordan and Puri, 2001). The large thermal stresses placed on the solid vessels may be associated with thermal microstructure effects and hence there is certainly a need for a well structured theory for a rigid solid which allows for microtemperature effects.

The theory we describe is based on the type II thermodynamics of (Green and Naghdi, 1991) and was explicitly developed by (Iesan and Nappa, 2005). For a three-dimensional rigid solid this theory involves four equations representing the balance of entropy per unit mass, but also the first entropy moment vector, η_i, $i = 1, 2$ or 3. Thus, the model consists of two balance equations, namely

$$\rho\frac{\partial\eta}{\partial t} = \frac{\partial S_k}{\partial x_k} + \rho s + \rho\xi, \qquad (1.132)$$

and

$$\rho\frac{\partial\eta_i}{\partial t} = \frac{\partial S_{ki}}{\partial x_k} + S_i - H_i + \rho s_i + \rho\xi_i\,. \qquad (1.133)$$

In these equations ρ, s, ξ, s_i and ξ_i are the density, the external rate of supply of entropy per unit mass, the internal rate of production of entropy per unit mass, the first moment of the external rate of supply of entropy, and the first moment of the internal rate of production of entropy. The terms S_i and S_{ki} are the entropy flux vector, and the first entropy moment tensor, while H_i is a so called mean entropy flux vector.

(Iesan and Nappa, 2005) assume that at a given point $\mathbf{x}$ there is a set of microcoordinates Σ_i and the absolute temperature θ' at $\mathbf{x}$ may be written as a linear combination of a temperature function $\theta(\mathbf{x}, t)$ and three microtemperatures $T_i(\mathbf{x}, t)$ such that

$$\theta'(\mathbf{x}, t) = \theta(\mathbf{x}, t) + T_i(\mathbf{x}, t)\Sigma_i\,. \qquad (1.134)$$

By introducing an internal energy function ϵ and a Helmholtz free energy function $\psi = \epsilon - \theta\eta - T_i\eta_i$, (Iesan and Nappa, 2005) propose an energy balance law and are able to deduce restrictions on the constitutive functions.

They start by assuming that the functions

$$\psi, \eta, \eta_i, S_i, S_{ij}, H_i, \xi \text{ and } \xi_i$$

depend on the variables

$$\theta, T_i, \alpha_{,i}, \beta_{i,j} \equiv \chi \tag{1.135}$$

where α and β_i are thermal displacement variables with

$$\alpha = \int_{t_0}^{t} \theta(\mathbf{x}, s)ds + \alpha_0, \qquad \beta_i = \int_{t_0}^{t} T_i(\mathbf{x}, s)ds + \beta_i^0.$$

Hence $\dot{\alpha} = \alpha_t = \theta$ and $\dot{\beta}_i = \beta_{i,t} = T_i$. Since the list (1.135) involves $\dot{\alpha}, \alpha_{,i}, \dot{\beta}_i$ and $\beta_{i,j}$ this is analogous to a type II theory.

The constitutive theory deduced by (Iesan and Nappa, 2005) has form

$$\begin{aligned} &\eta = -\frac{\partial\psi}{\partial\theta}, \quad \eta_i = -\frac{\partial\psi}{\partial T_i}, \quad S_i = \rho\frac{\partial\psi}{\partial\alpha_{,i}}, \quad S_{ij} = \rho\frac{\partial\psi}{\partial\beta_{j,i}}, \\ &\xi = -\frac{1}{\rho\theta}\Big(\rho\frac{\partial\psi}{\partial\alpha_{,i}} - H_i + \rho\xi_i\Big)T_i, \end{aligned} \tag{1.136}$$

where ψ, H_i and ξ_i depend on the list χ given by (1.135).

Thus, the governing set of nonlinear equations is (1.132) and (1.133) together with (1.135) and (1.136).

(Iesan and Nappa, 2005) further develop a linearized theory for a rigid heat conductor involving microtemperatures. This they do for both a fully anisotropic theory and an isotropic one.

The anisotropic equations may be written

$$\begin{aligned} &a\ddot{\alpha} - K_{ij}\alpha_{,ij} + M_{ij}\dot{\beta}_{i,j} = \rho s, \\ &B_{ij}\ddot{\beta}_j - D_{ijrs}\beta_{r,sj} + M_{ij}\dot{\alpha}_{,j} = \rho s_i\,, \end{aligned} \tag{1.137}$$

where a is a constant and K_{ij}, M_{ij}, B_{ij} and D_{ijrs} are constant tensors which satisfy the symmetries

$$K_{ij} = K_{ji}, \quad B_{ij} = B_{ji}, \quad D_{ijrs} = D_{rsij}\,. \tag{1.138}$$

The tensor M_{ij} is non-zero but not necessarily symmetric. We shall suppose B_{ij} and K_{ij} are also positive-definite.

In the isotropic case the relevant equations are

$$\begin{aligned} &a\ddot{\alpha} - K\Delta\alpha + m\dot{\beta}_{i,i} = \rho s, \\ &b\ddot{\beta}_i - d_2\Delta\beta_i - (d_1 + d_3)\beta_{j,ji} + m\dot{\alpha}_{,i} = \rho s_i\,. \end{aligned} \tag{1.139}$$

Structural stability and convergence results for a solution to equations (1.137) are given by (Ciarletta et al., 2010).

1.13 Exercises

Exercise 1.13.1 *Show that using q as given by* (1.12) *together with equation* $(1.6)_1$ *leads to θ satisfying* (1.11).

Exercise 1.13.2 *Show that* (1.14) *defines a hyperbolic system and verify that the characteristic equations are given by* (1.15).

Exercise 1.13.3 *Use the Fourier transform,*

$$F(s) = \frac{1}{\sqrt{2\pi}} \int_{-\infty}^{\infty} f(x) e^{-isx} dx,$$

and the inverse transform,

$$f(x) = \frac{1}{\sqrt{2\pi}} \int_{-\infty}^{\infty} F(s) e^{isx} ds,$$

to show (1.5) *is the solution to equation* (1.1) *together with the initial data* (1.3).

Exercise 1.13.4 *From the viscoelastic model*

$$\frac{\partial \theta}{\partial t} = D \int_0^t \phi(t-s) \frac{\partial^2 \theta}{\partial x^2}(x,s)\, ds$$

take $\phi(t) = \alpha e^{-\alpha t}$ and show $\theta(x,t)$ satisfies the damped wave equation

$$\frac{\partial^2 \theta}{\partial t^2} + \alpha \frac{\partial \theta}{\partial t} = D\alpha \frac{\partial^2 \theta}{\partial x^2}. \tag{1.140}$$

Suppose (1.140) *holds on $\{x \in (0,1)\} \times \{t > 0\}$ with*

$$\theta(0,t) = \theta(1,t) = 0.$$

Develop a Fourier series solution of form

$$\theta(x,t) = \sum_{j=1}^{\infty} b_j \sin k_j x\, e^{\sigma_j t} + \sum_{j=0}^{\infty} a_j \cos k_j x\, e^{\sigma_j t}.$$

Show that the solution will always decay in time and that $j^2 > \alpha/(4\pi^2 D)$ produces oscillatory damped modes.

Exercise 1.13.5 *Verify that* (1.18) *is a solution to* (1.16) *and* (1.17).

Exercise 1.13.6 *Let Ω be a bounded domain in $\mathbb{R}^3$ with boundary Γ. Consider a solution to the Graffi - Fabrizio system of equations* (1.83) *and* (1.84) *on $\Omega \times \{t > 0\}$ with θ prescribed on Γ, and θ, q_i given for $t = 0$. Show that a solution to the boundary-initial value problem so defined is unique, (cf. (Graffi, 1936a)).*

Exercise 1.13.7 *Heat transport in a bar is governed by the equation*

$$\frac{\partial \theta}{\partial t} = \frac{D_0}{\theta_0} \frac{\partial}{\partial x}\left(\theta \frac{\partial \theta}{\partial x}\right), \tag{1.141}$$

where θ *is the temperature of the bar. Suppose the temperature is* T *at the point* $x = 0$ *at time* $t = 0$. *By differentiation show that*

$$\theta(x,t) = \frac{a}{t^{1/3}} - b\frac{x^2}{t} \tag{1.142}$$

solves (1.141) *and thereby determine the constant* b.

Sketch the curve $\theta(x,t)$ *and suppose that the total temperature remains constant to determine the constant* a *and hence find the solution to* (1.141). *(By the total temperature we mean the value of the integral of* $\theta(\mathbf{x},t)$ *over its* x *range for a fixed* t*, and this is equal to the constant* T.*)*

Exercise 1.13.8 *The classical diffusion equation*

$$\frac{\partial \theta}{\partial t} = D_0 \frac{\partial^2 \theta}{\partial x^2} \tag{1.143}$$

has been used to model the heat distribution in a bar. In this equation $\theta(> 0)$ *represents the temperature of the bar. However, the solution to equation* (1.143) *has a major defect in this context. Explain the defect and suggest a method to remedy this.*

The distribution of heat in a bar has been described by the equation

$$\frac{\partial \theta}{\partial t} = \frac{D_0}{\theta_0^{1/3}} \frac{\partial}{\partial x}\left(\theta^{1/3} \frac{\partial \theta}{\partial x}\right). \tag{1.144}$$

Suppose the temperature is T *at the point* $x = 0$ *at time* $t = 0$. *By differentiation show that*

$$\theta(x,t) = \left(\frac{a}{t^{1/7}} - b\frac{x^2}{t}\right)^3 \tag{1.145}$$

solves (1.144) *and thereby determine the constant* b.

Sketch the curve $\theta(x,t)$ *and suppose the total temperature remains constant to determine the constant* a *and hence find the solution to* (1.145). *(The total temperature is as defined in exercise* (1.13.7).*)*

Equation (1.144) *is restricted to one spatial dimension whereas in practice heat travels in more than one - dimension. Suggest a modification of* (1.144) *which will apply in the two spatial dimensional case. Do you think the equation you have suggested is solvable?*

Exercise 1.13.9 *Professor A proposes that heat in a bar is distributed according to the equation*

$$\frac{\partial \theta}{\partial t} = D\frac{\partial}{\partial x}\left[\left(\frac{\theta}{\theta_0}\right)^m \frac{\partial \theta}{\partial x}\right] \tag{1.146}$$

together with the inital condition

$$\theta(x,0) = T\delta(x). \tag{1.147}$$

You may assume the solution to (1.146), (1.147) *is*

$$\theta(x,t) = \begin{cases} \dfrac{\theta_0}{\lambda(t)}\left[1 - \left(\dfrac{x}{r_0\lambda(t)}\right)^2\right]^{1/m}, & |x| \le r_0\lambda(t), \\ 0, & |x| > r_0\lambda(t), \end{cases}$$

where $\lambda(t)$ *and the constants* r_0 *and* t_0 *are given by*

$$\lambda(t) = \left(\frac{t}{t_0}\right)^{1/(2+m)}, \quad r_0 = \frac{N\Gamma(m^{-1}+3/2)}{\sqrt{\pi}\,\theta_0\Gamma(m^{-1}+1)}, \quad t_0 = \frac{r_0^2 m}{2D_0(m+2)},$$

Sketch this solution.

Professor A has two long bars of different material. He heats bar 1 in the point $x = 0$ *with temperature* T *at time* $t = 0$ *and supposes it satisfies his model with* $m = 1$*. He heats bar 2 at the same point at time* $t = \mathcal{T} > 0$*, with this bar satisfying his model with* $m = 1/2$*. Which heating effect will reach an observer a long way away first? (You must explain your reasoning carefully using the above solution.)*

2

Interaction with elasticity

2.1 Cattaneo theories

2.1.1 Cattaneo-Lord-Shulman theory

In sections 1.2 - 1.12 we have seen various ways of modifying the classical diffusion equation in order to allow heat to be transported with a finite wavespeed. The assumption was that the body would remain rigid. However, in many cases this is too strong since the body itself deforms or vibrates. Thus, in this chapter we wish to look at ways of coupling heat propagation in the case where the body is an elastic solid. This is the domain of thermoelasticity and, in particular, we shall review theories of thermoelasticity which allow temperature to travel as a wave with finite speed.

It would appear that the first attempts to couple elasticity with a way in which temperature can travel with a finite wavespeed are due to (Lord and Shulman, 1967) and to (Popov, 1967), as is observed in the short but very informative review by (Jordan and Puri, 2001). (Jordan and Puri, 2001) also derive a very useful comparison of the classical theory of thermoelasticity with two theories capable of allowing temperature to travel with a finite wavespeed. Extensive reviews of the early literature on thermoelasticity with temperature waves are by (Chandrasekharaiah, 1986), (Chandrasekharaiah, 1998) and by (Hetnarski and Ignaczak, 1999), and the recent book by (Ignaczak and Ostoja-Starzewski, 2009) concentrates on thermoelasticity with temperature waves, although the overlap with the current book is minimal.

To understand the situation we commence, as do (Lord and Shulman, 1967), with the classical theory of *linear* thermoelasticity. (Lord and Shulman, 1967) consider the isotropic case, but it is no more difficult to begin with the anisotropic situation and this we do now. In terms of the elastic displacement, u_i, and the temperature field, θ, the equations of classical linear thermoelasticity for an anisotropic and inhomogeneous body may be written,

$$\begin{aligned} \rho\ddot{u}_i &= (c_{ijkh}u_{k,h})_{,j} - (a_{ij}\theta)_{,j} + \rho f_i\,, \\ c\dot{\theta} &= -a_{ij}\dot{u}_{i,j} + (k_{ik}\theta_{,k})_{,i} + \rho r, \end{aligned} \tag{2.1}$$

where $\dot{\theta} = \theta_{,t}$ and standard indicial notation is used. Here ρ, c, f_i and r are, respectively, the density, density multiplied by the specific heat, externally supplied body force, and external supply of heat. The coefficients $c_{ijkh}(\mathbf{x}, t)$ are the elastic coefficients, or elasticities, $k_{ij}(\mathbf{x}, t)$ denote the components of the thermal conductivity tensor, and $a_{ij}(\mathbf{x}, t)$ are the components of a coupling tensor connecting the equations of elasticity to those for heat transport in the solid. We observe immediately that if we set $a_{ij} \equiv 0$, $f_i = 0$ and $r = 0$ then system (2.1) decouples into the two linear equations

$$\rho\ddot{u}_i = (c_{ijkh}u_{k,h})_{,j} \tag{2.2}$$

and

$$c\dot{\theta} = (k_{ik}\theta_{,k})_{,i}\,. \tag{2.3}$$

Equation (2.2) represents the standard equations of linear elasticity which under appropriate conditions on the elasticities c_{ijkh} allow elastic wave propagation and define a hyperbolic system, cf. (Knops and Payne, 1971b), (Knops and Wilkes, 1973). On the other hand, equation (2.3) for $c > 0$ and k_{ik} a positive-definite tensor, is the classical parabolic equation for the diffusion of θ. Thus, θ effectively has an infinite wavespeed as we saw in section 1.2. Thus, for the combined system (2.1) we expect a coupled hyperbolic - parabolic system of partial differential equations with the temperature field diffusing with infinite wavespeed.

(Lord and Shulman, 1967) proposed combining the Cattaneo approach (Maxwell-Cattaneo theory of section 1.2) together with the standard development of elasticity to derive a Cattaneo - type theory of thermoelasticity as we now describe. The approach of (Lord and Shulman, 1967) begins with the full nonlinear equations but they are mainly interested in developing a linear theory since they begin with ... "*small strains and small temperature changes*". With $\epsilon, \eta, t_{ij}, q_i$ and $e_{ij} = (u_{i,j} + u_{j,i})/2$ being the internal energy, entropy, stress tensor, heat flux and strain tensor for the elastic body, respectively, (Lord and Shulman, 1967) write the energy balance law as

$$\rho\theta\dot{\eta} = -q_{i,i}\,, \tag{2.4}$$

where η and ϵ are connected by the equation

$$\rho\dot{\epsilon} = \rho\theta\dot{\eta} + t_{ij}\dot{e}_{ij}\,, \tag{2.5}$$

superposed dot being the partial time derivative, $\partial/\partial t$. (Lord and Shulman, 1967) propose the general anisotropic equation for q_i which generalizes Cattaneo's equation $(1.45)_2$, namely,

$$A_{ij}\dot{q}_j + a\dot{q}_i + q_i = b\theta_{,i} + B_{ij}\theta_{,j}\,, \tag{2.6}$$

where the coefficients A_{ij}, a, b and B_{ij} depend on the material comprising the elastic body. They are principally interested in deriving an isotropic version of their theory and so note that in the isotropic case equation (2.6) may be replaced by

$$\tau\dot{q}_i + q_i = -k\theta_{,i}\,. \tag{2.7}$$

(Lord and Shulman, 1967) call τ the relaxation time, and they say it "*represents the time-lag needed to establish steady - state heat conduction in an element of volume when a temperature gradient is suddenly imposed on that element*".

(Lord and Shulman, 1967) proceed to introduce the Helmholtz free energy function $\psi = \psi(e_{ij}, \theta) = \epsilon - \eta\theta$ and then note

$$\frac{\partial\psi}{\partial t} = \dot{\psi} = \frac{\partial\psi}{\partial e_{ij}}\,\dot{e}_{ij} + \frac{\partial\psi}{\partial\theta}\,\dot{\theta} \tag{2.8}$$

and

$$\dot{\psi} = \dot{\epsilon} - \eta\dot{\theta} - \dot{\eta}\theta\,. \tag{2.9}$$

Equations (2.8) and (2.9) are employed in equation (2.5) to see that

$$\begin{aligned} t_{ij}\dot{e}_{ij} &= \rho(\dot{\epsilon} - \theta\dot{\eta}) \\ &= \rho(\dot{\psi} + \eta\dot{\theta}) \\ &= \rho\Big(\frac{\partial\psi}{\partial e_{ij}}\,\dot{e}_{ij} + \frac{\partial\psi}{\partial\theta}\,\dot{\theta} + \eta\dot{\theta}\Big)\,. \end{aligned} \tag{2.10}$$

From equation (2.10) (Lord and Shulman, 1967) infer that the stress tensor and entropy have the forms

$$\eta = -\frac{\partial\psi}{\partial\theta}\,, \qquad t_{ij} = \rho\frac{\partial\psi}{\partial e_{ij}}\,. \tag{2.11}$$

(Lord and Shulman, 1967) then employ the relation $(2.11)_1$ in the energy balance law (2.4) to derive the equation

$$\rho\theta\Big(\frac{\partial^2\psi}{\partial e_{ij}\partial\theta}\,\dot{e}_{ij} + \frac{\partial^2\psi}{\partial\theta^2}\,\dot{\theta}\Big) = q_{i,i}\,. \tag{2.12}$$

Let us observe that equations (2.7) and (2.12) (with replacement of appropriate time derivatives) could form the basis for a nonlinear Cattaneo -

Lord - Shulman theory. (Lord and Shulman, 1967) do not pursue this line and proceed to combine equations (2.7) and (2.12) linearizing in the process. In this way they derive the *linearized* energy balance law

$$-\rho\theta\psi_{\theta\theta}(\dot{\theta}+\tau\ddot{\theta})-\rho\theta\,\frac{\partial^2\psi}{\partial e_{ij}\partial\theta}\,(\dot{e}_{ij}+\tau\ddot{e}_{ij})=k\Delta\theta. \qquad (2.13)$$

(Lord and Shulman, 1967) then proceed to develop their theory in the isotropic case and expand about a constant temperature θ_0 and expand in terms of the strain invariants of elasticity theory. In this way they produce their famous system of equations for isotropic thermoelasticity. Their equation for the displacement u_i is the isotropic equivalent of equation $(2.1)_1$ and coupled to the isotropic equation which arises from (2.13) the Lord-Shulman system of equations is

$$\begin{aligned}&\rho\ddot{u}_i=(\lambda+\mu)u_{j,ij}+\mu\Delta u_i-(3\lambda+2\mu)\alpha\theta_{,i}\,,\\&\rho c(\tau\ddot{\theta}+\dot{\theta})+(3\lambda+2\mu)\alpha\theta_0(\tau\ddot{e}_{rr}+\dot{e}_{rr})=k\Delta\theta\,.\end{aligned} \qquad (2.14)$$

In equations (2.14), c is the specific heat and λ,μ are the coefficients which arise in isotropic elasticity, the Lamé moduli, the connection with the elastic coefficients c_{ijkh} being

$$c_{ijkh}=\lambda\delta_{ij}\delta_{kh}+\mu(\delta_{ik}\delta_{jh}+\delta_{ih}\delta_{jk})\,.$$

(Lord and Shulman, 1967) write their equations in non-dimensional form and then solve a one-dimensional problem which corresponds to zero initial conditions in a half space with the velocity $\partial u/\partial t$ experiencing a discontinuous input at time $t=0$ along the half space boundary, i.e. a displacement shock problem.

2.1.2 Cattaneo-Fox theory

The first development of a fully nonlinear thermoelastic theory which employs a Cattaneo equation for the heat flux is that of (Fox, 1969a). Fox begins with the momentum and continuity equations written in the *current* configuration as

$$\begin{aligned}&\rho\dot{v}_i=t_{ji,j}+\rho f_i\,,\\&\dot{\rho}+\rho d_{rr}=0,\end{aligned} \qquad (2.15)$$

where t_{ij} is the symmetric Cauchy stress tensor, a superposed dot denotes the *material* derivative, e.g. $\dot{\rho}=\partial\rho/\partial t+v_i\partial\rho/\partial x_i$, f_i is the externally supplied body force, and $d_{ij}=(v_{i,j}+v_{j,i})/2$, $v_i(\mathbf{x},t)$ being the velocity in the current reference frame. (Fox, 1969a) begins with a balance of energy equation and an entropy inequality postulated for arbitrary sub-bodies of an elastic body, and reduces these to pointwise form. In terms of the internal energy ϵ, entropy η, heat flux q_i, and temperature θ these are

$$\begin{aligned} \rho\dot{\epsilon} - \rho r + q_{i,i} - t_{ij}d_{ij} &= 0, \\ \rho\theta\dot{\eta} - \rho r + q_{i,i} - \frac{q_i\theta_{,i}}{\theta} &\geq 0, \end{aligned} \tag{2.16}$$

where r is the externally supplied source of heat. The entropy inequality $(2.16)_2$ is rewritten in terms of the Helmholtz free energy function $\psi = \epsilon - \eta\theta$ as

$$\rho\dot{\psi} + \rho\eta\dot{\theta} - t_{ij}d_{ij} + \frac{q_i\theta_{,i}}{\theta} \leq 0. \tag{2.17}$$

The constitutive theory of (Fox, 1969a) requires that

$$\psi, \eta, t_{ij}$$

depend on the independent variables

$$F_{iA} = \frac{\partial x_i}{\partial X_A} = x_{i,A} \qquad \text{and} \qquad \theta,$$

where $x_i = x_i(\mathbf{X}, t)$ is the mapping of points in the reference configuration to equivalent points in the current configuration. Upon introducing the right Cauchy - Green tensor $\mathbf{C} = \mathbf{F}^T\mathbf{F}$ (i.e. $C_{AB} = x_{i,A}x_{i,B}$) (Fox, 1969a) notes

$$\dot{C}_{AB} = 2d_{ij}x_{i,A}x_{j,B}$$

and rewrites inequality (2.17) in the form

$$-\rho\Big(\frac{\partial\psi}{\partial\theta} + \eta\Big)\dot{\theta} + \Big(t_{ij} - 2\rho\frac{\partial\psi}{\partial C_{AB}}\, x_{i,A}x_{j,B}\Big)d_{ij} - \frac{q_i\theta_{,i}}{\theta} \geq 0. \tag{2.18}$$

Using the fact that r and f_i may be selected at will it is now deduced from (2.18) that

$$\eta = -\frac{\partial\psi}{\partial\theta} \qquad \text{and} \qquad t_{ij} = 2\rho x_{i,A}x_{j,B}\,\frac{\partial\psi}{\partial C_{AB}}\,. \tag{2.19}$$

The residual of the entropy inequality (2.18) is

$$-q_i\theta_{,i} \geq 0, \tag{2.20}$$

and the energy balance law becomes

$$\theta\dot{\eta} = -q_{i,i} + \rho r. \tag{2.21}$$

(Fox, 1969a) uses superposed rigid body arguments and requests that the nonlinear time derivative of q_i in a Cattaneo law should be an objective derivative. This leads him to propose the general equation generalizing Cattaneo's one,

$$\dot{q}_i - \omega_{ij}q_j = \alpha q_i + \beta\theta_{,i}\,, \tag{2.22}$$

where $\omega_{ij} = (v_{i,j} - v_{j,i})/2$, $\dot{q}_i = q_{i,t} + v_k q_{i,k}$, and α, β depend on θ and the scalar invariants $q_i q_i, q_i \theta_{,i}$ and $\theta_{,i}\theta_{,i}$. (Fox, 1969a) specializes to the case where α and β are constants and introduces constants τ and κ by $\tau = -1/\alpha$, $\kappa = \beta/\alpha$ so that his equation (2.22) becomes

$$\tau(q_{i,t} + v_j q_{i,j} - \omega_{ij} q_j) = -q_i - \kappa\theta_{,i} \,. \tag{2.23}$$

Thus, the full nonlinear system of equations derived by (Fox, 1969a) to describe motion in a thermoelastic body generalizing the (Lord and Shulman, 1967) approach comprise equations (2.15), (2.21) and (2.23).

For easy reference we collect these here recalling the forms for η and t_{ij} given in equations (2.19),

$$\begin{aligned} &\dot{\rho} + \rho d_{ii} = 0, \\ &\rho \dot{v}_i = 2\frac{\partial}{\partial x_j}\Big(\rho x_{i,A} x_{j,B} \frac{\partial \psi}{\partial C_{AB}}\Big) + \rho f_i \,, \\ &-\theta \frac{d}{dt}\Big(\frac{\partial \psi}{\partial \theta}\Big) = -\frac{\partial q_i}{\partial x_i} + \rho r, \\ &\tau(\dot{q}_i - \omega_{ij} q_j) = -q_i - \kappa \theta_{,i} \,, \end{aligned} \tag{2.24}$$

where d/dt denotes the material derivative.

I am not aware of further use of the nonlinear system (2.24) apart from the solutions derived by (Fox, 1969a) himself. However, (Fox, 1969a) deserves full credit for producing a nonlinear invariant system of thermoelastic equations using a Cattaneo theory. The solutions given by (Fox, 1969a) involve a static deformation where he shows the heat flux decays exponentially in time, and one where the deformation is

$$x = 2ktY, \; y = Y, \; z = Z.$$

For this definition he solves his equation for q_i, $(2.24)_4$, exactly.

2.1.3 Hidden variables

(Caviglia et al., 1992) begin with the idea of introducing an internal vector variable ξ_i; an internal variable is sometimes also referred to as a hidden variable, and an extensive description of such variables may be found in (Maugin, 1990), (Maugin and Muschik, 1994a; Maugin and Muschik, 1994b). The vector $\boldsymbol{\xi}$ refers to a current configuration $\mathcal{R}$ which has deformed from a reference configuration $\mathcal{R}_0$ by a mapping $\mathbf{x} = \mathbf{x}(\mathbf{X}, t)$ or $x_i = x_i(X_A, t)$. They define the Cauchy stress tensor t_{ij}, the second Piola-Kirchoff stress tensor Y_{AB} and the first Piola-Kirchoff stress tensor S_{Ai}, where $\mathbf{Y} = J\mathbf{F}^{-1}\mathbf{t}(\mathbf{F}^{-1})^T$, $\mathbf{F}$ being the deformation gradient defined by $F_{iA} = \partial x_i/\partial X_A$ and $J = \det(F_{iA})$. They also introduce the heat flux q_i, the Helmholtz free energy ψ, the temperature θ, and temperature gradients $g_i = \theta_{,i}$ and $G_A = \theta_{,A}$ where $\theta_{,i} \equiv \partial\theta/\partial x_i$ whereas $\theta_{,A} \equiv \partial\theta/\partial X_A$. In terms of the displacement $u_i = x_i - X_i$, (Caviglia et al., 1992) have the balance of momentum equation

$$\rho_0 \ddot{u}_i = \frac{\partial}{\partial X_A} S_{Ai} + \rho_0 b_i \tag{2.25}$$

where ρ_0 is the density in $\mathcal{R}_0$, b_i is the body force and a superposed dot denotes $\partial/\partial t$ holding $\mathbf{X}$ fixed.

The thermodynamic procedure of (Caviglia et al., 1992) introduces the Cauchy-Green right tensor $\mathbf{C} = \mathbf{F}^T\mathbf{F}$ and requires the equivalent of the internal variable $\boldsymbol{\xi}$ referred to the reference configuration, namely $\boldsymbol{\Xi} = \mathbf{F}^T\boldsymbol{\xi}$. Then, (Caviglia et al., 1992) define their thermoelastic body to be one for which

$$\begin{aligned} \mathbf{t} &= \mathbf{F}\tilde{\mathbf{t}}(\mathbf{C}, \theta, \mathbf{G}, \boldsymbol{\Xi})\mathbf{F}^T\,, \\ \mathbf{q} &= \mathbf{F}\tilde{\mathbf{q}}(\mathbf{C}, \theta, \mathbf{G}, \boldsymbol{\Xi})\,, \\ \psi &= \tilde{\psi}(\mathbf{C}, \theta, \mathbf{G}, \boldsymbol{\Xi})\,, \end{aligned} \tag{2.26}$$

where $\tilde{\mathbf{t}}, \tilde{\mathbf{q}}, \tilde{\psi}$ are the functional forms of the indicated variables. The entropy inequality

$$-\rho_0(\dot{\psi} + \eta\dot{\theta}) + \frac{1}{2}\mathbf{Y}\cdot\dot{\mathbf{C}} - \frac{1}{\theta}\mathbf{Q}\cdot\mathbf{G} \geq 0 \tag{2.27}$$

is posed where η is the entropy. (Caviglia et al., 1992) show that inequality (2.27) leads to the deductions

$$\frac{\partial\tilde{\psi}}{\partial\mathbf{G}} = 0\,, \qquad \eta = -\frac{\partial\tilde{\psi}}{\partial\theta}\,, \qquad \mathbf{Y} = 2\rho_0\frac{\partial\tilde{\psi}}{\partial\mathbf{C}}\,, \tag{2.28}$$

and the residual entropy inequality is

$$\rho_0\theta\frac{\partial\tilde{\psi}}{\partial\boldsymbol{\Xi}}\cdot\dot{\boldsymbol{\Xi}} + \mathbf{Q}\cdot\mathbf{G} \leq 0. \tag{2.29}$$

Then, from the first of (2.28), the Helmholtz free energy function reduces to the form $\psi = \psi(\mathbf{C}, \theta, \boldsymbol{\Xi})$.

For the internal variable $\boldsymbol{\xi}$, (Caviglia et al., 1992) propose that $\boldsymbol{\Xi}$ satisfies an evolution equation of form

$$\dot{\boldsymbol{\Xi}} = -m\mathbf{G} - n\boldsymbol{\Xi} \tag{2.30}$$

where m, n are functions of the variables θ and $\mathbf{C}$ with $n > 0$. Upon employing $\dot{\boldsymbol{\Xi}}$ as given by (2.30) in the inequality (2.29) they deduce that

$$\mathbf{Q} = \rho_0 m\theta\frac{\partial\tilde{\psi}}{\partial\boldsymbol{\Xi}}\,. \tag{2.31}$$

Further, from (2.29), there remains the restriction

$$\boldsymbol{\Xi}\cdot\frac{\partial\tilde{\psi}}{\partial\boldsymbol{\Xi}} \geq 0\,. \tag{2.32}$$

To progress further (Caviglia et al., 1992) require that under stationary conditions $\mathbf{Q}$ satisfies Fourier's law so that

$$\mathbf{Q} = -\mathbf{K}\mathbf{G}$$

for $\mathbf{K}$ a positive-definite tensor which depends on θ and $\mathbf{C}$. Under stationary conditions equation (2.30) yields the connection

$$\mathbf{G} = -\frac{n}{m}\boldsymbol{\Xi}$$

and the last two relations lead (Caviglia et al., 1992) to propose the relationship

$$\mathbf{Q} = \frac{n}{m}\mathbf{K}\boldsymbol{\Xi}\,. \tag{2.33}$$

Then, from (2.31) they deduce that

$$\frac{\partial\tilde{\psi}}{\partial\boldsymbol{\Xi}} = \frac{n}{\rho_0\theta m^2}\mathbf{K}\boldsymbol{\Xi}$$

whence

$$\psi = \hat{\psi}(\theta, \mathbf{C}) + \frac{n}{2\rho_0\theta m^2}\,\Xi_A K_{AB}\Xi_B\,, \tag{2.34}$$

where $\hat{\psi}$ denotes a functional relationship of the indicated variables. Upon introducing the internal energy $\epsilon = \psi + \theta\eta$ one then sees that

$$\epsilon = \hat{\psi} - \theta\frac{\partial\hat{\psi}}{\partial\theta} + \frac{1}{2}\left(\frac{nK_{AB}}{\rho_0\theta m^2} - \theta\frac{\partial}{\partial\theta}\left[\frac{nK_{AB}}{\rho_0\theta m^2}\right]\right)\Xi_A\Xi_B\,. \tag{2.35}$$

(Caviglia et al., 1992) then require that ϵ be independent of $\boldsymbol{\Xi}$ and hence of $\mathbf{Q}$ and so

$$\frac{n}{\rho_o\theta m^2}\mathbf{K} = n\hat{\mathbf{K}}(\mathbf{C})$$

where $\hat{\mathbf{K}}$ denotes the functional form, $\hat{\mathbf{K}}$ also being a positive-definite tensor.

The constitutive theory of (Caviglia et al., 1992) may be summarized as

$$\begin{aligned}
\psi &= \hat{\psi}(\theta, \mathbf{C}) + \frac{\theta}{2}\hat{K}_{AB}\Xi_A\Xi_B\,,\\
Y_{AB} &= 2\rho_0\frac{\partial\tilde{\psi}}{\partial C_{AB}} + \frac{\theta}{2}\frac{\partial\hat{K}_{RS}}{\partial C_{AB}}\Xi_R\Xi_S\,,\\
\eta &= -\frac{\partial\hat{\psi}}{\partial\theta} - \frac{1}{2}\hat{K}_{RS}\Xi_R\Xi_S\,.
\end{aligned} \tag{2.36}$$

The (fully nonlinear) evolution equations for the model then follow from (2.25), the energy balance equation, equation (2.30), and may then be written as,

$$\rho_0 \ddot{u}_i = \frac{\partial}{\partial X_A} S_{Ai} + \rho_0 b_i\,,$$
$$-\rho_0 \theta\Big(\frac{\partial^2 \tilde{\psi}}{\partial \theta^2}\,\dot{\theta} + \frac{\partial^2 \tilde{\psi}}{\partial \theta \partial C_{AB}}\,\dot{C}_{AB}\Big) = -\frac{\partial}{\partial X_A}\big(\rho_0 m \theta^2 \hat{K}_{AB} \Xi_B\big), \tag{2.37}$$
$$\dot{\Xi}_A = -m\theta_{,A} - n\Xi_A\,,$$

where

$$S_{Ai} = \rho_0 \frac{\partial \psi}{\partial F_{iA}} = \rho_0 \frac{\partial \psi}{\partial C_{RS}} \frac{\partial C_{RS}}{\partial F_{iA}} = Y_{RS} \frac{\partial C_{RS}}{\partial F_{iA}}$$

and so

$$S_{Ai} = (\delta_{AR} F_{iS} + \delta_{SA} F_{iR}) Y_{RS} = F_{iS} Y_{AS} + F_{iR} Y_{RA}\,.$$

(Caviglia et al., 1992) then develop a linearized version of their theory. It is, however, important to note that they do this by considering a potentially large deformation from $\mathcal{R}_0$ to $\mathcal{R}$ followed by a "small" deformation to a new current configuration $\mathcal{R}^*$. In this way they are not simply developing a linear theory by suitably restricting $\hat{\psi}$ and $\hat{\mathbf{K}}$, they are producing a linearized theory which allows for linearization about a (nonlinear) state of pre-stress and possibly non-uniform temperature.

Let $\mathbf{X}$ denote the position of a particle in the reference configuration $\mathcal{R}_0$, let $\mathbf{x}$ be its position in $\mathcal{R}$, and let $\mathbf{x}^*$ be the corresponding position in $\mathcal{R}^*$. (Caviglia et al., 1992) assume that in $\mathcal{R}$ the temperature θ is constant so that $\mathbf{G} = \mathbf{0}$ and $\boldsymbol{\Xi} = \mathbf{0}$. The values of these variables in $\mathcal{R}^*$ are denoted by θ^*, $\mathbf{G}^*$ and $\boldsymbol{\Xi}^*$, with $\mathbf{C}$ and $\mathbf{C}^*$ denoting the values of the Cauchy-Green right tensor in $\mathcal{R}$ and $\mathcal{R}^*$. The perturbations to $\mathbf{x}, \theta$ and $\boldsymbol{\Xi}$ in $\mathcal{R}$ are written as $\mathbf{u}, \phi$ and $\boldsymbol{\Lambda}$, i.e.

$$x_i^* = x_i + u_i\,, \qquad \theta^* = \theta + \phi, \qquad \Xi_i^* = \Xi_i + \Lambda_i = \Lambda_i\,.$$

Then, equations (2.37) are linearized keeping only quantities linear in u_i, ϕ, Λ_i and their derivatives, in the equations which result. Full details are given in (Caviglia et al., 1992), we here only record the equations. However, we point out that (Caviglia et al., 1992) take $\mathbf{F} = \mathbf{I}$ so that in $\mathcal{R}$ the right Cauchy-Green tensor satisfies $\mathbf{C} = \mathbf{I}$, where θ is uniform. The pre-stress in $\mathcal{R}$ is maintained through the body force b_i and in equilibrium equation (2.37) is

$$\frac{\partial}{\partial X_A}(F_{iB} Y_{BA}) + \rho_0 b_i = 0. \tag{2.38}$$

The linearization of equations (2.37) relies on the fact that this procedure is performed about the solution of (2.38). It is important to note that the steady state deformation given by (2.38) is, in general, not homogeneous and represents a true nonlinear deformation before linearization.

The linearized equations of (Caviglia et al., 1992) are

$$\begin{aligned} \rho_0 \ddot{u}_i &= \big[(\delta_{ij} t_{kh} + A_{ihjk}) u_{j,k} - B_{ih}\phi\big]_{,h} \,, \\ c\dot{\phi} + B_{ij}\dot{u}_{i,j} &= -\frac{1}{\theta}\, q_{i,i} \,, \\ \tau a_{ij}\dot{q}_i + a_{ij} q_j &= -\phi_{,i} \,, \end{aligned} \tag{2.39}$$

where the coefficients involve quantities evaluated in the configuration $\mathcal{R}$ in which θ is uniform and $\mathbf{C} = \mathbf{I}$. The quantity t_{ij} is the Cauchy pre-stress tensor, and

$$\begin{aligned} A_{ijkh} &= 2\rho_0 \frac{\partial^2 \hat{\psi}}{\partial C_{ij} \partial C_{kh}} \,, \qquad B_{ij} = -2\rho_0 \frac{\partial^2 \hat{\psi}}{\partial \theta \partial C_{ij}} \,, \\ c &= -\rho_0 \frac{\partial^2 \hat{\psi}}{\partial \theta^2} \,, \qquad \tau = \frac{1}{n} \,, \qquad a_{ij} = \frac{n}{\rho_0 m^2 \theta^2} (\hat{\mathbf{K}}^{-1})_{ij} \,. \end{aligned} \tag{2.40}$$

We point out that the coefficients in (2.40) are all evaluated in $\mathcal{R}$.

When the body is isotropic, the coefficients become

$$\begin{aligned} A_{ijkh} &= \lambda \delta_{ij}\delta_{kh} + \mu(\delta_{ik}\delta_{jh} + \delta_{ih}\delta_{jk}), \\ t_{ij} &= \alpha \delta_{ij} \,, \qquad B_{ij} = \beta \delta_{ij} \,, \qquad a_{ij} = \frac{1}{\kappa}\, \delta_{ij} \end{aligned}$$

where κ is a constant and then equations (2.39) become

$$\begin{aligned} \rho_0 \ddot{u}_i &= (\alpha u_{i,j})_{,j} + (\lambda u_{j,j})_{,i} + (\mu u_{i,j})_{,j} + (\mu u_{j,i})_{,j} - (\beta\phi)_{,i} \,, \\ c\dot{\phi} + \beta \dot{u}_{i,i} &= -\frac{1}{\theta}\, q_{i,i} \,, \\ \tau \dot{q}_i + q_i &= -\kappa \phi_{,i} \,. \end{aligned} \tag{2.41}$$

In equations (2.39) and (2.41) the pre-stress is provided by the body force in equation (2.38). We could follow the procedure of (Iesan, 1980; Iesan, 1988) and allow a deformation from $\mathcal{R}_0$ to $\mathcal{R}$ which is induced by non-homogeneous boundary conditions in both x_i and θ. This would lead to coefficients which have pre-stress present due to the deformation but also due to non-uniform temperature in $\mathcal{R}$. The linearized equations which then arise contain extra terms to those in (2.39) and (2.41).

(Chandrasekharaiah, 1998), p. 722, remarks that the linearized theory of (Caviglia et al., 1992) closely resembles the Lord - Shulman theory. We point out that there is a resemblance, but equations (2.39) and (2.41) are different from those of Lord - Shulman. Firstly, in (2.39) the equations are for the anisotropic case. However, importantly both sets of equations (2.39) and (2.41) contain the effects of pre-stress. This is evident in (2.39) via the t_{kh} term but also in the equation for q_i through the coefficient a_{ij} which contains the pre-stress via $\hat{\mathbf{K}}$, see (2.40). In particular, due to the presence of the Cauchy pre-stress t_{ij} it is not true that, in general, the elastic coefficients $c_{ijkh} = \delta_{ij} t_{kh} + A_{ijkh}$ would be sign-definite.

2.2 Green-Lindsay theory

When one develops the classical theory of nonlinear thermoelasticity it is usual to begin with a constitutive assumption which is equivalent to requiring

$$\psi, \eta, q_i \quad \text{and} \quad S_{Ai} \tag{2.42}$$

to depend on the variables

$$X_A, \rho_0, \theta, \theta_{,A} \quad \text{and} \quad e_{AB}\,. \tag{2.43}$$

Here ψ, η are the Helmholtz free energy function and the entropy function, q_i is the heat flux vector and S_{Ai} is the Piola-Kirchoff stress tensor. The independent variables are $\mathbf{X}$, the coordinates of a point in the reference configuration, ρ_0 the density in the reference configuration, the temperature $\theta(\mathbf{X}, t)$, the temperature gradient $\theta_{,A} = \partial\theta/\partial X_A$, and the strain tensor, $e_{AB} = (x_{i,A}x_{i,B} - \delta_{AB})/2$, acting at time t but referred to the reference configuration. The function $\mathbf{x} = \mathbf{x}(\mathbf{X}, t)$ denotes the map defining the deformation (motion) of the elastic body.

The above prescription leads to a coupled set of nonlinear partial differential equations for the displacement $u_i = x_i - X_i$ and the temperature field θ. The balance of momentum equation which results may be thought of as yielding a hyperbolic equation but the corresponding balance of energy equation contains $\partial\theta/\partial t$ as the highest time derivative of θ and is effectively a parabolic equation. Thus, the system may be thought of as one of coupled parabolic - hyperbolic type. This has the undesirable feature that the temperature field essentially travels with infinite speed, cf. section 1.2. (This argument generalizes the analogous one from linear thermoelasticity as explained in section 2.1.1.) An appealing way to overcome this was suggested by (Müller, 1971a). His idea is to include $\dot{\theta}$ in the list of independent constitutive variables in (2.43). He develops a complete theory of thermoelasticity beginning with the balance laws for conservation of mass, momentum, and energy. In the balances of momentum and energy (Müller, 1971a) does not include a body force or external supply of heat. The thermodynamics of (Müller, 1971a) is based on his entropy inequality, (Müller, 1967a),

$$\rho\dot{\eta} + \frac{\partial\Phi_i}{\partial x_i} \geq 0 \tag{2.44}$$

where Φ_i is his entropy flux vector, see (Müller, 1967a). (Müller, 1971a) expands inequality (2.44) using the extended constitutive list, and he then argues that the balance equations which arise must hold in such a way that he is able to deduce relations between the functions $\psi, \eta, \Phi_i, S_{Ai}$, and q_i. In this way (Müller, 1971a) develops a fully nonlinear theory for thermoelasticity which, unlike the classical theory, allows θ to travel with a finite wavespeed. (Müller, 1971a) develops complete expressions for the stress, heat flux, and his entropy flux for an isotropic solid and deduces restrictions in equilibrium. He also shows the heat conduction tensor must be symmetric.

We here describe a theory due to (Green and Lindsay, 1972) which also employs $\dot{\theta}$ in the constitutive list. (Green and Lindsay, 1972) commence with the balance laws of mass, momentum, angular momentum, and energy, which are

$$\begin{aligned} &\rho_0 = \rho \det (x_{i,A}), \\ &\rho_0 \dot{v}_i = \rho_0 F_i + S_{Ai,A}\,, \\ &Y_{AB} = Y_{BA} \qquad \text{where} \qquad S_{Ai} = F_{iR} Y_{RA}\,, \\ &\rho_0 \dot{\epsilon} = \rho_0 r + Y_{AB} \dot{e}_{AB} - Q_{A,A}\,. \end{aligned} \tag{2.45}$$

Here ρ and ρ_0 denote the density in the current and reference configurations, v_i is the velocity, S_{Ai} is the Piola-Kirchoff stress tensor, ϵ the internal energy, Q_A the heat flux vector per unit area in the X_K frame but acting over the corresponding surface at time t, and e_{AB} (defined by (Green and Lindsay, 1972) as $e_{AB} = (x_{i,A}x_{i,B} - \delta_{AB})$) is the strain tensor referred to the reference configuration. The quantities $\mathbf{F}$ and r are an external body force and an external supply of heat, respectively. The Cauchy stress tensor, t_{ij}, (in the current frame) and the equivalent heat flux vector, q_i, are given in terms of Y_{AB} and Q_A as

$$\begin{aligned} (\det x_{r,K})\, t_{ij} &= x_{i,A} x_{j,B} Y_{AB} \\ (\det x_{r,K})\, q_i &= x_{i,A} Q_A\,. \end{aligned}$$

(Green and Lindsay, 1972) employ a general entropy inequality over any sub-body, this being based on the entropy inequality of (Green and Laws, 1972). However, they effectively reduce this to the following pointwise entropy inequality

$$\rho_0 \dot{\eta} - \frac{\rho_0 r}{\phi} + \left(\frac{Q_A}{\phi}\right)_{,A} \geq 0. \tag{2.46}$$

This inequality resembles the Clausius-Duhem inequality but the function ϕ is a generalized temperature which will be specified by constitutive theory. If one introduces the Helmholtz free energy function in terms of the generalized temperature ϕ, i.e.

$$\psi = \epsilon - \eta\phi \tag{2.47}$$

then inequality (2.46) may be rearranged with the aid of the energy conservation equation $(2.45)_4$, noting $\phi > 0$, as

$$-\rho_0(\dot{\psi} + \eta\dot{\phi}) + Y_{AB}\dot{e}_{AB} - \frac{Q_A \phi_{,A}}{\phi} \geq 0. \tag{2.48}$$

(Green and Lindsay, 1972) essentially use as constitutive theory the assertion that

$$\psi, \eta, \phi, Q_A \quad \text{and} \quad Y_{AB} \tag{2.49}$$

depend on the independent variables

$$X_A, \rho_0, \theta, \theta_{,A}, \dot{\theta} \quad \text{and} \quad e_{AB}\,. \tag{2.50}$$

Upon using (2.49) and (2.50) in inequality (2.48) (Green and Lindsay, 1972) deduce that

$$\begin{aligned}
&- \rho_0(\psi_\theta + \eta\phi_\theta)\dot{\theta} - \rho_0(\psi_{\dot{\theta}} + \eta\phi_{\dot{\theta}})\ddot{\theta} - \rho_0(\psi_{\theta_{,A}} + \eta\phi_{\theta_{,A}})\dot{\theta}_{,A}\\
&\quad + \left[Y_{AB} - \frac{\rho_0}{2}\Big(\frac{\partial\psi}{\partial e_{AB}} + \frac{\partial\psi}{\partial e_{BA}}\Big) - \frac{\rho_0}{2}\eta\Big(\frac{\partial\phi}{\partial e_{AB}} + \frac{\partial\phi}{\partial e_{BA}}\Big)\right]\dot{e}_{AB}\\
&\quad - \frac{Q_A}{\phi}\left[\phi_\theta\theta_{,A} + \frac{\partial\phi}{\partial\theta_{,B}}\,\theta_{,BA} + \phi_{\dot{\theta}}\dot{\theta}_{,A}\right.\\
&\quad \left. + \frac{1}{2}\Big(\frac{\partial\phi}{\partial e_{RS}} + \frac{\partial\phi}{\partial e_{SR}}\Big)e_{RS,A} + \frac{\partial\phi}{\partial\rho_0}\,\rho_{,A} + \frac{\partial\phi}{\partial X_A}\right] \geq 0.
\end{aligned} \tag{2.51}$$

(Green and Lindsay, 1972) then argue that $\ddot{\theta}, \dot{\theta}_{,A}\,, \dot{e}_{AB}\,, e_{RS,A}\,, \theta_{,AB}\,, \rho_{0,A}\,,$ may be selected independently in inequality (2.51) balancing the momentum and energy equations $(2.45)_2$ and $(2.45)_4$ by a suitable choice of F_i and r. In this manner they deduce the relations

$$\begin{aligned}
&\eta = -\frac{\partial\psi/\partial\dot{\theta}}{\partial\phi/\partial\dot{\theta}}\,,\\
&Y_{AB} = \frac{\rho_0}{2}\Big(\frac{\partial\psi}{\partial e_{AB}} + \frac{\partial\psi}{\partial e_{BA}}\Big) + \frac{\rho_0}{2}\eta\Big(\frac{\partial\phi}{\partial e_{AB}} + \frac{\partial\phi}{\partial e_{BA}}\Big),\\
&\rho_0\Big(\frac{\partial\psi}{\partial\theta_{,A}} + \eta\frac{\partial\phi}{\partial\theta_{,A}}\Big) + \frac{Q_A}{\phi}\,\phi_{\dot{\theta}} = 0\,,\\
&Q_A\,\frac{\partial\phi}{\partial\theta_{,B}} + Q_B\,\frac{\partial\phi}{\partial\theta_{,A}} = 0\,,\\
&Q_A\,\frac{\partial\phi}{\partial\rho_0} = 0, \qquad Q_A\Big(\frac{\partial\phi}{\partial e_{RS}} + \frac{\partial\phi}{\partial e_{SR}}\Big) = 0\,.
\end{aligned} \tag{2.52}$$

The residual entropy inequality follows from (2.51). However, (Green and Lindsay, 1972) then restrict attention to the case where the reference body is homogeneous (i.e. does not depend on $\mathbf{X}$) and then upon use of $(2.52)_{4,5,6}$ one finds

$$\phi = \phi(\theta, \dot{\theta}). \tag{2.53}$$

The residual entropy inequality then has form

$$-\rho_0\big(\psi_\theta + \eta\phi_\theta\big)\dot{\theta} - \frac{Q_A}{\phi}\,\phi_\theta\theta_{,A} \geq 0. \tag{2.54}$$

Since ρ_0 is non constant one employs (2.53) in equations $(2.52)_{2,3}$ to derive the forms for the stress tensor and heat flux, namely

$$\begin{aligned} Y_{AB} &= \frac{\rho_0}{2}\Big(\frac{\partial\psi}{\partial e_{AB}} + \frac{\partial\psi}{\partial e_{BA}}\Big) \\ Q_A &= -\rho_0\phi\,\frac{\dfrac{\partial\psi}{\partial\theta_{,A}}}{\dfrac{\partial\phi}{\partial\dot\theta}}\,. \end{aligned} \tag{2.55}$$

(Green and Lindsay, 1972) further reduce the energy equation $(2.45)_4$. One may then show that the full nonlinear system of equations for thermoelasticity of (Green and Lindsay, 1972) are given by

$$\begin{aligned} \rho_0\ddot x_i &= \rho_0 F_i + \frac{\partial S_{Ai}}{\partial X_A}\,, \\ \rho_0\phi\dot\eta &= \rho_0 r - \frac{\partial Q_A}{\partial X_A} - \rho_0\big(\psi_\theta + \eta\phi_\theta\big)\dot\theta - \rho_0\psi_{\theta,A}\,\dot\theta_{,A}\,, \end{aligned} \tag{2.56}$$

where S_{Ai} and Q_A are given by equations (2.55) with $S_{Ai} = F_{iR}Y_{RA}$.

A detailed analysis of acceleration waves, including curved waves, for system (2.56) is given by (Lindsay and Straughan, 1979).

(Green and Lindsay, 1972) write down expressions for ψ and ϕ which are quadratic in the variables $\theta, \dot\theta, \theta_{,i}, e_{ij}$ to develop a linearized theory of thermoelasticity from (2.56). They linearize about an initial body with zero stress and heat flux. The complete system of equations for linearized thermoelasticity derived by (Green and Lindsay, 1972) for an anisotropic thermoelastic body then have form

$$\begin{aligned} \rho\ddot u_i &= \rho F_i + (c_{ijkh}u_{k,h})_{,j} + \big[a_{ij}(\theta+\alpha\dot\theta)\big]_{,j}\,, \\ \rho(h\ddot\theta + d\dot\theta - a_{ij}\dot u_{i,j} - b_i\dot\theta_{,i}) &= \frac{\rho r}{\theta_0} + (b_i\dot\theta + k_{ij}\theta_{,j})_{,i}\,. \end{aligned} \tag{2.57}$$

Here u_i is the displacement about a reference state with positions denoted by X_i, ρ is the density, $h, d, b_i, c_{ijkh}, a_{ij}, k_{ij}$ are coefficients which have the symmetries

$$c_{ijkh} = c_{khij} = c_{jikh}\,,\quad a_{ij} = a_{ji}\,,\quad k_{ij} = k_{ji}\,. \tag{2.58}$$

(Green, 1972) has shown that the boundary-initial value problem for (2.57) is unique requiring only symmetry of the elastic coefficients c_{ijkh}. His proof employs a Lagrange identity technique. Uniqueness and continuous dependence on the initial data for a solution to the boundary-initial value problem for (2.57) requiring only symmetry of the elastic coefficients c_{ijkh} was established by (Straughan, 1974). His proof introduced a natural logarithmic convexity functional into thermoelasticity.

A very interesting study comparing the solutions to the equations of classical thermoelasticity, Cattaneo-Lord-Shulman theory, cf. section 2.1.1, and the (Green and Lindsay, 1972) theory is provided by (Jordan and Puri,

2001). These writers investigate the propagation of a thermal pulse in a thermoelastic shell employing each of the linearized equations for the three thermoelastic theories, classical, Lord-Shulman, and Green-Lindsay. Their numerical results are very revealing. They typically demonstrate that the classical theory leads to a smooth pulse while that of Lord-Shulman is less smooth showing discontinuities in derivatives. The theory of (Green and Lindsay, 1972) leads to strong pulse behaviour displaying distinct jumps. For the applications they have in mind, such as to the behaviour of stainless steel run tanks which hold cryogenic liquids for rocket fuel at NASA's John C. Stennis Space Center, the strong pulse solution is definitely of interest.

2.3 Green-Naghdi type II theory

(Green and Naghdi, 1993) adopt a different approach to thermoelasticity to other writers, this approach being based on an extension of the type II theory of heat propagation in a rigid solid developed by (Green and Naghdi, 1991), see section 1.10. The idea is to define a temperature θ, an empirical temperature T, and a thermal displacement α, such that θ depends on T and the properties of the material with $\theta > 0, \partial\theta/\partial T > 0$, and

$$\alpha(\mathbf{X}, t) = \int_{t_0}^{t} T(\mathbf{X}, s)ds + \alpha_0 \,. \tag{2.59}$$

Here t_0 is a "start time" at our disposal and α_0 is the value of α at $t = t_0$. (Although (Green and Naghdi, 1993) define T and θ in this way at the outset they later show that there is no loss in generality if one identifies T with θ.)

As usual, $x_i = x_i(X_A, t)$ denotes the motion of a body with positions $\mathbf{X}$ in the reference configuration, $\mathbf{x}$ being their counterparts in the current configuration. (Green and Naghdi, 1993) observe that

$$\dot{\alpha} = T \tag{2.60}$$

and they introduce the variables β_A and γ_i as

$$\beta_A = \frac{\partial \alpha}{\partial X_A}, \qquad \gamma_i = \frac{\partial T}{\partial x_i} = \dot{\alpha}_{,i} \,. \tag{2.61}$$

The variables $\dot{\boldsymbol{\beta}}$ and $\boldsymbol{\gamma}$ are connected by the equation

$$\dot{\beta}_A = F_{Ai}\gamma_i \,.$$

(Green and Naghdi, 1993) define t_{ij} to be the Cauchy stress tensor, $p_i = q_i/\theta$ to be the entropy flux vector, ψ, η, to be the Helmholtz free energy and entropy, respectively. Their momentum equation is

$$\rho\dot{v}_i = \rho b_i + t_{ji,j} \tag{2.62}$$

where ρ, v_i, b_i are the density, velocity, and body force. They work with an entropy balance *equation* rather than an entropy inequality and this requires them to introduce an intrinsic supply of entropy ξ in order to postulate their entropy balance equation as

$$\rho\dot{\eta} = \rho s + \rho\xi - p_{i,i}\,. \tag{2.63}$$

Here s is the external supply of entropy given by $s = r/\theta$, where r is the external supply of heat. The balance of energy equation employed by (Green and Naghdi, 1993) has form

$$t_{ij}d_{ij} - p_i\theta_{,i} - \rho(\dot{\psi} + \eta\dot{\theta}) - \rho\theta\xi = 0, \tag{2.64}$$

where $d_{ij} = (v_{i,j} + v_{j,i})/2$.

(Green and Naghdi, 1993) define a classical thermoelastic body to be one for which

$$t_{ij}, p_i, \psi, \eta, \theta \quad \text{and} \quad \xi$$

depend on the variables

$$T, \gamma_i = T_{,i} = \dot{\alpha}_{,i}\,, \quad \text{and} \quad F_{iA} = x_{i,A}\,. \tag{2.65}$$

This leads to the usual "hyperbolic-parabolic" system of nonlinear equations of thermoelasticity. The goal of (Green and Naghdi, 1993) is to introduce a new class of thermoelasticity equations by requiring

$$t_{ij}, p_i, \psi, \eta, \theta \quad \text{and} \quad \xi \tag{2.66}$$

to depend on

$$T, \alpha_{,A} \quad \text{and} \quad F_{iA}\,. \tag{2.67}$$

(Green and Naghdi, 1993) call this type of thermoelasticity, thermoelasticity of type II. They remark that ... "*it involves no dissipation of energy*" ... "*is perhaps a more natural candidate for its identification as thermoelasticity than the usual theory*".

(Green and Naghdi, 1993) employ relations (2.66) together with (2.67) in equation (2.64). They show that one may deduce from this the relations

$$\frac{\partial\theta}{\partial\beta_A} = 0, \qquad \frac{\partial\theta}{\partial F_{iA}} = 0 \tag{2.68}$$

whence

$$\theta = \theta(T).$$

They then argue that they may write $T = \theta - \theta_0$ and henceforth replace T by θ in the ensuing development. Thus,

$$\psi = \psi(\theta, \beta_A, F_{iA}) = \psi(\theta, \alpha_{,A}, F_{iA}).$$

They further show that the expanded energy equation leads to the results

$$\eta = -\frac{\partial \psi}{\partial \theta}, \qquad t_{ij} = \rho \frac{\partial \psi}{\partial F_{iA}} F_{Aj}, \qquad p_i = -\rho F_{iA} \frac{\partial \psi}{\partial \alpha_{,A}} \qquad \text{and} \qquad \xi = 0. \tag{2.69}$$

(An equivalent reduction employing the Piola-Kirchoff stress tensor S_{Ai} is given in section 4.4 where the forms more suitable for an acceleration wave analysis are derived.) (Green and Naghdi, 1993) then replace F_{iA} by the right Cauchy-Green tensor $C_{AB} = F_{Ai}F_{iB}$ to deduce

$$t_{ij} = \rho F_{iA} F_{Bj} \left(\frac{\partial \psi}{\partial C_{AB}} + \frac{\partial \psi}{\partial C_{BA}} \right). \tag{2.70}$$

The complete nonlinear equations of thermoelasticity of type II are then given in the current frame by equations (2.62) and (2.63) with η, p_i, ξ and t_{ij} given by (2.69) and (2.70). For ease of reference these are collected here as

$$\begin{aligned} \rho \ddot{x}_i &= \rho b_i + \frac{\partial}{\partial x_j} \left\{ \rho F_{iA} F_{Bj} \left(\frac{\partial \psi}{\partial C_{AB}} + \frac{\partial \psi}{\partial C_{BA}} \right) \right\}, \\ -\rho \frac{d}{dt} \left(\frac{\partial \psi}{\partial \theta} \right) &= \rho s + \frac{\partial}{\partial x_i} \left(\rho F_{iA} \frac{\partial \psi}{\partial \alpha_{,A}} \right), \end{aligned} \tag{2.71}$$

where b_i and s are externally supplied and d/dt denotes the material derivative. Once a prescription of the functional form of $\psi = \psi(\theta, \alpha_{,A}, F_{iA})$ is known, equations (2.71) yield a nonlinear system of partial differential equations for x_i and θ.

Linearized forms of the equations for type II thermoelasticity are given in the isotropic case by (Green and Naghdi, 1993) and in the anisotropic case by (Quintanilla, 1999; Quintanilla, 2002b). In terms of the displacement u_i and temperature perturbation θ these may be written for the isotropic case as

$$\begin{aligned} \rho_0 \ddot{u}_i &= \rho_0 b_i - E_1 \theta_{,i} + \mu \Delta u_i + (\lambda + \mu) u_{j,ij}, \\ c\ddot{\theta} &= \rho_0 r + \kappa \Delta \theta + \theta_0 E_1 \ddot{u}_{i,i}, \end{aligned} \tag{2.72}$$

where $\rho_0, E_1, \kappa, \theta_0$ are constants and μ, λ are the Lamé coefficients. In the anisotropic case for a body with a centre of symmetry the respective linear equations are

$$\begin{aligned} \rho \ddot{u}_i &= (c_{ijkh} u_{k,h})_{,j} - (a_{ij}\theta)_{,j} + \rho f_i, \\ c\ddot{\theta} &= -a_{ij} \ddot{u}_{i,j} + (k_{ij} \theta_{,j})_{,i} + \rho r, \end{aligned} \tag{2.73}$$

where f_i, r are the externally supplied body force and heat supply, ρ, c are positive constants, c_{ijkh} are the elastic coefficients, a_{ij} define a coupling tensor, and k_{ij} defines the thermal conductivity tensor.

A general uniqueness theorem for a solution to equations (2.73) requiring no definiteness of the elastic coefficients c_{ijkh} is given by (Quintanilla and

Straughan, 2000). Their proof relies on a logarithmic convexity argument. These writers also derive a variety of growth estimates for the solution depending on the elastic coefficients and the initial energy, see sections 6.2 and 6.3 of this book. Reciprocal theorems and variational principles for type II linear thermoelasticity are given by (Chirita and Ciarletta, 2010a).

As we mentioned in the introduction, section 1.1, the paper of (Green and Naghdi, 1991), and their companion papers (Green and Naghdi, 1992; Green and Naghdi, 1993) on type II and type III thermoelasticity (discussed in the next section), brought a new way of thinking to the area of heat wave propagation and their articles have influenced many subsequent developments. In fact, work since 1991 in this area has definitely increased as may be witnessed for example from the papers, and the references therein, of (Abd-Alla and Abo-Dahab, 2009), (Alvarez-Ramirez et al., 2006; Alvarez-Ramirez et al., 2008), (Anile and Romano, 2001), (Bargmann et al., 2008b), (Bargmann et al., 2008a), (Brusov et al., 2003), (Buishvili et al., 2002), (Caviglia and Morro, 2005), (Chandrasekharaiah, 1998), (Cai et al., 2006), (Christov and Jordan, 2005), (Christov, 2008), (Cimmelli and Frischmuth, 2007), (Ciancio and Quintanilla, 2007), (De Cicco and Diaco, 2002), (Duhamel, 2001), (Fabrizio et al., 1998), (Fabrizio et al., 2008), (Fichera, 1992), (Green and Naghdi, 1991; Green and Naghdi, 1992; Green and Naghdi, 1993; Green and Naghdi, 1995b; Green and Naghdi, 1995a; Green and Naghdi, 1996), (Han et al., 2006), (Hetnarski and Ignaczak, 1999), (Horgan and Quintanilla, 2005), (Iesan, 2002; Iesan, 2004; Iesan, 2008), (Iesan and Nappa, 2005), (Jaisaardsuetrong and Straughan, 2007), (Johnson et al., 1994), (Jordan and Puri, 2001), (Jou and Criado-Sancho, 1998), (Kalpakides and Maugin, 2004), (Lin and Payne, 2004a), (Linton-Johnson et al., 1994), (Loh et al., 2007), (Messaoudi and Said-Houari, 2008), (Metzler and Compte, 1999), (Meyer, 2006), (Mitra et al., 1995), (Morro, 2006), (Payne and Song, 2002; Payne and Song, 2004b), (Puri and Jordan, 1999b; Puri and Jordan, 1999a; Puri and Jordan, 2004; Puri and Jordan, 2006), (Puri and Kythe, 1997; Puri and Kythe, 1998), (Quintanilla, 2001b; Quintanilla, 2002a; Quintanilla, 2007b), (Quintanilla and Racke, 2003; Quintanilla and Racke, 2006a; Quintanilla and Racke, 2007; Quintanilla and Racke, 2008), (Quintanilla and Straughan, 2000; Quintanilla and Straughan, 2002; Quintanilla and Straughan, 2004; Quintanilla and Straughan, 2005b; Quintanilla and Straughan, 2005a; Quintanilla and Straughan, 2008), (Roy et al., 2009), (Ruggeri, 2001), (Saleh and Al-Nimr, 2008), (Sanderson et al., 1995), (Serdyukov, 2001), (Serdyukov et al., 2003), (Shnaid, 2003), (Straughan, 2004; Straughan, 2008), (Su et al., 2005), (Su and Dai, 2006), (Tzou, 1995b; Tzou, 1995a), (Vadasz, 2005), (Vadasz et al., 2005), (Vedavarz et al., 1992), (Zhang and Zuazua, 2003).

2.4 Green-Naghdi type III theory

The theory of type III thermoelasticity was formulated by (Green and Naghdi, 1992). The development starts very much like that for type II in section 2.3. Hence, the governing equations are (2.62) and (2.63) with the energy balance law (2.64) being used to reduce the constitutive theory. Again, the temperatures θ and T are introduced as is the thermal displacement α. The difference between type II and type III is in the constitutive list (2.67). The theory of type III adds the variable $\dot{\alpha}_{,i} = T_{,i}$ to the list (2.67). Thus, a thermoelastic material of type III is defined as one for which

$$t_{ij}, p_i, \psi, \eta, \theta \quad \text{and} \quad \xi \tag{2.74}$$

depend on

$$T, \alpha_{,A}, \dot{\alpha}_{,i} \quad \text{and} \quad F_{iA}\,. \tag{2.75}$$

(In a sense, type III combines the classical theory with that of type II as the list (2.75) is the union of the lists (2.65) and (2.67).)

(Green and Naghdi, 1992) employ (2.74) and (2.75) in the energy balance equation (2.64). After expanding the derivatives $\dot{\psi}$ and $\dot{\theta}$ in terms of the variables (2.75) the expanded energy equation is reduced. (Green and Naghdi, 1992) deduce that

$$\frac{\partial \theta}{\partial \dot{\alpha}_{,i}} = 0, \qquad \frac{\partial \theta}{\partial \alpha_{,A}} = 0, \qquad \frac{\partial \theta}{\partial F_{iA}} = 0 \qquad \text{and} \qquad \frac{\partial \psi}{\partial \dot{\alpha}_{,i}} = 0\,. \tag{2.76}$$

Thus,

$$\theta = \theta(T)$$

and (Green and Naghdi, 1992) show that T may be replaced by θ. Then, $(2.76)_4$ yields

$$\psi = \psi(\theta, \theta_{,A}, F_{iA}). \tag{2.77}$$

Further, (Green and Naghdi, 1992) show that

$$\eta = -\frac{\partial \psi}{\partial \theta}, \qquad t_{ij} = \rho \frac{\partial \psi}{\partial F_{iA}} F_{Aj} \tag{2.78}$$

but, unlike (2.69) for a type II material they cannot deduce an explicit expression for p_i, nor is ξ zero. Instead, the residual of the energy balance equation yields

$$p_i \dot{\alpha}_{,i} + \rho \frac{\partial \psi}{\partial \alpha_{,A}} F_{Ai} \dot{\alpha}_{,i} + \rho \theta \xi = 0. \tag{2.79}$$

We might think of equation (2.79) as defining the variable ξ.

To complete the theory of a type III thermoelastic material one needs, therefore, to specify the functional form of

$$p_i = p_i(\theta, \alpha_{,A}, \dot{\alpha}_{,i}, F_{iA}), \tag{2.80}$$

or equivalently, one needs to specify the heat flux $q_i = \theta p_i$. Clearly, one can write a general expression for p_i as a function of the vector terms which arise from combinations of $\alpha_{,A}, \dot{\alpha}_{,i}$ and F_{iA}. I am not aware of where this has been done, although (Quintanilla and Straughan, 2004) do study acceleration waves in the complete nonlinear theory.

The nonlinear theory for a thermoelastic body of type III then consists of equations (2.62) and (2.63) combined with (2.78), (2.79) and an explicit representation for p_i from (2.80). The general equations have form

$$\begin{aligned} \rho\ddot{x}_i &= \rho b_i + \frac{\partial}{\partial x_j}\Big(\rho\frac{\partial\psi}{\partial F_{iA}}F_{Aj}\Big)\,, \\ -\rho\frac{d}{dt}\Big(\frac{\partial\psi}{\partial\theta}\Big) &= \rho s - \frac{\partial p_i}{\partial x_i} - \frac{1}{\theta}p_i\dot{\alpha}_{,i} - \frac{\rho}{\theta}\frac{\partial\psi}{\partial\alpha_{,A}}F_{Ai}\dot{\alpha}_{,i}\,. \end{aligned} \tag{2.81}$$

Linearized forms of the equations for type III thermoelasticity are given by (Green and Naghdi, 1992) in the isotropic case and by (Quintanilla, 2001c) in the anisotropic case. In the isotropic case they are

$$\begin{aligned} \rho_0\ddot{u}_i &= \rho_0 b_i - E_1\theta_{,i} + \mu\Delta u_i + (\lambda+\mu)u_{j,ij}\,, \\ \rho_0 c\ddot{\theta} + E_1\theta_0\ddot{u}_i &= \rho_0\dot{r} + \kappa\Delta\dot{\theta} + \kappa^*\Delta\theta\,, \end{aligned} \tag{2.82}$$

where $\rho_0, E_1, c, \kappa, \kappa^*$ are constants, μ, λ are the Lamé constants, and b_i, r are the externally supplied body force and heat supply. In the anisotropic case when the body has a centre of symmetry the relevant equations are

$$\begin{aligned} \rho\ddot{u}_i &= (c_{ijkh}u_{k,h})_{,j} - (a_{ij}\theta)_{,j} + \rho f_i\,, \\ c\ddot{\theta} &= -a_{ij}\ddot{u}_{i,j} + (k_{ij}\theta_{,j})_{,i} + (b_{ij}\dot{\theta}_{,j})_{,i} + \rho r, \end{aligned} \tag{2.83}$$

where ρ, c are positive functions which may depend on $\mathbf{x}$, c_{ijkh} are the elastic coefficients, a_{ij} are coupling coefficients, and k_{ij}, b_{ij} represent the coefficients of thermal tensors. The terms f_i and r represent the body force and heat supply.

A general uniqueness theorem for a solution to (2.83) requiring only symmetry of the elastic coefficients c_{ijkh} is provided by (Quintanilla and Straughan, 2000). Their proof employs a Lagrange identity method, see section 6.4. Non-standard problems for thermoelasticity of type II or type III are considered by (Quintanilla and Straughan, 2005b), see also section 6.6.

2.5 Thermoelasticity with Voids

A class of theories which may be thought of as describing certain properties of porous media were derived by (Nunziato and Cowin, 1979). The key idea is to suppose there is an elastic body which has a distribution of voids throughout. The voids are gaps full of air, water, or some other fluid. This theory provides equations for the displacement of the elastic matrix of the porous medium and the void fraction occupied by the fluid. We believe the voids theory has a large potential, especially in wave propagation problems.

The theory of an elastic body containing voids essentially generalizes the classical theory of nonlinear elasticity by adding a function $\nu(\mathbf{X}, t)$ to describe the void fraction within the body. Here $\mathbf{X}$ denotes a point in the reference configuration of the body. Thus, in addition to the momentum equation for the motion $x_i = x_i(\mathbf{X}, t)$ as time evolves, one needs to prescribe an evolution equation for the void fraction ν. For a non-isothermal situation one also needs an energy balance law which effectively serves to determine the temperature field $T(\mathbf{X}, t)$. The original theory is due to (Nunziato and Cowin, 1979) and the temperature field development was largely due to D. Iesan, see details in chapter 1 of (Iesan, 2004). This theory has much in common with the continuum theory for granular materials, cf. (Massoudi, 2005; Massoudi, 2006a; Massoudi, 2006b).

In this chapter we wish to examine theories of thermoelastic materials containing voids. Such theories are particularly useful to describe nonlinear wave motion and account well for the elastic behaviour of the matrix, being a generalisation of nonlinear elasticity theory. Interestingly, while there are many studies involving the linearised theory of elastic materials with voids, see e.g. (Ciarletta and Iesan, 1993) or (Iesan, 2004), analysis of the fully nonlinear equations is only beginning, see e.g. (Iesan, 2005; Iesan, 2006).

The basic idea of including voids in a continuous body is due to (Goodman and Cowin, 1972), although they developed constitutive theory appropriate to a fluid. This they claim is more appropriate to flow of a granular medium. Acceleration waves in the Goodman-Cowin theory of granular media were studied by (Nunziato and Walsh, 1977; Nunziato and Walsh, 1978). For a reader interested in the theory of voids I would suggest first reading the article of (Goodman and Cowin, 1972), and then progressing to the theory of elastic materials with voids as given by (Nunziato and Cowin, 1979). General descriptions of the theory of elastic materials with voids and various applications are given in the books of (Ciarletta and Iesan, 1993) and (Iesan, 2004). Continuous dependence on the coupling coefficients of the voids theory (a structural stability problem) is studied by (Chirita et al., 2006).

The potential application area for the theory of elastic materials with voids is huge. In particular, wave motion in elastic materials with voids has many applications. (Ciarletta et al., 2007) mention four application areas of immediate interest. To appreciate the potential uses we briefly describe these areas. (Ouellette, 2004) is a beautiful and inspiring article which deals with many applications of acoustic microscopy. We are all aware of optical microscopy, but the potential uses of acoustic microscopy are enormous. (Ouellette, 2004) points out that the presence of voids presents a serious problem for acoustic microscopy, and a study of wave motion in an elastic material with voids is likely to be very helpful here. She observes that, "acoustic microscopy remains a niche technology and is especially sensitive to variations in the elastic properties of semiconductor materials, such as air gaps, known as delaminations or voids ..." In particular, (Ouellette, 2004) draws attention to several novel applications of acoustic microscopy in diagnostic medicine. She notes that one may, "apply a special ultrasound scanner to deliver pathological assessments of skin tumours or lesions, non-invasively," and especially there is, "no need to kill the specimen as is usually needed in optical microscopy." (Diebold, 2005) further emphasizes these and other applications.

Wave motion is important in the production of ceramics, or certainly in ceramic behaviour. (Saggio-Woyansky et al., 1992) observe that porous ceramics are either reticulate or foam and are made up of a porous network which has relatively low mass, low thermal conductivity, and low density, and (Raiser et al., 1994) report experimental results where microcracking along grain boundaries in ceramics is caused by compressive waves. Since reticulate porous ceramics are used for molten metal filters, diesel engine exhaust filters, as catalyst supports, and industrial hot-gas filters, and both reticulate and foam porous ceramics are used as light-structure plates, in gas combustion burners, and in fire - protection and thermal insulation materials, a study of wave motion in such materials is clearly useful.

A further important application area for elastic materials with voids is in the production of building materials such as bricks. Modern buildings are usually made with lighter, thinner bricks, often with many voids in the building materials. In seismic areas lighter materials are necessary and much applied research activity is taking place. However, the use of lighter materials, especially those with voids is creating an environmental problem because noise transmission through such objects is considerably greater. Consequently, there is much applied research ongoing in the area of acoustic materials with voids, cf. (Garai and Pompoli, 2005), (Maysenhölder et al., 2004), (Wilson, 1997), and any theoretical model for acoustic wave propagation in an elastic material with voids which yields useful results is desirable.

2.5.1 Basic theory of elastic materials with voids

To present ideas clearly we begin with the classical theory of thermoelasticity with voids, where the energy balance equation is essentially parabolic, so temperature is not transported as a wave. The balance equations for a continuous body containing voids are given by (Goodman and Cowin, 1972). We use the equations as given by (Nunziato and Cowin, 1979) since these are appropriate for an elastic body.

The key thing is to assume that there is a distribution of voids throughout the body B. If $\gamma(\mathbf{X}, t)$ denotes the density of the elastic matrix, then the mass density $\rho(\mathbf{X},t)$ of B has form

$$\rho = \nu\gamma \tag{2.84}$$

where $0 < \nu \le 1$ is a volume distribution function with $\nu = \nu(\mathbf{X},t)$. Since the density or void distribution in the reference configuration can be different we also have

$$\rho_0 = \nu_0\gamma_0$$

where ρ_0, γ_0, ν_0 are the equivalent functions to ρ, γ, ν, but in the reference configuration.

The first balance law is the balance of mass

$$\rho |\det \mathbf{F}| = \rho_0 \,.$$

With π_{Ai} being the Piola-Kirchoff stress tensor and $F_{iA} = x_{i,A}$ as before, the balance of angular momentum states

$$\boldsymbol{\pi}\mathbf{F}^T = \mathbf{F}\boldsymbol{\pi}^T.$$

The balance of linear momentum has form

$$\rho_0 \ddot{x}_i = \pi_{Ai,A} + \rho_0 f_i, \tag{2.85}$$

f_i being an external body force. The balance law for the voids distribution is

$$\rho_0 k\ddot{\nu} = h_{A,A} + g + \rho_0 \ell, \tag{2.86}$$

where k is an inertia coefficient, h_A is a stress vector, g is an intrinsic body force (giving rise to void creation/extinction inside the body), and ℓ is an external void body force. Actually, (Nunziato and Cowin, 1979) allow the inertia coefficient k to depend on $\mathbf{X}$ and/or t, but, for simplicity, we follow (Goodman and Cowin, 1972) and assume it to be constant.

The energy balance in the body may be expressed as

$$\rho_0 \dot{\epsilon} = \pi_{Ai}\dot{F}_{iA} + h_A \dot{\nu}_{,A} - g\dot{\nu} - q_{A,A} + \rho_0 r, \tag{2.87}$$

where ϵ, q_A and r are, respectively, the internal energy function, the heat flux vector, and the externally supplied heat supply function. To understand equation (2.87) we may integrate it over a fixed body B, integrate by parts, and use the divergence theorem to see that

$$\frac{d}{dt}\int_B \rho_0 \epsilon dV + \int_B (g\dot{\nu} + h_{A,A}\dot{\nu}) dV = \int_B \pi_{Ai}\dot{F}_{iA} dV - \oint_{\partial B} q_A N_A dS + \int_B \rho_0 r dV,$$

where ∂B is the boundary of B. Employing (2.86) with $\ell = 0$ we may rewrite the above as

$$\frac{d}{dt}\int_B (\rho_0 \epsilon + \frac{\rho_0 k}{2}\dot{\nu}^2) dV = \int_B \pi_{Ai}\dot{F}_{iA} dV - \oint_{\partial B} q_A N_A dS + \int_B \rho_0 r dV.$$

In this form we recognise the equation as an energy balance equation with a term added due to the kinetic energy of the voids. In fact, (Iesan, 2004), pp. 3–5, shows how one may begin with a conservation of energy law for an arbitrary sub-body of a continuous medium with voids, and then derive equations (2.85), (2.86) and (2.87) from the initial energy balance equation.

It is usual in continuum thermodynamics to also introduce an entropy inequality. We use the Clausius-Duhem inequality

$$\rho_0\dot{\eta} \geq -\left(\frac{q_A}{\theta}\right)_{,A} + \frac{\rho_0 r}{\theta}, \tag{2.88}$$

where η is the specific entropy function. Observe that the sign of the first term on the right of (2.88) is different from that of (Nunziato and Cowin, 1979). (One could use a more sophisticated entropy inequality where q_A/θ is replaced by a general entropy flux $\mathbf{k}$, as in (Goodman and Cowin, 1972), but the above is sufficient for our purpose.)

2.5.2 Thermodynamic restrictions

We consider an elastic body containing voids to be one which has as constitutive variables the set

$$\Sigma = \{\nu_0, \nu, F_{iA}, \theta, \theta_{,A}, \nu_{,A}\} \tag{2.89}$$

supplemented with $\dot{\nu}$. Thus, the constitutive theory assumes

$$\begin{aligned} \epsilon &= \epsilon(\Sigma, \dot{\nu}), & \pi_{Ai} &= \pi_{Ai}(\Sigma, \dot{\nu}), & q_A &= q_A(\Sigma, \dot{\nu}), \\ \eta &= \eta(\Sigma, \dot{\nu}), & h_A &= h_A(\Sigma, \dot{\nu}), & g &= g(\Sigma, \dot{\nu}). \end{aligned} \tag{2.90}$$

This is different from (Nunziato and Cowin, 1979) who regard η as the independent variable rather than θ and they also assume $q_A = 0$.

To proceed we introduce the Helmholtz free energy function ψ in the manner

$$\epsilon = \psi + \eta\theta. \tag{2.91}$$

Next, (2.87) is employed to remove the terms $-q_{A,A} + \rho_0 r$ from inequality (2.88) and then utilize (2.91) to rewrite (2.88) as

$$-\rho_0(\dot{\psi} + \eta\dot{\theta}) - \frac{q_A\theta_{,A}}{\theta} + \pi_{Ai}\dot{F}_{iA} + h_A\dot{\nu}_{,A} - g\dot{\nu} \geq 0. \tag{2.92}$$

The chain rule is used together with (2.90) to expand $\dot{\psi}$ and then (2.92) may be written as

$$\begin{aligned} &-\left(\rho_0\frac{\partial\psi}{\partial\nu} + g\right)\dot{\nu} - \frac{q_A\theta_{,A}}{\theta} - \left(\rho_0\frac{\partial\psi}{\partial F_{iA}} - \pi_{Ai}\right)\dot{F}_{iA} \\ &-\left(\rho_0\frac{\partial\psi}{\partial\theta} + \rho_0\eta\right)\dot{\theta} - \left(\rho_0\frac{\partial\psi}{\partial\nu_{,A}} - h_A\right)\dot{\nu}_{,A} \\ &-\rho_0\frac{\partial\psi}{\partial\theta_{,A}}\dot{\theta}_{,A} - \rho_0\frac{\partial\psi}{\partial\dot{\nu}}\ddot{\nu} \geq 0. \end{aligned} \tag{2.93}$$

The next step is to observe that $\dot{F}_{iA}, \dot{\theta}, \dot{\theta}_{,A}, \dot{\nu}_{,A}$ and $\ddot{\nu}$ appear linearly in inequality (2.93). We may then follow the procedure of (Coleman and Noll, 1963) and assign an arbitrary value to each of these quantities in turn, balancing equations (2.85), (2.86) and (2.87) by a suitable choice of the externally supplied functions f_i, ℓ and r. We may in this manner violate inequality (2.93) unless the coefficients of $\dot{F}_{iA}, \dot{\theta}, \dot{\theta}_{,A}, \dot{\nu}_{,A}$ and $\ddot{\nu}$ are each identically zero. Hence, we deduce that

$$\psi \neq \psi(\dot{\nu}, \theta_{,A}),$$

$$h_A = \rho_0 \frac{\partial \psi}{\partial \nu_{,A}} \quad \Rightarrow \quad h_A \neq h_A(\dot{\nu}, \theta_{,A}), \tag{2.94}$$

$$\pi_{Ai} = \rho_0 \frac{\partial \psi}{\partial F_{iA}} \quad \Rightarrow \quad \pi_{Ai} \neq \pi_{Ai}(\dot{\nu}, \theta_{,A}), \tag{2.95}$$

$$\eta = -\frac{\partial \psi}{\partial \theta} \quad \Rightarrow \quad \eta \neq \eta(\dot{\nu}, \theta_{,A}),$$

and further

$$\epsilon \neq \epsilon(\dot{\nu}, \theta_{,A}).$$

The residual entropy inequality, left over from (2.93), which must hold for all motions is

$$-\left(\rho_0 \frac{\partial \psi}{\partial \nu} + g\right)\dot{\nu} - \frac{q_A \theta_{,A}}{\theta} \geq 0.$$

Thus, to specify a material for an elastic body containing voids we have to postulate a suitable functional form for $\psi = \psi(\nu_0, \nu, F_{iA}, \theta, \nu_{,A})$. Such a form is usually constructed with the aid of experiments. The functions g and q_A still involve $\dot{\nu}$ and this can lead to behaviour almost viscoelastic-like, see (Nunziato and Cowin, 1979). Other writers, e.g. (Iesan, 2004), (Ciarletta and Iesan, 1993), omit $\dot{\nu}$ from the constitutive list at the outset. In this manner one deduces that g may be given as a derivative of the Helmholtz free energy, (Iesan, 2004), p. 7, although some of the possibly desirable features of viscoelasticity are lost. The wavespeeds of acceleration waves in this case are derived in (Iesan, 2004), (Ciarletta and Iesan, 1993).

2.5.3 Voids and Green - Lindsay thermoelasticity

In this section we consider a theory of voids as developed by (Nunziato and Cowin, 1979) but we allow for the possibility of propagation of a temperature wave, by generalizing the voids theory in the thermodynamic framework of (Green and Laws, 1972). In addition to allowing us to explicitly examine the important effects of temperature this allows us to study the propagation of a temperature wave in a porous material. In this section we concentrate on the theory of (Green and Laws, 1972) where a generalized temperature $\phi(\theta, \dot{\theta})$, θ being absolute temperature, is introduced. The theory was originally developed by (Ciarletta and Scarpetta, 1989).

The current literature increasingly recognises the importance thermal waves have in the theory of porous media. A very clever way to dry a saturated porous material via second sound is due to (Meyer, 2006) and (Johnson et al., 1994) show how second sound may be employed to calculate physical properties of water saturated porous media. Both of these cover highly important and useful topics. (Kaminski, 1990) reports experimental results for materials with non-homogeneous inner structures which indicate relaxation times of order 11 – 54 seconds rather than order picoseconds as was previously thought. There is evidence that second sound may be a key mechanism for heat transfer in some biological tissues as the experiments of (Mitra et al., 1995) and the work of (Vedavarz et al., 1992) indicate. Thus, we believe a theory of elastic materials with voids coupled to a suitable thermodynamic theory capable of admitting second sound has a place in modern engineering. One has to be careful how the theory of voids is married to the thermodynamics, however. The incorporation of time derivatives does present a serious problem. The thermodynamics of Green and his co-workers were specifically developed to incorporate into other areas of continuum mechanics and thus we believe these are natural approaches to use.

In this section we describe a thermo-poroacoustic theory which allows for nonlinear elastic effects and for the presence of voids, by using the thermodynamics of (Green and Laws, 1972). This thermodynamics utilises a generalized temperature $\phi(\theta, \dot{\theta})$ rather than just the standard absolute temperature θ.

The starting point is to commence with the standard balance equations for an elastic material containing voids, cf. (Nunziato and Cowin, 1979), or equations (2.85), (2.86), (2.87), and we follow the approach of (Ciarletta and Scarpetta, 1989), see also (Ciarletta and Straughan, 2007b),

$$\rho \ddot{x}_i = \pi_{Ai,A} + \rho F_i, \tag{2.96}$$

$$\rho k \ddot{\nu} = h_{A,A} + g + \rho \ell, \tag{2.97}$$

$$\rho \dot{\epsilon} = -q_{A,A} + \pi_{Ai}\dot{x}_{i,A} + h_A \dot{\nu}_{,A} - g\dot{\nu} + \rho r. \tag{2.98}$$

Here X_A denote reference coordinates, x_i denote spatial coordinates, a superposed dot denotes material time differentiation holding $\mathbf{X}$ fixed, and $_{,A}$ signifies $\partial/\partial X_A$. The variable ρ is the reference density, and we use ρ rather than ρ_0 henceforth, for simplicity. Furthermore, ν is the void fraction, ϵ is the specific internal energy, k is the inertia coefficient, F_i, ℓ and r are externally supplied body force, extrinsic equilibrated body force, and externally supplied heat. The tensor π_{Ai} is the stress per unit area of the X_A−plane in the reference configuration acting over corresponding surfaces at time t (the Piola-Kirchoff stress tensor), q_A is the heat flux vector, and h_A and g are a vector and a scalar function arising in the conservation law for void evolution. (Nunziato and Cowin, 1979) refer to h_A as the equilibrated stress and they call g the intrinsic equilibrated body force.

The thermodynamic development commences with the entropy inequality of (Green and Laws, 1972), and this is

$$\rho\dot{\eta} - \frac{\rho r}{\phi} + \left(\frac{q_A}{\phi}\right)_{,A} \geq 0. \tag{2.99}$$

In this inequality η is the specific entropy and $\phi(>0)$ is a generalised temperature function which reduces to θ in the equilibrium state. Next, introduce the Helmholtz free energy function ψ by $\psi = \epsilon - \eta\phi$ and rewrite inequality (2.99) using the energy equation (2.98) to obtain

$$-\rho\dot{\psi} - \rho\dot{\phi}\eta + \pi_{Ai}\dot{x}_{i,A} - \frac{q_A\phi_{,A}}{\phi} - g\dot{\nu} + h_A\dot{\nu}_{,A} \geq 0. \tag{2.100}$$

Now, we assume that the constitutive functions

$$\psi, \phi, \eta, \pi_{Ai}, q_A, h_A, g \tag{2.101}$$

depend on the variables

$$x_{i,A}, \nu, \nu_{,A}, \theta, \dot{\theta}, \theta_{,A}. \tag{2.102}$$

Note that we do not include $\dot{\nu}$ in the constitutive list and are so effectively following the voids approach of (Iesan, 2004), (Ciarletta and Iesan, 1993). One then expands $\dot{\psi}$ and $\dot{\phi}$ in (2.100) to reduce the constitutive equations. Inequality (2.100) expanded is

$$\begin{aligned}
&\dot{x}_{i,A}\Big(\pi_{Ai} - \rho\frac{\partial\psi}{\partial x_{i,A}} - \rho\eta\frac{\partial\phi}{\partial x_{i,A}}\Big) - \dot{\nu}\Big(\rho\frac{\partial\psi}{\partial\nu} + g + \rho\eta\frac{\partial\phi}{\partial\nu}\Big)\\
&- \dot{\theta}\Big(\rho\frac{\partial\psi}{\partial\theta} + \rho\eta\frac{\partial\phi}{\partial\theta}\Big) - \ddot{\theta}\Big(\rho\frac{\partial\psi}{\partial\dot{\theta}} + \rho\eta\frac{\partial\phi}{\partial\dot{\theta}}\Big)\\
&- \dot{\theta}_{,A}\Big(\rho\frac{\partial\psi}{\partial\theta_{,A}} + \rho\eta\frac{\partial\phi}{\partial\theta_{,A}} + \frac{q_A}{\phi}\frac{\partial\phi}{\partial\dot{\theta}}\Big) - \dot{\nu}_{,A}\Big(\rho\eta\frac{\partial\phi}{\partial\nu_{,A}} + \rho\frac{\partial\psi}{\partial\nu_{,A}} - h_A\Big)\\
&- \frac{q_A}{\phi}x_{i,AB}\frac{\partial\phi}{\partial x_{i,AB}} - \frac{q_A}{\phi}\frac{\partial\phi}{\partial\nu_{,J}}\nu_{,JA} - \frac{q_A}{\phi}\frac{\partial\phi}{\partial\theta_{,J}}\theta_{,JA}\\
&- \frac{q_A}{\phi}\Big(\frac{\partial\phi}{\partial\nu}\nu_{,A} + \frac{\partial\phi}{\partial\theta}\theta_{,A}\Big) \geq 0.
\end{aligned} \tag{2.103}$$

The terms in $x_{i,AB}, \nu_{,JA}$ and $\theta_{,JA}$ appear linearly and so using the fact that ℓ, r and F_i may be selected as we like to balance (2.96) – (2.98), we find

$$\frac{\partial\phi}{\partial x_{i,A}} = 0, \qquad \frac{\partial\phi}{\partial\nu_{,A}} = 0, \qquad \frac{\partial\phi}{\partial\theta_{,A}} = 0. \tag{2.104}$$

Thus

$$\phi = \phi(\theta, \dot{\theta}, \nu). \tag{2.105}$$

It is important to observe that the generalized temperature depends on ν in addition to θ and $\dot{\theta}$. Hence, the void fraction ν directly influences ϕ.

Furthermore, the linearity of $\dot{x}_{i,A}, \dot{\nu}, \ddot{\theta}, \dot{\theta}_{,A}$ and $\dot{\nu}_{,A}$ in (2.103) then allows us to deduce that

$$\begin{aligned} \pi_{Ai} &= \rho\frac{\partial\psi}{\partial x_{i,A}}, & q_A &= -\rho\frac{\partial\psi}{\partial\theta_{,A}}\Big/\frac{1}{\phi}\frac{\partial\phi}{\partial\dot{\theta}}, \\ h_A &= \rho\frac{\partial\psi}{\partial\nu_{,A}}, & g &= -\rho\Big(\frac{\partial\psi}{\partial\nu} + \eta\frac{\partial\phi}{\partial\nu}\Big), \end{aligned} \tag{2.106}$$

and

$$\eta = -\frac{\partial\psi}{\partial\dot{\theta}}\Big/\frac{\partial\phi}{\partial\dot{\theta}}\,. \tag{2.107}$$

The residual entropy inequality which remains from (2.103) after this procedure, has form

$$-\dot{\theta}\Big(\rho\frac{\partial\psi}{\partial\theta} + \rho\eta\frac{\partial\phi}{\partial\theta}\Big) - \frac{q_A}{\phi}\Big(\frac{\partial\phi}{\partial\nu}\nu_{,A} + \frac{\partial\phi}{\partial\theta}\theta_{,A}\Big) \geq 0. \tag{2.108}$$

This inequality places a further restriction on all constitutive equations and motions.

Thus, the complete nonlinear theory of Green - Lindsay thermoelasticity with voids as derived by (Ciarletta and Scarpetta, 1989) consists of equations (2.96) - (2.98) together with the constitutive equations (2.105) - (2.107). One needs functional forms for ψ and ϕ and then $\pi_{Ai}, h_A, g, \epsilon$ and q_A follow and the balance equations (2.96) - (2.98) are, in principle, determinate.

2.5.4 Voids and type II thermoelasticity

In this section we describe the theory of (De Cicco and Diaco, 2002). These writers generalize the thermodynamic procedure of (Green and Naghdi, 1993) and use a thermal displacement variable

$$\alpha = \int_{t_0}^{t} \theta(\mathbf{X}, s)ds + \alpha_0, \tag{2.109}$$

where $\mathbf{X}$ is the spatial coordinate in the reference configuration of the body with θ being the absolute temperature. A general procedure for deriving the equations for a continuous body from a single balance of energy equation is developed by (Green and Naghdi, 1995b). These writers derive the conservation equations for balance of mass, momentum, and entropy. The work of (De Cicco and Diaco, 2002), like that of (Green and Naghdi, 1993) starts with an entropy balance equation. (De Cicco and Diaco, 2002) extend the (Green and Naghdi, 1993) thermoelasticity theory to include voids in the manner of (Nunziato and Cowin, 1979). The full nonlinear equations are derived by (De Cicco and Diaco, 2002), although they only utilize a linearized version. We follow (Ciarletta et al., 2007) and rederive the (De Cicco and Diaco, 2002) theory referring to a reference configuration and employing a first Piola-Kirchoff stress tensor, as opposed to the symmetric stress tensor formulation of (De Cicco and Diaco, 2002).

It is worth observing that (Green and Naghdi, 1993) write, ... "*This type of theory, ... thermoelasticity type II, since it involves no dissipation of energy is perhaps a more natural candidate for its identification as thermoelasticity than the usual theory.*" Moreover, (Green and Naghdi, 1993) observe that, ... "*This suggests that a full thermoelasticity theory - along with the usual mechanical aspects - should more logically include the present type of heat flow (type II) instead of the heat flow by conduction (classical theory, type I).*" (The words in brackets have been added for clarity.) We would argue that it is beneficial to develop a fully nonlinear acceleration wave analysis for a Green - Naghdi type II thermoelastic theory of voids.

The starting point in the development of the theory is to consider the momentum and balance of voids equations for an elastic material containing voids, see (2.85), (2.86),

$$\rho \ddot{x}_i = \pi_{Ai,A} + \rho F_i, \tag{2.110}$$

$$\rho k \ddot{\nu} = h_{A,A} + g + \rho \ell. \tag{2.111}$$

One needs a balance of energy and from (De Cicco and Diaco, 2002) this is

$$\rho \dot{\epsilon} = \pi_{Ai} \dot{x}_{i,A} + h_A \dot{\nu}_{,A} - g\dot{\nu} + \rho s \theta + (\theta \Phi_A)_{,A}. \tag{2.112}$$

In these equations X_A denote reference coordinates, x_i denote spatial coordinates, a superposed dot denotes material time differentiation and $_{,A}$ stands for $\partial/\partial X_A$. The variables ρ, ν, ϵ, k, are the reference density, the void fraction, the specific internal energy, and the inertia coefficient. The terms F_i, ℓ and s denote externally supplied body force, extrinsic equilibrated body force, and externally supplied heat. The tensor π_{Ai} is the stress per unit area of the X_A-plane in the reference configuration acting over corresponding surfaces at time t (the Piola-Kirchoff stress tensor), Φ_A is the entropy flux vector, and h_A and g are a vector and a scalar function arising in the conservation law for void evolution. These are referred to by (Nunziato and Cowin, 1979) as the equilibrated stress and the intrinsic equilibrated body force, respectively.

The next step is to use the entropy balance equation, see (Green and Naghdi, 1993), (De Cicco and Diaco, 2002),

$$\rho \theta \dot{\eta} = \rho \theta s + \rho \theta \xi + (\theta \Phi_A)_{,A} - \Phi_A \theta_{,A} \tag{2.113}$$

where ξ is the internal rate of production of entropy per unit mass, and η, θ are the specific entropy and the absolute temperature. Introduce the Helmholtz free energy function $\psi = \epsilon - \eta\theta$ and then equation (2.112) is rewritten with the aid of (2.113) as

$$\rho \dot{\psi} + \rho \eta \dot{\theta} = \pi_{Ai} \dot{x}_{i,A} + h_A \dot{\nu}_{,A} - g\dot{\nu} + \Phi_A \theta_{,A} - \rho \theta \xi. \tag{2.114}$$

The constitutive theory of (De Cicco and Diaco, 2002) writes the functions

$$\psi, \eta, \pi_{Ai}, \Phi_A, h_A, g, \xi, \tag{2.115}$$

as depending on

$$x_{i,A}, \nu, \nu_{,A}, \dot{\alpha}, \alpha_{,A}. \tag{2.116}$$

The function $\dot{\psi}$ is expanded using the chain rule, and rearranging terms, recollecting $\dot{\alpha} = \theta$, equation (2.114) may be written as

$$\begin{aligned} &\dot{x}_{i,A}\Big(\rho\frac{\partial\psi}{\partial x_{i,A}} - \pi_{Ai}\Big) + \dot{\nu}_{,A}\Big(\rho\frac{\partial\psi}{\partial\nu_{,A}} - h_{Ai}\Big) + \dot{\alpha}_{,A}\Big(\rho\frac{\partial\psi}{\partial\alpha_{,A}} - \Phi_A\Big) \\ &\quad + \rho\ddot{\alpha}\Big(\frac{\partial\psi}{\partial\dot{\alpha}} - \eta\Big) + \dot{\nu}\Big(\rho\frac{\partial\psi}{\partial\nu} + g\Big) + \rho\theta\xi = 0. \end{aligned} \tag{2.117}$$

We now use the fact that $\dot{x}_{i,A}, \dot{\nu}_{,A}, \dot{\alpha}_{,A}, \ddot{\alpha}$ and $\dot{\nu}$ appear linearly in (2.117) and so one derives the forms, cf. (De Cicco and Diaco, 2002), equations (19),

$$\begin{aligned} &\pi_{Ai} = \rho\frac{\partial\psi}{\partial x_{i,A}}, \qquad \Phi_A = \rho\frac{\partial\psi}{\partial\alpha_{,A}}, \qquad h_A = \rho\frac{\partial\psi}{\partial\nu_{,A}}, \\ &g = -\rho\frac{\partial\psi}{\partial\nu}, \qquad \eta = -\frac{\partial\psi}{\partial\theta} = -\frac{\partial\psi}{\partial\dot{\alpha}}, \qquad \xi = 0. \end{aligned} \tag{2.118}$$

A theory of type II thermoelasticity containing voids is then given by equations (2.110) - (2.112) with the constitutive theory prescribed by equations (2.118).

2.5.5 Voids and type III thermoelasticity

As we have seen in section 2.5.4, (De Cicco and Diaco, 2002) have developed a theory of thermoelasticity with voids which is a generalization of the dissipationless theory of thermoelasticity of (Green and Naghdi, 1993). The latter writers refer to this as thermoelasticity of type II, type I being the classical theory where the equation governing the temperature field is effectively parabolic as opposed to hyperbolic in type II theory. The theory of a thermoelastic body with voids corresponding to type I thermoelasticity was developed by D. Iesan, see e.g. (Iesan, 2004). However, as shown in section 2.4 (Green and Naghdi, 1992) have developed a further theory of thermoelasticity which employs the thermal displacement variable α and the thermodynamics of (Green and Naghdi, 1991; Green and Naghdi, 1995b). This theory leads to what is essentially a second order in time equation for the thermal displacement field, but differently from the type II theory of (Green and Naghdi, 1993) the theory of (Green and Naghdi, 1992) does have damping and hence dissipation. (Green and Naghdi, 1991; Green and Naghdi, 1992) refer to this theory as being of type III, cf. section 2.4.

The goal of this section is to develop a type III theory of thermoelasticity, but allowing for the accommodation of a distribution of voids throughout the body. The essential difference between type II and type III thermoelasticity is that the variable $\dot{\alpha}_{,A}$ is added to the constitutive list (2.116), whereas it is absent in section 2.5.4, cf. section 2.4. The presentation follows (Straughan, 2008), chapter 7.

We commence with the balance laws for a thermoelastic body with voids, equations (2.85), (2.86) and (2.87). With ρ denoting the density in the reference configuration and referring everything to this configuration, we have the equation of momentum balance

$$\rho \ddot{x}_i = \pi_{Ai,A} + \rho f_i. \tag{2.119}$$

The equation of voids distribution is

$$\rho k \ddot{\nu} = h_{A,A} + g + \rho \ell. \tag{2.120}$$

The equation of energy balance is

$$\rho \dot{\epsilon} = \pi_{Ai} \dot{x}_{i,A} + h_A \dot{\nu}_{,A} - g \dot{\nu} + \rho s \theta - (\theta p_A)_{,A}. \tag{2.121}$$

We let s be the heat supply and $p_A = q_A/\theta$ is the entropy flux vector. We choose this representation to keep in line with (Green and Naghdi, 1991; Green and Naghdi, 1992), and observe that $p_A = -\Phi_A$ where Φ_A is the entropy flux vector of (De Cicco and Diaco, 2002). We follow (Green and Naghdi, 1992) and postulate an entropy balance equation

$$\rho \dot{\eta} = \rho s + \rho \xi - p_{A,A}, \tag{2.122}$$

where ξ is the internal rate of production of entropy per unit mass. The variable θ is the absolute temperature and $\alpha(\mathbf{X}, t)$ is the thermal displacement.

We next introduce the Helmholtz free energy function ψ in terms of the internal energy ϵ, entropy η and temperature θ, by $\psi = \epsilon - \eta\theta$. Then, from (2.121) and (2.122) it is a straightforward matter to derive the reduced energy equation, cf. (Green and Naghdi, 1992), equation (2.5),

$$\rho \dot{\psi} + \rho \eta \dot{\theta} = \pi_{Ai} \dot{x}_{i,A} + h_A \dot{\nu}_{,A} - g \dot{\nu} - \rho \xi \theta - \theta_{,A} p_A. \tag{2.123}$$

A thermoelastic body of type III which contains a distribution of voids is defined to be one for which the functions

$$\psi, \eta, \pi_{Ai}, p_A, h_A, g \text{ and } \xi \tag{2.124}$$

depend on the independent variables

$$F_{iA} = x_{i,A}, \nu, \nu_{,A}, \dot{\alpha}, \alpha_{,A}, \dot{\alpha}_{,A}\,. \tag{2.125}$$

We do not consider the inhomogeneous situation which would also require inclusion of X_A in the list (2.125), cf. (Iesan, 2004). Observe that we do not include $\dot{\nu}$ in the list (2.125). This follows (Iesan, 2004) and allows us to determine g from ψ.

The procedure now is to expand ψ in terms of the variables in the list (2.125), and recalling $\dot{\alpha} = \theta$, we obtain from (2.123),

$$(\rho\psi_{F_{iA}} - \pi_{Ai})\dot{F}_{iA} + \dot{\nu}(\rho\psi_\nu + g) + \dot{\nu}_{,A}(\rho\psi_{\nu,A} - h_A) \\ + \ddot{\alpha}(\rho\psi_{\dot{\alpha}} + \rho\eta) + \rho\psi_{\dot{\alpha}_{,A}}\ddot{\alpha}_{,A} + \dot{\alpha}_{,A}(p_A + \rho\psi_{\alpha,A}) + \rho\xi\dot{\alpha} = 0. \tag{2.126}$$

We observe that $\dot{F}_{iA}, \dot{\nu}_{,A}, \ddot{\alpha}, \ddot{\alpha}_{,A}, \dot{\nu}$, appear linearly in (2.126). Thus, we may deduce that the coefficients of these terms in (2.126) must be zero. The process is akin to that described in Appendix A of (Green and Naghdi, 1992). Thus, we find that

$$\pi_{Ai} = \rho\psi_{F_{iA}}, \qquad g = -\rho\psi_\nu, \qquad h_A = \rho\psi_{\nu,A}, \\ \eta = -\psi_{\dot{\alpha}}, \qquad \psi \neq \psi(\dot{\alpha}_{,A}). \tag{2.127}$$

Hence, once we prescribe a functional form for the Helmholtz free energy function ψ we also know the stress tensor, entropy, and the voids functions h_A and g. What remains from (2.126) is

$$\rho\xi\dot{\alpha} + \dot{\alpha}_{,A}(\rho\psi_{\alpha,A} + p_A) = 0. \tag{2.128}$$

This leads to further restrictions on constitutive functions. We now also have that

$$\psi = \psi(x_{i,A}, \nu, \nu_{,A}, \dot{\alpha}, \alpha_{,A}), \\ p_A = p_A(x_{i,A}, \nu, \nu_{,A}, \dot{\alpha}, \alpha_{,A}, \dot{\alpha}_{,A}), \\ \xi = \xi(x_{i,A}, \nu, \nu_{,A}, \dot{\alpha}, \alpha_{,A}, \dot{\alpha}_{,A}). \tag{2.129}$$

Thus, once we have a form for the functional dependence of ψ on its variables, and a form for p_A, equations (2.119) - (2.121) yield the complete nonlinear theory of type III thermoelasticity with voids, the function ξ being determined by equation (2.128).

2.5.6 Linear voids type III thermoelasticity

One may study acceleration waves in the nonlinear theory of section 2.5.5. The acceleration waves in this case do not have a separately propagating temperature wave. The reason is that in some sense type III thermoelasticity behaves more like type I thermoelasticity. For acceleration wave motion in thermoelasticity without voids this is explained in detail by (Quintanilla and Straughan, 2004), and a similar explanation holds here. Nevertheless, the extra damping present in the current theory may be useful in practical problems and with this in mind we now develop the equations for a linear theory. Let the body have a centre of symmetry although we allow it to be anisotropic. We denote the displacement in this section as u_i. We then write ψ as a quadratic function of the variables in the list (2.129). Thus,

$$\begin{aligned}\rho\psi =&\frac{1}{2}a_{iAjB}u_{i,A}u_{j,B} - \frac{a_1}{2}\theta^2 - \frac{a_2}{2}\nu^2 + A_{iA}\theta u_{i,A} + B_{iA}\nu u_{i,A} \\ &+ \frac{R_{AB}}{2}\nu_{,A}\nu_{,B} + S_{AB}\nu_{,A}\alpha_{,B} + \frac{T_{AB}}{2}\alpha_{,A}\alpha_{,B}\,,\end{aligned} \tag{2.130}$$

where a_{iAjB}, R_{AB}, T_{AB} have the following symmetries,

$$a_{iAjB} = a_{jBiA}, \qquad R_{AB} = R_{BA}, \qquad T_{AB} = T_{BA}.$$

From (2.127) we now see that

$$\begin{aligned}&\pi_{Ai} = a_{iAjB}u_{j,B} + A_{iA}\theta + B_{iA}\nu, \qquad h_A = R_{AB}\nu_{,B} + S_{AB}\alpha_{,B}, \\ &\rho\eta = a_1\theta - A_{iA}u_{i,A}\,, \qquad g = a_2\nu - B_{iA}u_{i,A}\,.\end{aligned} \tag{2.131}$$

We also write

$$\begin{aligned}\rho\xi &= \phi_1\nu + \phi_2\dot{\alpha}, \\ p_A &= -K_{AB}\nu_{,B} - L_{AB}\alpha_{,B} - M_{AB}\dot{\alpha}_{,B}\,.\end{aligned}$$

From (2.128) one may use the cyclic thermomechanical process argument of (Green and Naghdi, 1991), section 9, to infer that L_{AB}, M_{AB}, R_{AB} are non-negative tensor forms, $\phi_2 \leq 0$, $\phi_1 = 0$, and $S_{AB} = K_{AB}$, $T_{AB} = L_{AB}$.

In this manner, equations (2.119), (2.120) and (2.122) lead to the linear equations

$$\begin{aligned}\rho\ddot{u}_i &= (a_{iAjB}u_{j,B})_{,A} + (A_{iA}\theta)_{,A} + (B_{iA}\nu)_{,A}, \\ \rho k\ddot{\nu} &= (R_{AB}\nu_{,B})_{,A} + (K_{AB}\alpha_{,B})_{,A} + a_2\nu - B_{iA}u_{i,A}\,, \\ a_1\ddot{\alpha} &= A_{iA}\dot{u}_{i,A} + \phi_2\dot{\alpha} + (K_{AB}\nu_{,B})_{,A} + (T_{AB}\alpha_{,B})_{,A} + (M_{AB}\dot{\alpha}_{,B})_{,A}\,.\end{aligned} \tag{2.132}$$

One may study the boundary - initial value problem for (2.132). For example, uniqueness and stability are easily investigated either by using an energy method, or if definiteness of the elastic coefficients a_{iAjB} is not imposed, by a logarithmic convexity argument. For the latter one will be better employing a time integrated version of α as done by (Ames and Straughan, 1992; Ames and Straughan, 1997) and (Quintanilla and Straughan, 2000), these articles following the introduction of this method for the (Green and Laws, 1972), (Green, 1972), version of thermoelasticity in (Straughan, 1974). One may also study one-dimensional waves as in (Green and Naghdi, 1992) and then (2.132) essentially reduce to

$$\begin{aligned}\rho u_{tt} &= au_{xx} + A\theta_x + B\nu_x, \\ \rho k\nu_{tt} &= R\nu_{xx} + K\alpha_{xx} + a_2\nu - Bu_x, \\ a_1\alpha_{tt} &= Au_{tx} + \phi_2\alpha_t + K\nu_{xx} + T\alpha_{xx} + M\alpha_{txx}.\end{aligned} \tag{2.133}$$

The damped character of the temperature wave is evident from (2.133) as is observed in the non voids case by (Green and Naghdi, 1992), page 262. If the displacement and voids effects are absent from $(2.133)_3$, then we see that α satisfies the equation

$$a_1\frac{\partial^2\alpha}{\partial t^2} - M\frac{\partial^3\alpha}{\partial t\partial x^2} = \phi_2\frac{\partial\alpha}{\partial t} + T\frac{\partial^2\alpha}{\partial x^2}.$$

This equation clearly does not permit the possibility of undamped thermal waves, unless $M = \phi_2 = 0$. The damping evident in equations (2.133) may be useful for description of some practical situations.

(Eringen, 1990; Eringen, 2004) develops a voids theory which has a richer structure than the (Nunziato and Cowin, 1979) model. This is achieved by incorporating an equation for the spin at each point of the body. Again, this theory is likely to have rich application in wave propagation problems. (Straughan, 2008) describes this theory in connection with nonlinear wave motion in section 7.6. A general study of singular surface propagation in a continuous body formed of a thermo-microstretch material which has memory is given by (Iesan and Scalia, 2006).

The theory developed by (Eringen, 1990) includes temperature effects while (Eringen, 2004) also includes electromagnetic effects which could be important in wave motion in ceramics, for example. However, we here ignore electromagnetic effects. The basic variables of the theory of (Eringen, 1990; Eringen, 2004) are the displacement u_i, microstretch φ, and the microrotation vector ϕ_i. The microstretch theory of (Eringen, 1990; Eringen, 2004) is based on balance laws for these quantities. These are balance of momentum,

$$\rho_0\ddot{u}_i = \pi_{Ai,A} + \rho_0 f_i \tag{2.134}$$

and balance of microstretch

$$\rho_0\frac{j_0}{2}\ddot{\varphi} = m_{A,A} + T + \rho_0\ell, \tag{2.135}$$

in which we measure quantities in the current configuration but refer back to the reference configuration. Thus, π_{Ai} is a Piola-Kirchoff stress tensor, f_i is a prescribed body force, j_0 is the microinertia, m_A is a microstretch couple, ℓ is a prescribed microstretch source term and T (denoted by $t - s$ in (Eringen, 2004)) is the microstretch stress. Here $,A$ denotes $\partial/\partial X_A$. In addition to equations (2.134) and (2.135), the Eringen theory has a balance of spins equation of form

$$\rho_0 J\ddot{\phi}_i = m_{Ai,A} + \epsilon_{iAj}\pi_{Aj} + \rho_0\ell_i, \tag{2.136}$$

where ℓ_i is an applied body couple density, m_{Ai} is the couple stress tensor, and we have taken the microinertia tensor $J_{ik} = J\delta_{ik}$ for simplicity. The constitutive theory assumes that

$$\pi_{Ai}, m_A, T \text{ and } m_{Ai} \tag{2.137}$$

are functions of the variables

$$F_{iA} = u_{i,A},\ \phi_i,\ \phi_{i,A},\ \varphi \text{ and } \varphi_{,A}. \tag{2.138}$$

In fact, (Eringen, 2004) combines $u_{i,A}$ and ϕ_i into a single strain measure $e_{iA} = u_{i,A} + \epsilon_{Ami}\phi_m$.

(Straughan, 2008) addresses some new questions regarding singular surfaces for the (Eringen, 1990) theory.

A detailed account of many properties of elastic bodies containing voids may also be found in the book by (Iesan, 2004), chapters 1 to 3.

2.6 Generalized thermoelasticity with microstructure

2.6.1 Hetnarski-Ignaczak theory

(Ignaczak, 1990) and (Hetnarski and Ignaczak, 1996; Hetnarski and Ignaczak, 1997; Hetnarski and Ignaczak, 1999) present an interesting thermoelastic theory which is capable of describing soliton - like thermoelastic waves. The wave aspect is further analysed in (Hetnarski and Ignaczak, 2000) where a comparison is made with wave propagation in other thermoelastic models. The model described by (Hetnarski and Ignaczak, 1999) consists of equations for the displacement u_i, temperature θ, and an elastic heat flow field b_i. In the isotropic case these equations are given by (Hetnarski and Ignaczak, 1999) as

$$\begin{aligned} \zeta^2 \ddot{u}_i &= f_i - \epsilon\theta_{,i} + \frac{1}{2(1-\nu)}\, u_{j,ij} + \kappa\Delta u_i\,, \\ \dot{\theta} &= r - \theta\dot{u}_{i,i} + \Delta\theta + \frac{b_i\theta_{,i}}{\theta} - b_{i,i}\,, \\ \omega\dot{b}_i &= -\frac{\theta_{,i}}{\theta}\,, \end{aligned} \tag{2.139}$$

where θ is the absolute temperature, ζ, ϵ are constants, f_i and r are body force and heat supply, ν is Poisson's ratio and $\kappa = (1-2\nu)/(2-2\nu)$. The constant ω is much less than 1 although positive. (Hetnarski and Ignaczak, 1999) show how equations (2.139) lead to soliton - like thermoelastic waves which move with different wavespeeds.

2.6.2 Micropolar, dipolar, affine microstructure

A type II thermoelastic theory incorporating micropolar effects was developed by (Ciarletta, 1999). He concentrates on producing a linear theory. In addition to the type II thermoelasticity theory of section 2.3 (Ciarletta, 1999) introduces a microrotation vector ϕ_i which represents spin at a point.

His basic equations, in the current frame are

$$\begin{aligned}\rho_0\ddot{u}_i &= t_{ji,j} + \rho_0 f_i\,,\\ \rho_0\dot{\eta} &= \rho_0 s + \Phi_{i,i}\,,\\ I_{ij}\ddot{\phi}_j &= m_{ji,j} + \epsilon_{ijk}t_{jk} + \rho_0 g_i\,.\end{aligned} \tag{2.140}$$

Equation $(2.140)_1$ is the balance of linear momentum, ρ_0 being density, u_i displacement, t_{ij} Cauchy stress, and f_i body force. Equation $(2.140)_2$ is the balance of entropy equation, η being entropy, s entropy supply, Φ_i entropy flux, and we observe the intrinsic entropy supply ξ is shown by (Ciarletta, 1999) to be zero. In the equation $(2.140)_3$ I_{ij} represents the coeffcients of inertia, m_{ij} is the couple stress tensor, and g_i is the body couple density. (Ciarletta, 1999) introduces the variables e_{ij} and κ_{ij} by

$$e_{ij} = u_{j,i} + \epsilon_{jik}\phi_k\,, \qquad \kappa_{ij} = \phi_{j,i}\,, \tag{2.141}$$

and he shows the energy balance law may be written as

$$\rho_0\dot{\psi} - t_{ij}\dot{e}_{ij} - m_{ij}\dot{\kappa}_{ij} + \rho_0\eta\dot{\theta} - \Phi_i\theta_{,i} = 0, \tag{2.142}$$

where ψ is the Helmholtz free energy and θ is the temperature.

(Ciarletta, 1999) linearizes about a reference state in which $\theta = T_0, \alpha = \alpha_0$, T_0 and α_0 being constants, where α is the thermal displacement. By introducing a free energy ψ which is quadratic in e_{ij}, κ_{ij}, T and $\tau_{,i}$, where $T = \theta - T_0$, $\tau = \int_{t_0}^t T ds$, he shows the constitutive equations are

$$\begin{aligned}t_{ij} &= A_{ijrs}e_{rs} + B_{ijrs}\kappa_{rs} - D_{ij}T + G_{ijr}\tau_{,r}\,,\\ m_{ij} &= B_{rsij}e_{rs} + C_{ijrs}\kappa_{rs} - E_{ij}T + H_{ijr}\tau_{,r}\,,\\ \rho_0\eta &= D_{ij}e_{ij} + E_{ij}\kappa_{ij} + aT + b_i\tau_{,i}\,,\\ \Phi_i &= G_{rsi}e_{rs} + H_{rsi}\kappa_{rs} - b_iT + K_{ij}\tau_{,j}\,.\end{aligned}$$

(Ciarletta, 1999) principally works with the isotropic theory for a body with a centre of symmetry. For this case he shows the governing evolutionary equations become

$$\begin{aligned}\rho_0\ddot{u}_i &= (\mu+\kappa)\Delta u_i + (\lambda+\mu)u_{j,ji} + \kappa\epsilon_{irs}\phi_{s,r} - mT_{,i} + \rho_0 f_i\,,\\ I\ddot{\phi}_i &= \gamma\Delta\phi_i + (\alpha+\beta)\phi_{j,ji} + \kappa\epsilon_{irs}u_{s,r} - 2\kappa\phi_i + \rho_0 g_i\,,\\ aT_0\ddot{T} &= k\Delta T - mT_0\ddot{u}_{i,i} + \rho_0\dot{s}.\end{aligned} \tag{2.143}$$

(Ciarletta, 1999) solves a problem of a concentrated heat source and proves a continuous dependence result. (Passarella and Zampoli, 2011) derive reciprocal and variational principles.

(Quintanilla, 2002c) develops a theory for thermoelasticity of type II for a body which includes an affine microstructure term x_{iK}. He writes that this determines the homogeneous deformation of the particle with centre of mass at $\mathbf{X}$. He uses the equation of balance of linear momentum,

$$\rho\ddot{x}_i = t_{Ki,K} + \rho f_i\,, \tag{2.144}$$

where t_{Ki} is here the Piola-Kirchoff stress tensor. His balance of entropy is

$$\rho\dot{\eta} = \rho S + \rho\xi + \Phi_{A,A}\,. \tag{2.145}$$

He also needs an equation for micromotion,

$$\rho J_{KL}\ddot{x}_{iL} = S_{LiK,L} - S_{iK} + \rho f_{iK}\,, \tag{2.146}$$

where J_{KL} is an inertia tensor, S_{LiK} is the dipolar stress tensor, S_{iK} is a second order tensor defined below, and f_{iK} is a source term for the micromotion. The energy balance equation is

$$\rho(\dot{\psi} + \dot{\theta}\eta) - t_{Ki}\dot{x}_{i,K} - S_{LiK}\dot{x}_{iK,L} - S_{iK}\dot{x}_{iK} + \rho\theta\xi - \Phi_A\theta_{,A} = 0. \tag{2.147}$$

(Quintanilla, 2002c) postulates constitutive theory that

$$\psi, t_{Kj}, S_{LiK}, S_{iK}, \eta, \Phi_A \quad \text{and} \quad \xi$$

depend on the variables

$$x_{i,K}, x_{iK}, x_{iK,L}, \theta \quad \text{and} \quad \alpha_{,K}\,,$$

α being the thermal displacement. He shows that this leads to

$$\begin{aligned} &t_{Kj} = \rho\frac{\partial\psi}{\partial x_{j,K}}, \qquad S_{Kj} = \rho\frac{\partial\psi}{\partial x_{Kj}}, \qquad S_{KiJ} = \rho\frac{\partial\psi}{\partial x_{iJ,K}}, \\ &\Phi_A = \rho\frac{\partial\psi}{\partial\alpha_{,A}}, \qquad \eta = -\frac{\partial\psi}{\partial\theta} \qquad \text{and} \qquad \xi = 0. \end{aligned} \tag{2.148}$$

Then, a nonlinear theory for thermoelasticity of type II including affine microstructure consists of the differential equations (2.144) - (2.146) together with the constitutive equations (2.148).

(Quintanilla, 2002c) linearizes about a state in which $\alpha = \alpha_0$ and $\theta = T_0$. He puts $T = \theta - T_0$, $u_i = x_i - X_i$, $u_{iA} = x_{iA} - X_{iA}$, and postulates a Helmholtz free energy function ψ which is quadratic. In this way he derives the governing evolution equations

$$\begin{aligned} \rho\ddot{u}_i &= (A_{iJRs}u_{s,R} + B_{iJrS}u_{r,S} - \beta_{Ji}T)_{,J} + \rho f_i\,, \\ \rho J_{KL}\ddot{u}_{iL} &= (E_{KiLSjR}u_{jR,S} + M_{KiLR}\tau_{,R})_{,L} \\ &\qquad - (B_{rSiK}u_{r,S} + C_{SriK}u_{r,S} - \chi_{iK}T) + \rho f_{iK}\,, \\ a\ddot{\tau} &= -\beta_{Ki}\dot{u}_{i,K} - \chi_{iK}\dot{u}_{iK} + M_{LjKI}\,u_{jL,KI} + K_{IJ}\,\tau_{,IJ} + \frac{\rho}{T_0}R\,, \end{aligned} \tag{2.149}$$

where $\tau = \int_{t_0}^{t} T ds$ is a thermal displacement. (Quintanilla, 2002c) introduces an interesting functional to establish uniqueness via logarithmic convexity without assuming definiteness of the elastic coefficients. He also establishes an existence theorem using a semigroup approach.

Thermoelasticity theories based on Green-Naghdi type II and type III thermodynamics are also investigated with internal variables in the interesting article of (Ciancio and Quintanilla, 2007).

2.6.3 Piezoelectricity and thermoelasticity

Piezoelectricity is an interesting phenomenon. It is basically the ability of some materials to generate an electric field or an electric potential when a mechanical stress is applied. Some crystals and especially certain ceramics exhibit piezoelectric behaviour. In this section we briefly describe some work which has developed and employed theories for piezoelectricity in a thermoelastic body when the temperature wave behaviour arises from a Lord-Shulman, Green-Lindsay, or Green-Naghdi type II approach.

Since ceramics are porous materials it makes sense to develop a piezoelectric theory for thermoelasticity which also incorporates porosity. This is what (Ciarletta and Scalia, 1993) did. They derive a thermoeleastic theory which allows the body to have a distribution of voids. Their thermodynamics is based on the (Green and Laws, 1972) and (Green and Lindsay, 1972) θ and $\dot{\theta}$ theory. Let u_i denote the displacement and ν the void fraction. Then (Ciarletta and Scalia, 1993) begin with the balance of linear momentum and balance equation for the voids, i.e.

$$\begin{aligned} \rho_0 \ddot{u}_i &= t_{ji,j} + f_i, \\ \rho_0 \chi \ddot{\nu} &= H_{i,i} + g + \ell\,, \end{aligned} \tag{2.150}$$

where t_{ij}, H_i are the Cauchy stress tensor and the equilibrated stress vector, f_i and ℓ are externally supplied body forces, χ is an inertia coefficient, and g is an intrinsic equilibrated body force. They adopt Maxwell's equations in the form

$$D_{i,i} = f, \qquad E_i = -\phi_{,i}\,, \tag{2.151}$$

where $\mathbf{D}, \mathbf{E}$ are the electric displacement field and the electric field, f is the charge density and ϕ is the electric potential. Their equation of energy balance is

$$\rho_0 \dot{\epsilon} = t_{ij}\dot{e}_{ij} + H_i \dot{\nu}_{,i} - g\dot{\nu} - q_{i,i} + E_i \dot{D}_i + \rho_0 r, \tag{2.152}$$

in which ϵ is the internal energy, $e_{ij} = (u_{i,j} + u_{j,i})/2$, q_i is the heat flux and r is the heat supply.

(Ciarletta and Scalia, 1993) employ the entropy inequality of (Green and Laws, 1972)

$$\rho_0 \dot{\eta} \ge \frac{\rho_0 r}{\phi} - \left(\frac{q_i}{\phi}\right)_{,i},$$

with η being entropy and ϕ a function depending on the constitutive variables. They assume there is a constant temperature T_0 in the reference state and ν_0 is the distribution of ν in that state. They then put $\theta = T - T_0, \zeta = \nu - \nu_0$, and define a generalized Helmholtz free energy of form

$$G = \epsilon - \phi\eta - \frac{1}{\rho_0} D_i E_i\,.$$

(Ciarletta and Scalia, 1993) define a piezoelectric material to be one for which

$$G, t_{ij}, H_i, q_i, g, \eta, D_i \quad \text{and} \quad \phi$$

depend on the variables

$$e_{ij}, \theta, \dot{\theta}, \theta_{,i}\,, E_i, \zeta \quad \text{and} \quad \zeta_{,i}\,.$$

They then exploit the entropy inequality to show that

$$\begin{aligned} t_{ij} &= \rho_0 \frac{\partial G}{\partial e_{ij}}\,, \qquad D_i = -\rho_0 \frac{\partial G}{\partial E_i}\,, \qquad H_i = \rho_0 \frac{\partial G}{\partial \zeta_{,i}}\,, \\ q_i &= \rho_0 \phi \frac{\partial G}{\partial \theta_{,i}} \Big/ \frac{\partial \phi}{\partial \dot{\theta}}\,, \qquad \eta = -\frac{\partial G}{\partial \dot{\theta}} \Big/ \frac{\partial \phi}{\partial \dot{\theta}}\,, \\ g &= -\rho_0 \Big(\frac{\partial G}{\partial \zeta} + \eta \frac{\partial \phi}{\partial \zeta} \Big) \\ \text{and} \quad & \phi = \phi(\zeta, \theta, \dot{\theta}). \end{aligned} \tag{2.153}$$

They assume further that in thermodynamic equilibrium ϕ becomes $T_0 + \theta$, i.e. $\phi(\zeta, \theta, 0) = T_0 + \phi$.

Thus, the full system of nonlinear equations for piezoelectric behaviour in a thermoelastic body as derived by (Ciarletta and Scalia, 1993) are equations (2.150), (2.151), and (2.152) together with (2.153).

(Ciarletta and Scalia, 1993) futher develop a linear version of their theory and establish reciprocity relations and a uniqueness theorem.

The paper of (Iesan, 2008) proceeds along the lines of Green-Naghdi type II thermoelasticity to develop a theory of piezoelectricity in a microstretch continuous body. The idea of microstretch was introduced in section 2.5.6. As (Iesan, 2008) usefully points out a microstretch continuum is a dipolar one with a dipolar displacement u_{ij} where $u_{ij} = \phi\delta_{ij} + \epsilon_{ijk}\phi_k$. Here ϕ is a microstretch function (i.e. a porosity function) while ϕ_i is a microrotation vector. He remarks that ϕ may be thought of as a breathing motion whereas ϕ_i represents a rigid microrotation. He also notes that when ϕ is zero one obtains a Cosserat continuum.

The lucid paper of (Iesan, 2008) employs balance equations for entropy, linear momentum, moment of momentum, energy, microstretch, and Maxwell's equations. The full thermodynamic development is given in (Iesan, 2008). We simply present the relevant equations and constitutive theory. The form of Maxwell's equations are

$$D_{i,i} = f, \qquad E_i = -\psi_{,i}\,, \tag{2.154}$$

where D_i, E_i are the electric displacement field and the electric field, f is the charge density and ψ is the electric potential. The balance of entropy equation is

$$\rho_0 \dot{\eta} = \rho_0 s + \rho_0 \xi + \Phi_{i,i} \tag{2.155}$$

where ρ_0 is density, η entropy, Φ_i entropy flux, s is the external supply of entropy, and ξ is the internal rate of production of entropy. The balance of linear momentum is

$$\rho_0 \ddot{u}_i = \rho_0 f_i + t_{ji,j} \tag{2.156}$$

where u_i is the elastic displacement, f_i is the prescribed body force, and t_{ij} is the Cauchy stress tensor. The balance of moment of momentum equation is

$$I_{ij}\ddot{\phi}_j = \rho_0 g_i + \epsilon_{ijk}t_{jk} + m_{ji,j} \tag{2.157}$$

where I_{ij} is an inertia tensor, g_i is the external body couple, and m_{ij} is the couple stress tensor. Finally the equation for microstretch balance is

$$j_0\ddot{\phi} = \pi_{i,i} + \rho_0\ell - \sigma. \tag{2.158}$$

Here j_0 is a coefficient, π_i is the microstretch stress vector, ℓ is an extrenally supplied microstretch body force and σ is a function defined in terms of the electric enthalpy, see below.

(Iesan, 2008) introduces the electric enthalpy function A by

$$A = \epsilon - \eta\theta - \frac{1}{\rho_0} D_i E_i \tag{2.159}$$

where ϵ is the internal energy. His constitutive theory for a piezoelectric thermoelastic body requires that

$$A, t_{ij}, m_{ij}, \pi_i, \sigma, \eta, \Phi, \xi \qquad \text{and} \qquad D_i$$

depend on the variables

$$e_{ij}, \phi_{j,i}, \phi_{,i}, \phi, \theta \qquad \text{and} \qquad \alpha_{,i}$$

where

$$e_{ij} = u_{j,i} + \epsilon_{jik}\phi_k \qquad \text{and} \qquad \dot{\alpha} = \theta,$$

θ being the temperature. (Iesan, 2008) shows that

$$\begin{aligned}
&m_{ij} = \rho_0\frac{\partial A}{\partial \phi_{j,i}}, \qquad t_{ij} = \rho_0\frac{\partial A}{\partial e_{ij}}, \qquad \Phi_i = \rho_0\frac{\partial A}{\partial \alpha_{,i}},\\
&\eta = -\frac{\partial A}{\partial \theta}, \qquad D_i = -\rho_0\frac{\partial A}{\partial E_i}, \qquad \pi_i = \rho_0\frac{\partial A}{\partial \phi_{,i}},\\
&\sigma = \rho_0\frac{\partial A}{\partial \phi}, \qquad \text{and} \qquad \xi = 0.
\end{aligned} \tag{2.160}$$

The fully nonlinear theory of (Iesan, 2008) then consists of equations (2.154) - (2.158) with the forms (2.160). Once a form for functional dependence of A is prescribed this yields a complete set of equations.

(Iesan, 2008) further develops a linear theory. He linearizes about a reference state in which $\theta = T_0$ and $\alpha = \alpha_0$, T_0 and α_0 being constants. He defines $T = \theta - T_0$ and $\tau = \int_{t_0}^t T ds$ and then proposes a quadratic form for A. The complete form for the functions $t_{ij}, m_{ij}, \pi_i, \sigma, \Phi_i, \eta$ and D_i is then given in the general anisotropic case by (Iesan, 2008) in his equations (2.25) For an isotropic and homogeneous body (Iesan, 2008) develops the linear equations as

$$\begin{aligned}
\rho_0 \ddot{u}_i &= (\mu+\kappa)\Delta u_i + (\lambda+\mu)u_{j,ji} + \kappa\epsilon_{ijk}\phi_{k,j} \\
&\qquad + \lambda_0\phi_{,i} - \beta_0\dot{\tau}_{,i} + \rho_0 f_i \,, \\
I\ddot{\phi}_i &= \gamma\Delta\phi_i + (\alpha+\beta)\phi_{j,ji} + \kappa\epsilon_{ijk}u_{k,j} - 2\kappa\phi_{,i} + \rho_0 g_i \,, \\
j_0\ddot{\phi} &= (a_0\Delta - \xi_0)\phi - \lambda_2\Delta\psi + \nu_1\Delta\tau \\
&\qquad - \lambda_0 u_{j,j} + c_0\dot{\tau} + \rho_0\ell \,, \\
a\ddot{\tau} &= k\Delta\tau + \nu_1\Delta\phi - \nu_3\Delta\psi - \beta_0\dot{u}_{i,i} - c_0\dot{\phi} + \frac{\rho_0}{T_0}S \,, \\
\lambda_2\Delta\phi &+ \chi\Delta\psi + \nu_3\Delta\tau = -f,
\end{aligned} \tag{2.161}$$

where f_i, g_i, ℓ, S are external supplies. (Iesan, 2008) pointedly remarks that equation $(2.161)_5$ generalizes the classical equation $\chi\Delta\psi = -f$ for the electric potential. Here, the λ_2 term represents a porosity effect on the electric potential while the ν_3 term represents a thermal effect.

(Iesan, 2008) establishes a general uniqueness theorem and a continuous dependence result for his linear theory. He also obtains the solution for the problem of a concentrated heat source and for an impulsive body force. He also derives the solution for the problem of a thick-walled spherical shell where the shell surfaces are subject to different but constant pressures.

(Walia et al., 2009) study the propagation of Lamb waves in a transversely isotropic thermoelastic piezoelectric plate which is rotating about an axis orthogonal to the plate. They allow for finite speed thermal wave propagation by using both a Lord-Shulman type theory and a Green-Lindsay one, with the appropriate modifications to account for piezoelectric effects. Many numerical results are presented and their theory is applied specifically to a plate made of PZT-5A piezoelectric thermoelastic material. Other relevant references are provided by (Walia et al., 2009), see also (Ciarletta and Scarpetta, 1996).

2.6.4 Other theories

There are several other theories of thermoelasticity which cater for second sound effects which have been proposed and analysed in the literature. We briefly mention some.

(Iesan and Quintanilla, 2009) develop a type II thermoelasticity theory which includes microstretch effects and also allows for microtemperatures. Within the linearized theory they study uniqueness, existence, and instability of solutions. (Green and Naghdi, 1995c) present a general development of their entropy balance thermodynamics to Cosserat continua, Cosserat surfaces and to Cosserat curves. In (Green and Naghdi, 1995d) they present a similar development for the theory of mixtures of interacting continua. (Caviglia and Morro, 2005) present a general theory for a class of linear thermoviscoelastic materials and study this in detail when there is variation in a particular direction, the z-direction say. They also investigate the energy flux, and problems of reflection and transmission of waves.

Functionally graded elastic bodies are man made and have the property that elastic coefficients or other coefficients are not constant but change continuously throughout in a way that the material is designed for a specific purpose. Within second sound theory functionally graded thermoelastic bodies have been studied by (Ghosh and Kanoria, 2009) and (Mallik and Kanoria, 2007). The work of (Ghosh and Kanoria, 2009) is based on a Green-Lindsay type of thermoelasticity whereas that of (Mallik and Kanoria, 2007) proposes equations based on type II thermoelasticity. The effect of a magnetic field on the response of a thermoelastic body in the context of second sound theories has also been studied. (Aouadi, 2008) studies magnetic field effects within Green-Lindsay thermoelasticity. (Abd-Alla and Abo-Dahab, 2009) investigate a time-dependent problem with a magnetic field in type II thermoelasticity theory. (Sharma and Thakar, 2006) analyse the effect of rotation and a magnetic field for both Lord-Shulman and Green-Lindsay theories of thermoelasticity.

A thermoelasticity theory based on the two temperature approach, see section 1.7, was developed by (Chen et al., 1969). A variety of shock wave problems within the context of this theory were tackled by (Warren and Chen, 1973). (Puri and Jordan, 2006) also present an in-depth study of harmonic waves in the two-temperature thermoelastic theory. They investigate particularly the low and high frequency regimes and present detailed numerical results for both the elastic and temperature waves. Another study of wave propagation in the two temperature thermoelasticity theory is due to (Kumar and Mukhopadhyay, 2010). We also mention the study of (Othman and Singh, 2007) who study a rotating micropolar thermoelastic body. They present solutions for harmonic waves and compare the results within theories of classical thermoelasticity, Lord-Shulman theory, Green-Lindsay theory, type II theory, and a dual phase lag theory.

Analytical results for the solution to thermoelasticity of type III for a beam are given by (Zelati et al., 2010), while (Liu and Quintanilla, 2010a) establish analyticity results for a type III plate. Energy decay in a mixed thermoelastic system of type II and type III is studied by (Liu and Quintanilla, 2010b).

A novel result for a Timoshenko beam system is established by (Sare and Racke, 2009), who show that exponential decay of the solution is to be expected for a Timoshenko system with Fourier's law, but incorporation of a Cattaneo - like heat flux law does not lead to exponential decay.

2.7 Exercises

Exercise 2.7.1 *Consider the boundary - initial value problem, $\mathcal{P}$, for equations (2.132) with u_i, ν and α prescribed on the boundary Γ, of a bounded domain $\Omega \subset \mathbb{R}^3$. Let (u_i^1, ν_1, α_1) and (u_i^2, ν_2, α_2) be solutions to $\mathcal{P}$ for the same boundary and initial data. Write out the boundary initial value problem for the difference solution $u_i = u_i^1 - u_i^2, \nu = \nu_1 - \nu_2, \alpha = \alpha_1 - \alpha_2$ to $\mathcal{P}$. For appropriate symmetry conditions on the coefficients derive the energy equation*

$$\begin{aligned}\frac{d}{dt}\Big[\frac{1}{2}\int_\Omega \rho\dot{u}_i\dot{u}_i\,dx &+ \frac{1}{2}\int_\Omega a_{iAjB}u_{j,B}u_{i,A}\,dx + \frac{1}{2}\int_\Omega \rho k\dot{\nu}^2 dx\\ &+ \frac{1}{2}\int_\Omega R_{AB}\nu_{,A}\nu_{,B}dx + \frac{a_1}{2}\|\theta\|^2 + \frac{1}{2}\int_\Omega T_{AB}\alpha_{,A}\alpha_{,B}dx\\ &+ \int_\Omega K_{AB}\alpha_{,A}\nu_{,B}dx + \int_\Omega B_{iA}u_{i,A}\nu\,dx\Big]\\ &+ \int_\Omega M_{AB}\theta_{,A}\theta_{,B}dx - \phi_2\|\theta\|^2 = 0,\end{aligned} \tag{2.162}$$

where $\|\cdot\|$ is the norm on $L^2(\Omega)$. Use this equation to deduce uniqueness for appropriate signs on and relations between coefficients.

Exercise 2.7.2 *For the Hetnarsky - Ignazcak equations (2.139) with $f_i = 0$ and $r = 0$, show that*

$$\omega b_i\dot{b}_i = \frac{\partial}{\partial t}\frac{\omega}{2}|\mathbf{b}|^2 = -\frac{b_i\theta_{,i}}{\theta}\,.$$

Then show that

$$\dot{\theta} + \frac{\partial}{\partial t}\frac{\omega}{2}|\mathbf{b}|^2 = -\theta\dot{u}_{i,i} + \Delta\theta - b_{i,i}\,.$$

Show further that if Ω is a bounded domain in $\mathbb{R}^3$ with boundary Γ,

$$-\oint_\Gamma n_i\theta_{,i}dS = \omega\oint_\Gamma \theta\dot{b}_i n_i\,dS$$

and so $\partial\theta/\partial n = 0$ on Γ is consistent with $b_i n_i = 0$ on Γ.

Deduce also that with $u_i = 0$ on Γ,

$$\frac{d}{dt}\frac{\zeta^2}{2}\|\dot{\mathbf{u}}\|^2 = -\epsilon\int_\Omega \theta_{,i}\dot{u}_i dx - \frac{d}{dt}\frac{A}{2}\int_\Omega (u_{i,i})^2 dx - \frac{d}{dt}\frac{\kappa}{2}\|\nabla\mathbf{u}\|^2,$$

where $\|\cdot\|$ is the norm on $L^2(\Omega)$. Hence, conclude that with $u_i = 0, b_i n_i = 0$, and $\partial\theta/\partial n = 0$ on Γ,

$$F(t) = \frac{\zeta^2}{2\epsilon}\|\dot{\mathbf{u}}\|^2 + \frac{A}{2\epsilon}\|u_{i,i}\|^2 + \frac{\kappa}{2\epsilon}\|\nabla\mathbf{u}\|^2 + \frac{\omega}{2}\|\mathbf{b}\|^2 + \int_\Omega \theta\,dx$$

satisfies

$$F(t) = F(0) \qquad \text{for all } t > 0.$$

Exercise 2.7.3 *Prove that a solution to the boundary initial value problem* $\mathcal{P}$ *for* (2.143) *is unique.*
Hint. Let (2.143) *be defined on a bounded spatial domain* $\Omega \subset \mathbb{R}^3$, *for* $t > 0$. *Let* Γ *be the boundary of* Ω. *On* Γ *suppose* u_i, ϕ_i *and* T *are given. Also, initial values are given for* $u_i, u_{i,t}, \phi_i, \phi_{i,t}, T$ *and* T_t. *Let* (u_i^1, ϕ_i^1, T^1) *and* (u_i^2, ϕ_i^2, T^2) *be solutions which satisfy* $\mathcal{P}$ *for the same boundary and initial data. Define the difference solution* $u_i = u_i^1 - u_i^2, \phi_i = \phi_i^1 - \phi_i^2, T = T_1 - T_2$. *Integrate in time the equation which arises for* T *and set* $\tau = \int_{t_0}^t T\,ds$. *Show that one may find*

$$\begin{aligned}
\frac{d}{dt}\Bigg[\int_\Omega \frac{\rho_0}{2}\dot{u}_i\dot{u}_i dx &+ \Big(\frac{\mu+\kappa}{2}\Big)\|\nabla\mathbf{u}\|^2 + \Big(\frac{\mu+\lambda}{2}\Big)\|u_{i,i}\|^2 \\
&- \kappa\epsilon_{irs}\int_\Omega u_i\phi_{s,r}dx + \frac{I}{2}\int_\Omega \dot{\phi}_i\dot{\phi}_i dx + \frac{\gamma}{2}\|\nabla\boldsymbol{\phi}\|^2 + \Big(\frac{\alpha+\beta}{2}\Big)\|\phi_{i,i}\|^2 \\
&+ \frac{a}{2}\|T\|^2 + \frac{k}{T_0}\|\nabla\tau\|^2\Bigg] = -2\kappa\|\boldsymbol{\phi}\|^2.
\end{aligned}$$

(Note $\dot{\tau} = T$.*) Hence, deduce uniqueness when* κ *is suitably restricted (a restriction which does follow from thermodynamics).*

3
Interaction with fluids

3.1 Cattaneo theories

Within the field of fluid mechanics modifications of the Navier-Stokes equations to incorporate finite speed heat transport via a Cattaneo - like theory have not been as prevalent as they are in solid mechanics. The earliest approaches to doing this would appear to be those of (Müller, 1967b), of (Fox, 1969b) and of (Carrassi and Morro, 1972). Second sound in fluid mechanics has been known for a long time through heat waves in Helium II below the lambda point of about 2.2°K. (Peshkov, 1944) reports results of experiments on Helium II in which he detects a heat wave. (Peshkov, 1947) further analyses experimental results and relates these to Landau's theory. A review of the physics literature on this subject may be found in (Donnelly, 2009).

(Fox, 1969b) adopts a very general approach at the outset and writes the constitutive theory for the Helmholtz free energy function, ψ, stress tensor, t_{ij}, and entropy, η, as functions of the variables $F_{iA}, \theta, \theta_{,i}$ and q_i, these being the deformation gradient $F_{iA} = \partial x_i / \partial X_A$, temperature θ, and heat flux q_i. He proposes instead of a Fourier law for the heat flux $\mathbf{q}$, a general rate-type equation of form

$$h_k(\mathbf{F}, \theta, \theta_{,i}, q_i, \dot{\mathbf{F}}, \dot{\theta}, \dot{\theta}_{,i}, \dot{q}_i) = 0,$$

where the vector h_k is a linear function in each of the variables $\dot{\mathbf{F}}, \dot{\theta}, \dot{\theta}_{,i}$, and $\dot{q}_i$. In these expressions a superposed dot denotes the material derivative, e.g.

$$\dot{q}_i = \frac{\partial q_i}{\partial t} + v_j \frac{\partial q_i}{\partial x_j}\,.$$

He develops a general theory for what he calls a fluid phase, (Fox, 1969b), section 4. His full theory is totally nonlinear and involves a very general set of equations for a viscous fluid. However, he also develops a reduced theory for an inviscid fluid. (Fox, 1969b) stresses the use of an objective derivative rather than the material derivative $\dot{q}_i$ for the heat flux. His inviscid theory is based on the equations

$$\begin{aligned}
&\dot{\rho} + \rho d_{ii} = 0\,,\\
&\rho \dot{v}_i = \rho b_i - p_{,i}\,,\\
&\rho\theta\dot{\eta} + q_{i,i} = \rho r - 2\rho \frac{\partial \psi}{\partial \xi}\, q_i(\epsilon_1 \theta_{,i} + \epsilon_2 q_i),\\
&\dot{q}_i - \omega_{ij} q_j = \epsilon_1 \theta_{,i} + \epsilon_2 q_i\,,
\end{aligned} \tag{3.1}$$

where $\xi = q_i q_i$, $\omega_{ij} = (v_{i,j} - v_{j,i})/2$, and

$$p = \rho^2 \frac{\partial \psi}{\partial \rho}\,, \qquad \eta = -\frac{\partial \psi}{\partial \theta}\,.$$

The coefficients ϵ_1 and ϵ_2 are, in general, nonlinear functions of the scalar variables $\rho, \theta, \theta_{,i}\theta_{,i}, \xi$, and $\theta_{,i}q_i$. The derivative $\dot{q}_i - \omega_{ij}q_j$ is an objective (Jaumann) derivative. (Fox, 1969b) applies his theory to describe a fountain effect, and shows his theory is consistent with heat travelling as a wave.

(Müller, 1967b) adopts a different approach. He writes equations for $\dot{q}_i$, $\dot{t}_{ij}$ and couples these with the balances of mass, energy and momentum. This is effectively requiring the system of equations to form a hyperbolic system from the outset. The paper of (Müller, 1967b) has been very influential in that he developed the idea of an extended theory of thermodynamics. Theories of extended thermodynamics are described in detail in the books of (Müller and Ruggeri, 1998), (Jou et al., 2010a) and of (Lebon et al., 2008). We do not pursue this here, although the interested reader might wish to consult the article of (Muschik, 2007). For a gas, there is a connection with extended thermodynamics and the early work of (Grad, 1949), based on kinetic theory. We think it is worth drawing attention to the paper of (Truesdell, 1976) who writes, ... "*to claim that the kinetic theory can bear in any way upon the principle of material frame - indifference is presently ridiculous.*" (Truesdell, 1976) also writes, ... "*The kinetic theory of gases provides little support for continuum mechanics except in very special flows,*" and he writes, ... "*He who regards the kinetic theory as providing the one and only right approach to gas flows should discard all of continuum mechanics, not just one or another part of it.*" Whether one regards an equation like $(3.1)_4$ as a balance law or as a constitutive equation is a matter of some controversy in the literature. For the case of a balance law the material derivative, $\dot{q}_i$, is employed. When $(3.1)_4$ is regarded as a constitutive equation then an objective derivative is preferred for $\dot{q}_i$. Fourier's law, $q_i = -k\theta_{,i}$, is a constitutive equation and one viewpoint is to regard equation $(3.1)_4$ as a generalization of Fourier's law. Then, an objective derivative for $\dot{q}_i$ is natural. (Dauby et al., 2002) write, ...*"When the constitutive equations (like* $(3.1)_4$*) are used to describe heat transfer in a moving fluid as in the present work, it is important to recall that objective time derivatives (Jou et al., 2010a) must be introduced instead of the partial time derivatives."* (The words in brackets have been added.) (Carrassi and Morro, 1972) also adopt a different approach. While they are interested in acoustic waves they do develop a general theory for a viscous fluid. They have the standard equations for balance of mass, momentum, and energy, namely

$$\begin{aligned} \dot{\rho} + \rho d_{ii} &= 0, \\ \rho \dot{v}_i &= -p_{,i} + t_{ji,j}\,, \\ \rho \dot{\epsilon} &= -p d_{ii} + t_{ij} d_{ij} - q_{i,i}\,. \end{aligned}$$

However, in addition to adopting a relaxation law for q_i they adopt a similar relation for the (extra) stress tensor t_{ij}. Thus, (Carrassi and Morro, 1972) suggest employing the evolution equations

$$\tau \frac{\partial q_i}{\partial t} + q_i = -k\theta_{,i}\,,$$

and

$$\tau_v \frac{\partial t_{ij}}{\partial t} + t_{ij} = 2\mu d_{ij} + \lambda \delta_{ij} d_{rr}\,.$$

The constant τ_v is a relaxation time for the stress. The paper of (Carrassi and Morro, 1972) then focusses on acoustic waves in some detail.

(Morro, 1980) is also interested in describing wave motion in a heat conducting viscous fluid. His is an inspiring paper which involves the use of hidden variables. (Morro, 1980) uses the balance equations

$$\begin{aligned} \dot{\rho} + \rho d_{ii} &= 0, \\ \rho \dot{v}_i &= t_{ji,j} + \rho b_i\,, \\ \rho \dot{\epsilon} &= \rho r + t_{ij} d_{ij} - q_{i,i}\,. \end{aligned} \tag{3.2}$$

However, he works with hidden variables, and these are the vector, α_i^1, and a tensor, α_{ij}^2, in component form; in direct notation the hidden variables are $\boldsymbol{\alpha}^1$ and $\boldsymbol{\alpha}^2$. (These, in certain cases approach the heat flux and stress tensor, respectively.) The governing equations for $\boldsymbol{\alpha}^1$ and $\boldsymbol{\alpha}^2$ have form

$$\begin{aligned} \tau_1 \dot{\alpha}_i^1 + \alpha_i^1 &= \theta_{,i} \\ \tau_2 \dot{\alpha}_{ij}^2 + \alpha_{ij}^2 &= d_{ij} \end{aligned}$$

for constants $\tau_1, \tau_2 > 0$. (Morro, 1980) shows that thermodynamics requires

$$\eta = -\frac{\partial\psi}{\partial\theta}, \qquad t_{ij} = -\rho^2\psi_\rho\delta_{ij} + \frac{\rho}{\tau_2}\frac{\partial\psi}{\partial\alpha^2_{ij}}, \qquad q_i = -\frac{\rho\theta}{\tau_1}\frac{\partial\psi}{\partial\alpha^1_i},$$

and the free energy must have form

$$\psi = \Psi(\theta,\rho) + \frac{1}{\rho}\left[\frac{\kappa\tau_1}{2\theta}\,\alpha^1_i\alpha^1_i + \mu\tau_2\alpha^2_{ij}\alpha^2_{ij} + \frac{\lambda\tau_2}{2}(\alpha^2_{ii})^2\right].$$

The constitutive theory of (Morro, 1980) then becomes

$$\begin{aligned}
\eta &= -\Psi_\theta + \frac{\kappa\tau_1}{2\rho\theta^2}\,\alpha^1_i\alpha^1_i\,,\\
t_{ij} &= -p\delta_{ij} + 2\mu\alpha^2_{ij} + \lambda\alpha^2_{rr}\delta_{ij}\,,\\
q_i &= -\kappa\alpha^1_i\,.
\end{aligned}$$

(Morro, 1980) shows how one may develop an acceleration wave analysis in detail. It is important that he shows the free energy and the entropy depend on the variable α^1_i which is closely related to the heat flux q_i. (Morro, 1980) also considers objective derivatives for $\boldsymbol{\alpha}^1$ and $\boldsymbol{\alpha}^2$ which are generalizations of those of (Fox, 1969b).

3.1.1 Cattaneo-Fox theory

(Straughan and Franchi, 1984) adopted a specific form of incompressible thermoviscous fluid equations which uses a Boussinesq approximation in the buoyancy term in the momentum equation. They also employed the Jaumann derivative of (Fox, 1969b) for q_i in a Cattaneo model. Thus, the Cattaneo-Fox equations proposed by (Straughan and Franchi, 1984) have form

$$\begin{aligned}
\dot{v}_i &= -\frac{1}{\rho}p_{,i} + k_i g\alpha\theta + \nu\Delta v_i\,,\\
v_{i,i} &= 0,\\
\dot{\theta} &= -q_{i,i}\,,\\
\tau(\dot{q}_i - \epsilon_{ijk}\omega_j q_k) &= -q_i - \kappa\theta_{,i}\,.
\end{aligned}\tag{3.3}$$

Here $\mathbf{k} = (0,0,1)$ and $\boldsymbol{\omega} = \mathrm{curl}\,\mathbf{v}/2$. The quantities g, α, ν, τ and κ are, respectively, gravity, the thermal expansion coefficient of the fluid, kinematic viscosity, thermal relaxation time, and thermal conductivity of the fluid. In deriving equation $(3.3)_1$ one begins with the balance of momentum equation

$$\rho\dot{v}_i = t_{ji,j} + \rho f_i \tag{3.4}$$

where t_{ij} and f_i are the stress tensor and body force, respectively. For an incompressible, linear viscous fluid $t_{ij} = -p\delta_{ij} + 2\mu d_{ij}$, where μ is the dynamic viscosity and d_{ij} is the symmetric part of the velocity gradient, namely $d_{ij} = (v_{i,j} + v_{j,i})/2$. We note $\nu = \mu/\rho$ and suppose in the body force term $\mathbf{f} = -g\mathbf{k}$, and ρ is a linear function of temperature θ, i.e.

$$\rho = \rho_0\big(1 - \alpha(\theta - \theta_0)\big), \tag{3.5}$$

where ρ_0 is the value of ρ when $\theta = \theta_0$, and $\alpha(> 0)$ is the thermal expansion coefficient of the fluid. Then equation (3.4) becomes with ρ replaced by the constant ρ_0,

$$\rho_0\dot{v}_i = -p_{,i} + 2\mu d_{ij,j} - \rho_0\big(1 - \alpha(\theta - \theta_0)\big)gk_i. \tag{3.6}$$

We note $2d_{ij,j} = \Delta v_i$ since $v_{j,j} = 0$ and we incorporate the constant terms $\rho_0[1 + \alpha\theta_0]g$ into p, i.e. redefine

$$p \to p + \rho_0 g[1 + \alpha\theta_0]z.$$

Then upon division by ρ_0 and replacing ρ_0 by a constant ρ, equation (3.6) yields equation $(3.3)_1$.

(Lebon and Cloot, 1984) suggested modifying the Jaumann derivative in (3.3) and studied a thermal convection problem incorporating the effect of surface tension.

3.1.2 Cattaneo-Christov theory

(Christov, 2009) is an inspiring piece of work and he has suggested another objective derivative be employed for q_i. He suggests the following Lie derivative which is based on very sound physical principles,

$$\dot{q}_i - q_j v_{i,j} + q_i d_{rr} \equiv \frac{\partial q_i}{\partial t} + v_j q_{i,j} - q_j v_{i,j} + v_{r,r} q_i\,. \tag{3.7}$$

When the fluid is incompressible $d_{rr} = 0$ and then instead of equations (3.3) one may pose the Cattaneo-Christov equations for thermoviscous fluid motions, namely

$$\begin{aligned}
&\dot{v}_i = -\frac{1}{\rho}p_{,i} + k_i g\alpha\theta + \nu\Delta v_i\,,\\
&v_{i,i} = 0,\\
&\dot{\theta} = -q_{i,i}\,,\\
&\tau(q_{i,t} + v_j q_{i,j} - q_j v_{i,j}) = -q_i - \kappa\theta_{,i}\,.
\end{aligned} \tag{3.8}$$

Uniqueness and structural stability questions for a general Cattaneo-Christov fluid are presented by (Ciarletta and Straughan, 2010). These writers allow compressibility but they restrict attention to the case where the velocity field is *a priori* known. A uniqueness result for the incompressible heat conducting Cattaneo-Christov model is given by (Tibullo and Zampoli, 2011).

A general non-isothermal thermodynamic theory for a compressible gas which is based on the Cattaneo-Christov equations is derived by

(Straughan, 2010a). He shows how an acceleration wave may propagate and derives an explicit formula for the wavespeeds. The Cattaneo-Christov theory has been placed on a sound thermodynamic footing by (Morro, 2010). He derives objective evolution equations for both the heat flux and the stress which allow the body to deform and are completely compatible with thermodynamics.

3.1.3 Guyer-Krumhansl model

(Franchi and Straughan, 1994b) suggested modifying equation $(3.3)_4$ by adding Guyer-Krumhansl terms for q_i. In this way one derives instead of (3.3) the system

$$\begin{aligned}
&\dot{v}_i = -\frac{1}{\rho}p_{,i} + k_i g\alpha\theta + \nu\Delta v_i\,,\\
&v_{i,i} = 0,\\
&\dot{\theta} = -q_{i,i}\,,\\
&\tau(\dot{q}_i - \epsilon_{ijk}\omega_j q_k) = -q_i - \kappa\theta_{,i} + \hat{\tau}(\Delta q_i + 2q_{k,ki})\,,
\end{aligned} \tag{3.9}$$

where the relaxation time $\hat{\tau}$ is discussed in section 1.3. (Franchi and Straughan, 1994b) study thermal convection on the basis of these equations.

(Dauby et al., 2002) propose a similar set of equations to (3.9) and investigate thermal convection also incorporating surface tension effects at a free surface.

3.1.4 Alternative Guyer-Krumhansl model

In view of the findings of (Straughan, 2010d; Straughan, 2010c) on thermal convection employing the Cattaneo-Christov equatons (3.8), it may be also worth considering a Guyer - Krumhansl invariant. Then, one would modify equations (3.8) to

$$\begin{aligned}
&\dot{v}_i = -\frac{1}{\rho}p_{,i} + k_i g\alpha\theta + \nu\Delta v_i\,,\\
&v_{i,i} = 0,\\
&\dot{\theta} = -q_{i,i}\,,\\
&\tau(q_{i,t} + v_j q_{i,j} - q_j v_{i,j}) = -q_i - \kappa\theta_{,i} + \hat{\tau}(\Delta q_i + 2q_{k,ki})\,.
\end{aligned} \tag{3.10}$$

3.1.5 Further Cattaneo type fluid models

(Puri and Kythe, 1997) worked with system (3.3) and solved a problem of a plate moving in a Maxwell-Cattaneo fluid. This allowed them to simplify the equations and seek a solution $\mathbf{v} = (0, 0, u(x,t))$ with a temperature field $\theta(x,t)$, x being the one-dimensional spatial variable. The reduced (linear) system of equations they worked with is

$$\begin{aligned} \frac{\partial u}{\partial t} &= \frac{\partial^2 u}{\partial x^2} + G\theta, \\ \lambda P \frac{\partial^2 \theta}{\partial t^2} + P \frac{\partial \theta}{\partial t} &= \frac{\partial^2 \theta}{\partial x^2}, \end{aligned} \tag{3.11}$$

where λ, P, G are positive constants.

(Puri and Kythe, 1998) analysed a similar class of problem but when the stress tensor is allowed to have a non-Newtonian form. In this case instead of equations (3.11) they derived the system

$$\begin{aligned} \frac{\partial u}{\partial t} - k \frac{\partial^3 u}{\partial t \partial x^2} &= \frac{\partial^2 u}{\partial x^2} + G\theta, \\ \lambda P \frac{\partial^2 \theta}{\partial t^2} + P \frac{\partial \theta}{\partial t} &= \frac{\partial^2 \theta}{\partial x^2}, \end{aligned} \tag{3.12}$$

where k is another positive constant, this term representing the viscoelastic effect.

(Puri and Jordan, 1999b) [see also (Puri and Jordan, 1999a)] analysed a problem of an oscillating vertical plate which is periodically heated. They adopted a Maxwell-Cattaneo fluid but also assumed the fluid was of dipolar type. This led them to study the system of equations

$$\begin{aligned} \frac{\partial u}{\partial t} - \ell_1^2 \frac{\partial^3 u}{\partial t \partial x^2} &= \frac{\partial^2 u}{\partial x^2} - \ell_2^2 \frac{\partial^4 u}{\partial x^4} + G\theta, \\ \lambda P \frac{\partial^2 \theta}{\partial t^2} + P \frac{\partial \theta}{\partial t} &= \frac{\partial^2 \theta}{\partial x^2}. \end{aligned} \tag{3.13}$$

The coefficient ℓ_2^2 is a positive dipolar constant.

If we analyse a problem like that of (Puri and Kythe, 1997) but instead of using equations (3.3) we employ the GMC system (3.9) then we may arrive at the system of partial differential equations

$$\begin{aligned} &\frac{\partial u}{\partial t} - \nu \Delta u = G\theta, \\ &c \frac{\partial \theta}{\partial t} = -\frac{\partial q_i}{\partial x_i}, \\ &\tau \frac{\partial^2 q_i}{\partial t \partial x_i} = -\frac{\partial q_i}{\partial x_i} - \kappa \Delta \theta + 3\hat{\tau}\, \Delta \frac{\partial q_i}{\partial x_i}. \end{aligned} \tag{3.14}$$

Upon elimination of $q_{i,i}$ we find

$$\begin{aligned} &\frac{\partial u}{\partial t} - \nu \Delta u = G\theta, \\ &\tau_1 \frac{\partial^2 \theta}{\partial t^2} + c \frac{\partial \theta}{\partial t} - \kappa \Delta \theta - \tau_2\, \Delta \frac{\partial \theta}{\partial t} = 0, \end{aligned} \tag{3.15}$$

where $\tau_1 = \tau c > 0$ and $\tau_2 = 3\hat{\tau} c > 0$.

Thus, equations (3.11), (3.12) and (3.15) represent interaction of a fluid with an MC or GMC thermodynamic law in a suitable linear sense.

The simplified systems (3.11), (3.12) or (3.15) are certainly much more amenable to analysis then the original systems (3.3) or (3.9).

3.2 Green-Laws theory

(Müller, 1971b) begins with the equations of balance of mass, balance of linear momentum, and balance of energy without body force and without heat supply, namely

$$\begin{aligned} \dot{\rho} + \rho v_{i,i} &= 0, \\ \rho \dot{v}_i - t_{ji,j} &= 0, \\ \rho \dot{\epsilon} + q_{i,i} &= t_{ij} v_{i,j}\,, \end{aligned}$$

where $\rho, v_i, t_{ij}, \epsilon, q_i$ are density, velocity, stress tensor, internal energy, and heat flux, respectively. He assumes constitutive theory of form

$$\begin{aligned} t_{ij} &= t_{ij}(\rho, \theta, \theta_{,i}, \dot{\theta}, d_{rs}) \\ q_i &= q_i(\rho, \theta, \theta_{,i}, \dot{\theta}, d_{rs}) \\ \epsilon &= \epsilon(\rho, \theta, \theta_{,i}, \dot{\theta}, d_{rs}) \end{aligned} \tag{3.16}$$

where θ is the temperature, $d_{ij} = (v_{i,j} + v_{j,i})/2$. He exploits his entropy inequality

$$\rho\dot{\eta} + \Phi_{i,i} \geq 0,$$

for an entropy flux vector $\mathbf{\Phi}$ which like the entropy, η, depends on the constitutive list (3.16). (Müller, 1971b) derives equations for a viscous fluid and for an inviscid fluid. He also shows how one may include a body force and a heat supply and use the classical arguments of Coleman and Noll to reduce the constitutive theory.

In this section we describe the equations for an inviscid fluid derived using the thermodynamic arguments of (Green and Laws, 1972). The details may be found in (Lindsay and Straughan, 1978).

The equations presented by (Lindsay and Straughan, 1978) are conservation of mass, linear momentum, angular momentum, and energy and have form

$$\begin{aligned} \dot{\rho} + \rho v_{i,i} &= 0, \\ \rho \dot{v}_i &= \rho f_i + t_{ji,j}, \\ t_{ij} &= t_{ji}\,, \\ \rho \dot{\epsilon} &= \rho r - q_{i,i} + t_{ij} d_{ij} \end{aligned} \tag{3.17}$$

where f_i and r are the body force and externally supplied heat supply, respectively.

The entropy inequality employed is that of (Green and Laws, 1972),

$$\frac{d}{dt}\int_V \rho\eta\, dV - \int_V \frac{\rho r}{\phi}\, dV + \oint_{\partial V} \frac{q_i n_i}{\phi}\, dS \geq 0 \tag{3.18}$$

where V is any subbody in a continuous body B. The notation ∂V denotes the boundary of V, η is the entropy function and ϕ is a scalar function to be more precisely identified. In terms of the Helmholtz free energy function ψ,

$$\psi = \epsilon - \eta\phi \tag{3.19}$$

one may reduce (3.18) to a pointwise form and rewrite it with the aid of equation $(3.17)_4$ as

$$-\rho(\dot{\psi} + \eta\dot{\phi}) + t_{ij}d_{ij} - \frac{q_i\phi_{,i}}{\phi} \geq 0. \tag{3.20}$$

To describe an inviscid (perfect) fluid (Lindsay and Straughan, 1978) suppose that the functions

$$\psi, \phi, \eta, q_i, t_{ij} \tag{3.21}$$

depend on the independent variables

$$\rho, \theta, \dot{\theta}, \theta_{,i} \tag{3.22}$$

where θ is the temperature in the body. For the scalars ψ, ϕ, η (Lindsay and Straughan, 1978) show that the list (3.22) may be replaced by

$$\rho, \theta, \dot{\theta}, \lambda \tag{3.23}$$

where $\lambda = \theta_{,i}\theta_{,i}/2$. The forms (3.21) - (3.22) are now inserted into the entropy inequality (3.20) and (Lindsay and Straughan, 1978) deduce that

$$\phi = \phi(\theta, \dot{\theta}), \qquad \psi = \psi(\rho, \theta, \dot{\theta}, \lambda), \tag{3.24}$$

$$\eta = -\frac{\partial\psi/\partial\dot{\theta}}{\partial\phi/\partial\dot{\theta}} = \eta(\rho, \theta, \dot{\theta}, \lambda), \tag{3.25}$$

$$q_i = -K\theta_{,i}\,, \tag{3.26}$$

$$K = \frac{\rho\phi\partial\psi/\partial\lambda}{\partial\phi/\partial\dot{\theta}} = K(\rho, \theta, \dot{\theta}, \lambda), \tag{3.27}$$

$$t_{ij} = -p\delta_{ij} - \rho\frac{\partial\psi}{\partial\lambda}\theta_{,i}\theta_{,j}\,, \tag{3.28}$$

$$p = \rho^2\frac{\partial\psi}{\partial\rho} = p(\rho, \theta, \dot{\theta}, \lambda). \tag{3.29}$$

What remains of the entropy inequality (3.20) is

$$-\rho\Big(\frac{\partial\psi}{\partial\theta} + \eta\frac{\partial\phi}{\partial\theta}\Big)\dot{\theta} + 2K\frac{\partial\phi}{\partial\theta}\frac{\lambda}{\phi} \geq 0. \tag{3.30}$$

From this inequality (Lindsay and Straughan, 1978) deduce that in thermodynamic equilibrium (for which $\dot{\theta} = 0$, $\theta_{,i} = 0$ and is denoted by E) the following relations hold

$$\left(\frac{\partial\psi}{\partial\theta} + \eta\frac{\partial\phi}{\partial\theta}\right)\bigg|_E = 0, \tag{3.31}$$

$$\left(\frac{\partial\eta}{\partial\dot{\theta}}\frac{\partial\phi}{\partial\dot{\theta}}\right)\bigg|_E - \left(\frac{\partial\eta}{\partial\dot{\theta}}\right)\bigg|_E \geq 0, \tag{3.32}$$

$$K|_E \geq 0. \tag{3.33}$$

Thefunction $\phi(\theta, \dot{\theta})$ is usually called a generalized temperature. One may show that the system of equations (3.17) reduces to

$$\begin{aligned}
&\dot{\rho} + \rho v_{i,i} = 0,\\
&\rho\dot{v}_i = \rho f_i - p_{,i} - \left(\rho\frac{\partial\psi}{\partial\lambda}\theta_{,i}\theta_{,j}\right)_{,j},\\
&\rho\phi\eta_{\dot{\theta}}\ddot{\theta} + (\rho\psi_\theta + \eta\phi_\theta + \rho\phi\eta_\theta)\dot{\theta} + (\rho\psi_\lambda + \rho\phi\eta_\lambda - K_{\dot{\theta}})\dot{\lambda}\\
&\qquad - K_\lambda\theta_{,i}\theta_{,j}\theta_{,ij} - K\Delta\theta - K_\rho\rho_{,i}\theta_{,i} - 2\lambda K_\theta\\
&\qquad + \rho^2\phi\eta_\rho d_{ii} + (\rho\psi_\lambda - K_{\dot{\theta}})\theta_{,i}\theta_{,j}d_{ij} = \rho r.
\end{aligned} \tag{3.34}$$

Equations (3.34) represent the complete system of equations for thermodynamic motion in an inviscid fluid when one employs the thermodynamics of (Green and Laws, 1972).

(Lindsay and Straughan, 1978) develop a detailed analysis of acceleration wave behaviour for a solution to (3.34) including curved waves of arbitrary shape. Particular solutions are presented for a cylindrical shaped wave moving into a shear flow or for a spherical wave advancing into a radial flow.

3.3 Type II fluid

(Green and Naghdi, 1995a) used their thermal displacement variable α and their entropy balance equation to derive a new class of fluid theories. In this book we refer to their theories as being of a fluid of type II or type III. We believe that both of these theories may have application in the active area of research into heat transfer characteristics of nanofluids, cf. chapter 8. As we point out in chapter 8 nanofluids typically consist of a suspension of metals or their oxides, Cu, CuO, Al_2O_3, SiO_2, TiO_2, in water or a base fluid like ethylene glycol, see e.g. (Hwang et al., 2007), (Maiga et al., 2005), (Kim et al., 2007). An interesting article of (Vadasz et al., 2005) suggests that a mechanism for the increased heat transfer characteristics of a nanofluid may be through a hyperbolic equation for the temperature field. In view of the fact that the temperature displacement field essential to the type II theory of (Green and Naghdi, 1995a) satisfies what is effectively a hyperbolic equation it may be that the extension of the Green - Naghdi model developed by (Quintanilla and Straughan, 2008) which we now describe will be applicable to nanofluids.

(Quintanilla and Straughan, 2008) commence with the reduced energy balance equation

$$T_{ij}L_{ij} - p_i\gamma_i - \rho(\dot{\psi} + \eta\dot{\theta}) - \rho\theta\xi = 0, \tag{3.35}$$

written in the current configuration since we are dealing with a fluid. Here $T_{ij}, p_i, \rho, \psi, \eta, \theta$ and ξ are, respectively, the (symmetric) stress tensor, entropy flux vector, density, Helmholtz free energy function, entropy, absolute temperature, and the internal rate of production of entropy. Now, v_i denotes the velocity field, $L_{ij} = v_{i,j}$, $\gamma_i = (\dot{\alpha})_{,i}$, where $\alpha = \int_{t_0}^{t} \theta(\mathbf{X}, s)ds + \alpha_0$ is the thermal displacement field. We also require the (Green and Naghdi, 1995a) entropy balance law written in the current configuration

$$\rho\dot{\eta} = \rho s + \rho\xi - p_{i,i}, \tag{3.36}$$

where s is the external rate of supply of entropy per unit mass. Since we are now developing a fluid theory we also require the balance of mass,

$$\dot{\rho} + \rho v_{i,i} = 0, \tag{3.37}$$

and the balance of linear momentum,

$$\rho\dot{v}_i = T_{ji,j} + \rho b_i, \tag{3.38}$$

in which b_i is an externally supplied body force.

The development of (Quintanilla and Straughan, 2008) is different from that of (Green and Naghdi, 1995a). To understand this we observe that (Green and Naghdi, 1995a) commence with the assumption that ψ, η, T_{ij}, p_i and ξ depend on the variables $\rho, L_{ij}, \theta, \alpha_{,i}$ and γ_i. However, (Green and Naghdi, 1995a) p. 293 assume that p_i is linear in γ_i, T_{ij} is quadratic in d_{ij} ($d_{ij} = (v_{i,j} + v_{j,i})/2$), ξ is quadratic in d_{ij} and γ_i, and ψ has the form

$$\psi = \frac{1}{2} m\delta_i\delta_i + f(\rho, \theta) \tag{3.39}$$

where $\delta_i = \alpha_{,i}$ and m is a constant. After this they analyse a class of dissipationless flows by assuming the Reynolds, Peclet and m numbers are suitably large and this leads to a restricted class of dissipationless flows. (Quintanilla and Straughan, 2008) develop what is a more general dissipationless theory from the outset. To do this they omit $\gamma_i = \theta_{,i} = (\dot{\alpha})_{,i}$ as a variable in the constitutive theory from the outset. (This corresponds to the way (Green and Naghdi, 1993) develop their theory of thermoelasticity without energy dissipation, discussed in section 2.3). (Quintanilla and Straughan, 2008) are thus able to obtain a more complete nonlinear constitutive theory in which a variable such as the entropy flux vector, p_i, is defined naturally in terms of the Helmholtz free energy rather than having a preimposed form.

The work of (Quintanilla and Straughan, 2008) begins with the assumption that

$$T_{ij}, \psi, \eta, p_i \text{ and } \xi \tag{3.40}$$

are functions of the independent variables

$$\rho, L_{ij}, \theta, \alpha_{,i} \,. \tag{3.41}$$

Next, write $L_{ij} = d_{ij} + \omega_{ij}$, $\omega_{ij} = (v_{i,j} - v_{j,i})/2$, and use (3.41) together with (3.40) in the energy balance law (3.35) to see that

$$\begin{aligned}&\Big[T_{ij} + \delta_{ij}\rho^2\frac{\partial\psi}{\partial\rho} + \frac{\rho}{2}\Big(\frac{\partial\psi}{\partial\alpha_{,i}}\alpha_{,j} + \frac{\partial\psi}{\partial\alpha_{,j}}\alpha_{,i}\Big)\Big]d_{ij} + T_{ij}\omega_{ij}\\ &-\gamma_i\Big(p_i + \rho\frac{\partial\psi}{\partial\alpha_{,i}}\Big) - \rho\frac{\partial\psi}{\partial L_{ij}}\dot{\overline{L_{ij}}} - \dot\theta\rho\Big(\frac{\partial\psi}{\partial\theta} + \eta\Big)\\ &-\rho\theta\xi + \frac{\rho}{2}\omega_{ij}\Big(\frac{\partial\psi}{\partial\alpha_{,j}}\alpha_{,i} - \frac{\partial\psi}{\partial\alpha_{,i}}\alpha_{,j}\Big) = 0.\end{aligned} \tag{3.42}$$

(Quintanilla and Straughan, 2008) deduce from (3.42) that p_i, η and ψ reduce to the forms

$$p_i = -\frac{\partial\psi}{\partial\alpha_{,i}}, \qquad \eta = -\frac{\partial\psi}{\partial\theta} \qquad \text{and} \qquad \psi = \psi(\rho, \theta, \alpha_{,i}). \tag{3.43}$$

They then restrict attention to the situation where $\xi = \xi(\rho, \theta, \alpha_{,i})$ and equation (3.42) leaves

$$\begin{aligned}&\Big[T_{ij} + \delta_{ij}\rho^2\frac{\partial\psi}{\partial\rho} + \frac{\rho}{2}\Big(\frac{\partial\psi}{\partial\alpha_{,i}}\alpha_{,j} + \frac{\partial\psi}{\partial\alpha_{,j}}\alpha_{,i}\Big)\Big]d_{ij} - \rho\theta\xi\\ &+\frac{\rho}{2}\omega_{ij}\Big(\frac{\partial\psi}{\partial\alpha_{,j}}\alpha_{,i} - \frac{\partial\psi}{\partial\alpha_{,i}}\alpha_{,j}\Big) = 0.\end{aligned} \tag{3.44}$$

From (3.44) (Quintanilla and Straughan, 2008) show further that

$$\frac{\partial\psi}{\partial\alpha_{,i}}\alpha_{,j} = \frac{\partial\psi}{\partial\alpha_{,j}}\alpha_{,i}\,, \tag{3.45}$$

and

$$T_{ij} = -p\delta_{ij} - \frac{\rho}{2}\Big(\frac{\partial\psi}{\partial\alpha_{,i}}\alpha_{,j} + \frac{\partial\psi}{\partial\alpha_{,j}}\alpha_{,i}\Big) \tag{3.46}$$

where p is a pressure defined by $p = \rho^2\partial\psi/\partial\rho$. From the remainder of equation (3.44) it follows that $\xi = 0$, in agreement with (Green and Naghdi, 1995a).

In view of the above, the equations for a fluid of type II are given by the balance equations (3.36) - (3.38) with the constitutive theory (3.43),

(3.45) and (3.46) together with $\xi = 0$. If we recollect these explicitly then the balances of mass, linear momentum, and entropy become

$$\begin{aligned} &\dot{\rho} + \rho v_{i,i} = 0, \\ &\rho \dot{v}_i = -p_{,i} - \frac{1}{2}\big[\rho(\psi_{\alpha_{,i}}\alpha_{,j} + \psi_{\alpha_{,j}}\alpha_{,i})\big]_{,j} + \rho b_i, \\ &-\rho\frac{d}{dt}\Big(\frac{\partial\psi}{\partial\theta}\Big) = \rho s + \frac{\partial}{\partial x_i}\Big(\frac{\partial\psi}{\partial\alpha_{,i}}\Big), \end{aligned} \tag{3.47}$$

where d/dt, like the superposed dot, denotes the material derivative.

3.4 Type III fluid

(Green and Naghdi, 1995a) develop a further theory for a thermoviscous fluid which utilizes their thermal displacement variable α,

$$\alpha(\mathbf{x}, t) = \int_{t_0}^{t} \theta(\mathbf{x}, s)ds + \alpha_0 \tag{3.48}$$

where θ is the temperature field and $\mathbf{x}$ refers to the current configuration. They begin with the equations of balance of mass, balance of linear momentum, and balance of entropy in the form

$$\rho_t + v_i\rho_{,i} + \rho v_{i,i} = 0, \tag{3.49}$$

$$\rho(v_{i,t} + v_j v_{i,j}) = T_{ji,j} + \rho b_i, \tag{3.50}$$

$$\rho(\eta_t + v_i\eta_{,i}) = -p_{i,i} + \rho s + \rho\xi\,. \tag{3.51}$$

In these equations ρ, v_i and η are the density, velocity and entropy. Additionally T_{ji} and p_i are the (Cauchy) stress tensor and entropy flux vector, while b_i, s are the externally supplied body force and entropy supply, respectively. The variable ξ is an internal rate of production of entropy per unit mass.

(Green and Naghdi, 1995a) also employ the reduced energy equation

$$-\rho(\dot{\psi} + \eta\dot{\theta}) - \rho\theta\xi - p_i\gamma_i + T_{ji}v_{i,j} = 0 \tag{3.52}$$

where a superposed dot denotes the material derivative and $\gamma_i = \theta_{,i} = \partial\dot{\alpha}/\partial x_i$. They also define the variable

$$\delta_i = \alpha_{,i} = \frac{\partial\alpha}{\partial x_i}\,. \tag{3.53}$$

They then define a thermoviscous fluid to be one for which the Helmholtz free energy function ψ, the entropy, stress tensor, entropy flux vector, and the internal rate of production of entropy depend on the independent constitutive variables

$$\rho, v_{i,j}, \theta, \delta_i, \gamma_i \tag{3.54}$$

i.e.

$$\begin{aligned} \psi &= \psi(\rho, v_{i,j}, \theta, \delta_i, \gamma_i) \\ T_{ij} &= T_{ij}(\rho, v_{i,j}, \theta, \delta_i, \gamma_i) \\ \eta &= \eta(\rho, v_{i,j}, \theta, \delta_i, \gamma_i) \\ p_i &= p_i(\rho, v_{i,j}, \theta, \delta_i, \gamma_i) \\ \xi &= \xi(\rho, v_{i,j}, \theta, \delta_i, \gamma_i). \end{aligned} \tag{3.55}$$

Unlike the theory of section 3.3 the constitutive list (3.54) contains the variable $\gamma_i = \dot{\alpha}_{,i}$ which is in addition to those of (3.41). For this reason we refer to this as a fluid of type III, by analogy with thermoelasticity of type III as defined in section 2.4.

By manipulating the energy equation (3.52) (Green and Naghdi, 1995a) are able to reduce the constitutive list and indeed, they demonstrate that ψ does not depend on γ_i and $v_{i,j}$, so

$$\psi = \psi(\rho, \theta, \alpha_{,i}), \tag{3.56}$$

and additionally

$$\eta = -\frac{\partial \psi}{\partial \theta}, \qquad \alpha_{,i} \frac{\partial \psi}{\partial \alpha_{,j}} = \alpha_{,j} \frac{\partial \psi}{\partial \alpha_{,i}}, \tag{3.57}$$

while the energy equation assumes the form

$$\Big(T_{ij} + p\delta_{ij} + \rho \alpha_{,j} \frac{\partial \psi}{\partial \alpha_{,i}}\Big) d_{ij} - \Big(p_i + \rho \frac{\partial \psi}{\partial \alpha_{,i}}\Big) \gamma_i - \rho \theta \xi_i = 0, \tag{3.58}$$

where $d_{ij} = (v_{i,j} + v_{j,i})/2$ and p is a pressure given by

$$p = \rho^2 \frac{\partial \psi}{\partial \rho}. \tag{3.59}$$

At this point (Green and Naghdi, 1995a) specialize to the situation in which

$$\psi = \frac{m}{2} \alpha_{,i} \alpha_{,i} + f(\rho, \theta) \tag{3.60}$$

for $m > 0$ a constant and

$$\begin{aligned} p_i &= -\rho m \delta_i - \frac{\kappa}{\theta_0} \gamma_i \\ T_{ij} &= -p\delta_{ij} + \lambda d_{kk} \delta_{ij} + 2\mu d_{ij} - 2m \alpha_{,i} \alpha_{,j} \\ \rho \xi \theta &= \lambda d_{ii}^2 + 2\mu d_{ij} d_{ij} + \frac{\kappa}{\theta_0} \gamma_i \gamma_i , \end{aligned} \tag{3.61}$$

where $\theta_0, \kappa, \lambda$ and μ are constants.

3.4.1 Type III viscous fluid

We do not in this work adopt equations (3.60) and (3.61). Instead we leave things more general. We dot not impose a form for ψ and select

$$p_i = -\rho \frac{\partial \psi}{\partial \alpha_{,i}} - \frac{\kappa}{\theta} \gamma_i \,. \tag{3.62}$$

This is different to equation $(3.61)_1$ for two reasons. One, the first term is more general. Secondly, we employ θ rather than a constant θ_0. This we believe leads to a more natural energy equation which reduces to the classical energy equation in appropriate circumstances. In (3.62) κ may depend on the variables in the constitutive list. Our viscous theory is completed by specifying

$$\begin{aligned} T_{ij} &= -p\delta_{ij} + \hat{T}_{ij} \,, \\ \hat{T}_{ij} &= -\rho\alpha_{,i} \frac{\partial \psi}{\partial \alpha_{,j}} + \lambda d_{kk}\delta_{ij} + 2\mu d_{ij} \,, \\ \rho\theta\xi &= \frac{\kappa}{\theta} \gamma_i\gamma_i \,. \end{aligned} \tag{3.63}$$

The governing equations of motion for a type III fluid are then obtained upon employment of (3.62) and (3.63) in the conservation laws (3.49) - (3.51).

3.4.2 Type III inviscid fluid

Since the theme of this book is heat waves it is appropriate to develop a theory for an inviscid type III fluid. To this end we effectively neglect the dependence on $v_{i,j}$ in the constitutive list and drop the d_{ij} terms. Thus, our constitutive theory for an inviscid fluid of type III is

$$p_i = -\rho \frac{\partial \psi}{\partial \alpha_{,i}} - \frac{\kappa}{\theta} \gamma_i \,, \tag{3.64}$$

together with

$$\begin{aligned} T_{ij} &= -p\delta_{ij} - \rho\alpha_{,i} \frac{\partial \psi}{\partial \alpha_{,j}} \,, \\ \rho\theta\xi &= \frac{\kappa}{\theta} \gamma_i\gamma_i \,, \end{aligned} \tag{3.65}$$

where in its fullest generality ψ has the functional form (3.56) and κ depends on the constitutive variables $\rho, \theta, \alpha_{,i}, \gamma_i$.

The governing equations for an inviscid fluid of type III then become upon utilizing (3.64) and (3.65) in the evolution equations (3.49) - (3.51)

$$\dot{\rho} + \rho v_{i,i} = 0, \tag{3.66}$$

$$\rho \dot{v}_i = -p_{,i} - (\rho \alpha_{,j} \psi_{\alpha_{,i}})_{,j} + \rho b_i\,, \tag{3.67}$$

$$\rho \dot{\eta} = (\rho \psi_{\alpha_{,i}})_{,i} + \left(\frac{\kappa}{\theta}\gamma_i\right)_{,i} + \rho s + \frac{\kappa}{\theta^2}\gamma_i\gamma_i\,, \tag{3.68}$$

$$= (\rho \psi_{\alpha_{,i}})_{,i} + \frac{1}{\theta}(\kappa\gamma_i)_{,i} + \rho s\,. \tag{3.69}$$

3.5 Green-Naghdi extended theory

(Green and Naghdi, 1996) continue their development for describing the behaviour of a continuous body which relies on an entropy balance law rather than an entropy inequality. Again they introduce a quantity T which is the "empirical" temperature and the "thermal displacement variable"

$$\alpha = \int_{t_0}^{t} T(\mathbf{x}, s)ds + \alpha_0\,.$$

In fact, the full theory developed by (Green and Naghdi, 1996) is very general. They remark, (Green and Naghdi, 1996), p. 240, that ... "the theory ... leads to a set of differential equations ... which are rather unmanagable from the point of view of understanding turbulent or other flows." To produce a more tractable theory they restrict attention to a reduced version of their general theory which leaves only one equation as the mechanical differential equation. Precisely, the theory of (Green and Naghdi, 1996) develops a novel theory for fluids which involves vorticity and spin of vorticity. This introduces higher spatial gradients into the equations than those of Navier-Stokes theory and so is likely to be relevant where non-Newtonian fluid behaviour is expected. They work with two temperatures and are motivated by attempting to describe turbulence. In this respect, they are continuing the work of (Marshall and Naghdi, 1989a; Marshall and Naghdi, 1989b).

We simply describe the relevant differential equations for the model of (Green and Naghdi, 1996). Full details of the continuum thermodynamical development from the entropy balance law is given in (Green and Naghdi, 1996). The basic equations of (Green and Naghdi, 1996) are the balance of linear momentum, balance of mass, and balances of entropy for two temperatures θ_H and θ_T, which they regard as the usual temperature, and a turbulent temperature, respectively. However, other interpretations may be given to the different temperatures, see e.g. section 8.4 and (Straughan, 2010b). The balance of linear momentum, balance of mass, and balances of entropy as given by (Green and Naghdi, 1996) for their incompressible fluid may be written

$$\begin{aligned}
&\rho\left(\dot{v}_i - \frac{\mu_1}{\mu}\frac{d}{dt}\Delta v_i\right) = \rho b_i - p_{,i} + \mu\Delta v_i - 2\mu_1\Delta^2 v_i,\\
&v_{i,i} = 0,\\
&\rho\dot{\eta}_H = \rho s_H + \rho\xi_H - p^H_{i,i},\\
&\rho\dot{\eta}_T = \rho s_T + \rho\xi_T - p^T_{i,i}.
\end{aligned} \tag{3.70}$$

Here a superposed dot denotes the material derivative $d/dt = \partial/\partial t + v_i\partial/\partial x_i$. The variables ρ, v_i, b_i, p are the density, velocity, body force and pressure. The coefficient μ is the kinematic viscosity of the fluid while μ_1 is another constant reflecting the geometry of the particles and the interaction with the fluid. The appendices to (Bleustein and Green, 1967) and (Green and Rivlin, 1964) derive an expression for the kinetic energy of a system of particles as a function of the velocity of the centroid and the derivative of this velocity. While neither of the articles of (Green and Rivlin, 1964) nor (Bleustein and Green, 1967) has a direct bearing on the fluid theory of (Green and Naghdi, 1996), their procedure leads to a kinetic energy which can be equated to the kinetic energy of the fluid currently being described. The quantities η_H, η_T are the entropies corresponding to the temperatures θ_H and θ_T. The terms s_H, s_T, are external supplies of entropy, ξ_H, ξ_T, are intrinsic supplies of entropy which depend on the variables of the theory, and p^H_i, p^T_i are entropy flux vectors. (Green and Naghdi, 1996) assume that the Helmholtz free energy function ψ has form

$$\psi = c_H(\theta_H - \theta_H \ln\theta_H) + c_T(\theta_T - \theta_T \ln\theta_T), \tag{3.71}$$

with c_H, c_T positive constants, while the entropies and entropy fluxes assume the form

$$\eta_H = c_H \ln\theta_H, \qquad \eta_T = c_T \ln\theta_T, \tag{3.72}$$

and

$$p^H_i = -\frac{\kappa_H}{\theta_0}\frac{\partial\theta_H}{\partial x_i}, \qquad p^T_i = -\frac{\kappa_T}{\theta_0}\frac{\partial\theta_T}{\partial x_i}, \tag{3.73}$$

for positive constants $\kappa_H, \kappa_T, \theta_0$, with θ_0 being some reference temperature. The instrinsic entropy supply functions are given by

$$\rho\xi_H\theta_H = \frac{\kappa_H}{\theta_0}\frac{\partial\theta_H}{\partial x_i}\frac{\partial\theta_H}{\partial x_i} + 2\mu d_{ij}d_{ij} + \phi, \tag{3.74}$$

$$\rho\xi_T\theta_T = \frac{\kappa_T}{\theta_0}\frac{\partial\theta_T}{\partial x_i}\frac{\partial\theta_T}{\partial x_i} + 4\mu_1 d_{ij}P_{ij} + \frac{2\mu_1^2}{\mu}P_{ij}P_{ij} - \phi. \tag{3.75}$$

In these equations the variables d_{ij} and P_{ij} are defined by $d_{ij} = (v_{i,j} + v_{j,i})/2$, $P_{ij} = -\Delta v_{i,j}$ and ϕ is constant. It is very important to note, however, that (Green and Naghdi, 1996) observe that for some purposes ϕ could depend on temperatures, cf. section 8.4, and (Straughan, 2010b).

Thus, the complete system of equations for an incompressible viscous fluid in the (Green and Naghdi, 1996) extended theory are

$$\begin{aligned}
&\rho\frac{dv_i}{dt} - \frac{\rho\mu_1}{\mu}\frac{d}{dt}\Delta v_i = \rho b_i - p_{,i} + \mu\Delta v_i - 2\mu_1\Delta^2 v_i\,,\\
&v_{i,i} = 0,\\
&\rho c_H\frac{d\theta_H}{dt} = \rho s_H\theta_H + \frac{\kappa_H}{\theta_0}\frac{\partial\theta_H}{\partial x_i}\frac{\partial\theta_H}{\partial x_i} + 2\mu d_{ij}d_{ij} + \phi,\\
&\rho c_T\frac{d\theta_T}{dt} = \rho s_T\theta_T + \frac{\kappa_T}{\theta_0}\frac{\partial\theta_T}{\partial x_i}\frac{\partial\theta_T}{\partial x_i} + 4\mu d_{ij}P_{ij} + 2\frac{\mu_1^2}{\mu}P_{ij}P_{ij} - \phi,
\end{aligned} \tag{3.76}$$

where d/dt has been employed to denote the material derivative.

(Green and Naghdi, 1996) determine the basic solution to plane Poiseuille flow for their theory and show that it leads to a flattened profile rather than the parabolic one of classical Newtonian theory. They also address a similar basic solution to Poiseuille flow in a pipe. Additionally, (Green and Naghdi, 1996) address the problem of flow of a circular jet from a round hole. Finally, (Green and Naghdi, 1996) address two problems where the solution is time-dependent. All the problems addressed by (Green and Naghdi, 1996) are in an isothermal situation.

4
Acceleration waves

The general theory of acceleration waves in continuum mechanics is covered in detail in the research review articles of (Chen, 1973) and (McCarthy, 1972), see also the accounts in the books of (Fabrizio and Morro, 2003), (Iesan and Scalia, 1996), (Ogden, 1997) and (Straughan, 2008). (Truesdell and Toupin, 1960) and (Truesdell and Noll, 1992) cover many aspects of acceleration waves and singular surfaces in general. We now include an account of some recent studies employing acceleration waves in theories of heat transport associated with second sound.

4.1 Maxwell-Cattaneo theory

Suppose we have a rigid body occupying $\mathbb{R}^3$ and the temperature field and heat flux are governed by the Maxwell - Cattaneo equations when the thermal conductivity κ depends on temperature θ, cf. chapter 1, so the governing equations are

$$\begin{aligned} c\theta_{,t} &= -q_{i,i}, \\ \tau q_{i,t} + q_i &= -\kappa(\theta)\theta_{,i}\,, \end{aligned} \tag{4.1}$$

where q_i is the heat flux, and c, τ are positive constants. Recall that $,i$ denotes differentiation with respect to x_i, e.g. $\theta_{,i} = \partial\theta/\partial x_i$. An *acceleration wave* for a solution to equations (4.1) is a surface $\mathcal{S}$ across which $\theta_{,t}, \theta_{,i}, q_{i,t}, q_{i,j}$, suffer at most finite discontinuities, with the functions θ, q_i continuous everywhere. Even though the jump across $\mathcal{S}$ is in $\theta_{,t}$ and $q_{i,t}$ we call $\mathcal{S}$ an acceleration wave. Numerical solutions to the Maxwell - Cattaneo equations with κ a linear function of θ are presented by (Glass et al., 1986), (Cramer et al., 2001), (Christov and Jordan, 2010), and for κ a more general function of θ they are presented by (Reverberi et al., 2008), see section 5.3. An analytical solution of the Maxwell - Cattaneo equations with κ constant, for a step input at the boundary, is provided by means of a Laplace transform technique by (Al-Qahtani and Yilbas, 2010).

To illustrate the basic concepts of acceleration wave analysis, we shall for now restrict attention to a plane acceleration wave moving in the direction of the $x-$axis, with one-dimensional motion.

In one space dimension the heat flux has one component, q, and equations (4.1) become

$$\begin{aligned} c\theta_t &= -q_x, \\ \tau q_t + q &= -\kappa(\theta)\theta_x\,, \end{aligned} \tag{4.2}$$

where $\theta_t = \partial\theta/\partial t$, $q_x = \partial q/\partial x$, etc.

For a function $h(x,t)$ we define

$$h^+(x,t) = \lim_{x\to\mathcal{S}} h(x,t) \text{ from the right,}$$

$$h^-(x,t) = \lim_{x\to\mathcal{S}} h(x,t) \text{ from the left.}$$

In particular, h^+ is the value of h at $\mathcal{S}$ approaching from the region which $\mathcal{S}$ is about to enter. The jump of h at $\mathcal{S}$, written as $[h]$, is,

$$[h] = h^- - h^+\,. \tag{4.3}$$

We take the jump of each of equations (4.2), to find

$$\begin{aligned} c[\theta_t] + [q_x] &= 0, \\ \tau[q_t] + \kappa(\theta^+)[\theta_x] &= 0\,, \end{aligned} \tag{4.4}$$

since $\kappa(\theta)$ is continuous across $\mathcal{S}$. Next, employ the kinematic condition of compatibility, sometimes known as the Hadamard relation,

$$\frac{\delta}{\delta t}[f] = \Big[\frac{\partial f}{\partial t}\Big] + V\Big[\frac{\partial f}{\partial X}\Big] \tag{4.5}$$

where $\delta/\delta t$ denotes the time derivative at the wave. (The Hadamard relation is discussed in detail in (Chen, 1973), appendix 1, and also in (Truesdell and Toupin, 1960), section 180.)

Note, since $q \in C^0(\mathbb{R})$, $[q] = 0$ and so by using the Hadamard relation

$$0 = \frac{\delta}{\delta t}[q] = [q_t] + V[q_x]$$

so that

$$[q_t] = -V[q_x]\,. \tag{4.6}$$

Similarly,

$$[\theta_t] = -V[\theta_x]\,. \tag{4.7}$$

We use equations (4.6) and (4.7) in (4.4) and then obtain

$$\begin{pmatrix} -cV & 1 \\ \kappa & -\tau V \end{pmatrix} \begin{pmatrix} [\theta_x] \\ [q_x] \end{pmatrix} = \begin{pmatrix} 0 \\ 0 \end{pmatrix} \tag{4.8}$$

We require the amplitudes $[\theta_x], [q_x]$ to be non-zero and so from (4.8) we need

$$\begin{vmatrix} -cV & 1 \\ \kappa & -\tau V \end{vmatrix} = 0$$

and so

$$V^2 = \frac{\kappa(\theta^+)}{c\tau}\,. \tag{4.9}$$

Note that V depends on the value of θ at the wave (here $\theta^+ = \theta^-$).

Define the wave amplitudes $A(t)$ and $B(t)$ as

$$A = [\theta_x], \qquad B = [q_x], \tag{4.10}$$

and then observe that from $(4.8)_1$

$$cVA = B. \tag{4.11}$$

To find the equation governing the amplitude $A(t)$ (or $B(t)$) we differentiate equations (4.2) with respect to x and take the jumps of the results to find

$$\begin{aligned} c[\theta_{tx}] &= -[q_{xx}] \\ \tau[q_{tx}] + [q_x] &= -\kappa'(\theta^+)[\theta_x^2] - \kappa[\theta_{xx}]\,. \end{aligned} \tag{4.12}$$

From the definition of $[h]$ we may prove the relation for the jump of a product of functions g, h,

$$[gh] = g^+[h] + h^+[g] + [g][h]. \tag{4.13}$$

From the Hadamard relation we have that

$$\frac{\delta}{\delta t}[q_x] = [q_{xt}] + V[q_{xx}]\,, \tag{4.14}$$

$$\frac{\delta}{\delta t}[\theta_x] = [\theta_{xt}] + V[\theta_{xx}]\,. \tag{4.15}$$

Thus, recalling definitions (4.10) we eliminate $[\theta_{tx}]$ and $[q_{tx}]$ from (4.12) to find

$$c\Big(\frac{\delta A}{\delta t} - V[\theta_{xx}]\Big) + [q_{xx}] = 0, \tag{4.16}$$

$$\tau\Big(\frac{\delta B}{\delta t} - V[q_{xx}]\Big) + B + \kappa'(A^2 + 2\theta_x^+ A) + \kappa[\theta_{xx}] = 0. \tag{4.17}$$

We wish to remove the $[\theta_{xx}]$ and $[q_{xx}]$ terms from (4.16) and (4.17) using the wavespeed relation (4.9). Hence, form (4.16)+λ(4.17) to obtain

$$\begin{aligned} c\frac{\delta A}{\delta t} + (\lambda\kappa - cV)[\theta_{xx}] + [q_{xx}](1 - \lambda\tau V) + \lambda\tau\frac{\delta B}{\delta t} \\ + \lambda B + \lambda\kappa'(A^2 + 2\theta_x^+ A) = 0. \end{aligned} \tag{4.18}$$

We see that the correct choice of λ is

$$\lambda = \frac{cV}{\kappa} = \sqrt{\frac{c}{\kappa\tau}}. \tag{4.19}$$

Note that from (4.11) $B = cVA$ and so since

$$\frac{\delta B}{\delta t} = c\Big(A\frac{\delta V}{\delta t} + V\frac{\delta A}{\delta t}\Big)$$

we find from (4.18)

$$2c\frac{\delta A}{\delta t} + \Big(\frac{c}{\tau} + \frac{c}{V}\frac{\delta V}{\delta t} + \frac{2\theta_x^+\kappa'}{V\tau}\Big)A + \frac{\kappa'}{V\tau}A^2 = 0. \tag{4.20}$$

This is the equation governing the evolutionary behaviour of the amplitude $A(t)$ - the amplitude equation. It is a Bernoulli equation, which may be written in the form

$$\frac{\delta A}{\delta t} + \alpha(t)A + \beta(t)A^2 = 0.$$

It may be solved by the substitution $\gamma = 1/A$ to yield the general solution

$$A(t) = \frac{A(0)}{\exp\{\int_0^t \alpha(s)ds\} + \int_0^t \beta(s)\exp\{\int_s^t \alpha(\eta)d\eta\}ds}. \tag{4.21}$$

4.1.1 Wave into equilibrium

Suppose now the region ahead of the wave is such that

$$\theta = \text{constant} \qquad \text{and} \qquad \kappa(\theta) = \gamma\theta, \qquad \gamma > 0 \text{ (constant)}.$$

Then, $V =$ constant and $\theta_x^+ = 0$. Hence, equation (4.20) reduces to

$$\frac{\delta A}{\delta t} + \alpha A + \beta A^2 = 0, \tag{4.22}$$

where $\alpha = 1/2\tau$, $\beta = \gamma/2cV\tau$ are both constant. Then (4.22) is solved to find

$$A(t) = \frac{A(0)}{e^{\alpha t} + \dfrac{\beta}{\alpha}A(0)(e^{\alpha t} - 1)}. \tag{4.23}$$

From this equation we see that if $A(0) > 0$ then $A(t)$ decays to zero. If $A(0) < 0$ then $A(t)$ will blow-up in a finite time $\mathcal{T}$, with

$$\mathcal{T} = \frac{1}{\alpha}\log\Big(\frac{|A(0)|(\beta/\alpha)}{[|A(0)|(\beta/\alpha) - 1]}\Big). \tag{4.24}$$

Since here $A(t) = \theta_x^-(t)$ we see that when $A(0) < 0$ the temperature gradient steepens at the wave and $\theta_x^-(t)$ blows up. It is believed that a thermal shock forms, i.e. θ develops a discontinuity across $\mathcal{S}$ at $t = \mathcal{T}$. (Fu and Scott, 1991) have investigated this behaviour in elasticity in detail and very interesting computations of (Christov and Jordan, 2008), (Christov et al., 2006; Christov et al., 2007) and (Jordan, 2007) follow the acceleration wave development into a shock in a variety of situations, see also chapter 5, section 5.1, of this book.

4.1.2 Acceleration wave in three dimensions

An acceleration wave for the Maxwell-Cattaneo equations (4.1) is defined as in section 4.1. Namely, θ and q_i are C^0 everywhere and the first and higher derivatives of θ and q_i are allowed to have finite discontinuities across a surface $\mathcal{S}$. For simplicity we suppose now that we are dealing with a wave moving into an equilibrium region for which

$$\theta = \text{constant}, \quad q_i = 0.$$

Then $\dot{\theta}^+ = 0, \theta^+_{,i} = 0$.

General compatibility relations for a function $\psi(\mathbf{X}, t)$ are needed across $\mathcal{S}$. These are given in detail in (Truesdell and Toupin, 1960) or in (Chen, 1973). We simply quote those we need. If ψ is continuous in $\mathbb{R}^3$ but its derivative is discontinuous across $\mathcal{S}$ then

$$[\psi_{,i}] = n_i B, \qquad \text{where } B = [n^i \psi_{,i}]. \tag{4.25}$$

When $\psi \in C^1(\mathbb{R}^3)$ then

$$[\psi_{,ij}] = n_i n_j C, \qquad \text{where } C = [n^i n^j \psi_{,ij}]. \tag{4.26}$$

In (4.25) and (4.26), n_i refers to the unit normal to $\mathcal{S}$. Relations (4.25) and (4.26) are derived from (Chen, 1973), equations (4.13), (4.14). The relation corresponding to the Hadamard formula (4.5) in three dimensions is, cf. (Chen, 1973) (4.15),

$$\frac{\delta}{\delta t}[\psi] = [\dot{\psi}] + U_N B \tag{4.27}$$

where $\dot{\psi} = \partial\psi/\partial t$, U_N is the speed at the point on $\mathcal{S}$ with unit normal n_i and B is defined in (4.25).

We being by taking the jump of (4.1) to find

$$\begin{aligned} c[\theta_t] &= -[q_{i,i}]\,, \\ \tau[q_{i,t}] &= -\kappa[\theta_{,i}]\,. \end{aligned} \tag{4.28}$$

Define the three-dimensional wave amplitudes as

$$A(t) = [n^i \theta_{,i}], \quad B_i(t) = [n^j q_{i,j}]\,.$$

Then, since $q_i, \theta \in C^0(\mathbb{R}^3)$ we find using (4.28)

$$0 = \frac{\delta}{\delta t}[q_i] = [\dot{q}_i] + U_N[n^j q_{i,j}],$$
$$0 = \frac{\delta}{\delta t}[\theta] = [\dot{\theta}] + U_N[n^j \theta_{,j}].$$

Whence

$$[\dot{q}_i] = -U_N B_i, \qquad [\dot{\theta}] = -U_N A. \tag{4.29}$$

Using (4.29) in (4.28) we thus see that

$$\begin{aligned} - cU_N A + n_i B_i &= 0, \\ - \tau U_N B_i + \kappa n_i A &= 0. \end{aligned} \tag{4.30}$$

From equation $(4.30)_2$ we see that the wave must be longitudinal, i.e. $B_i = Bn_i$, where

$$B = [n^i n^j q_{i,j}]. \tag{4.31}$$

Then we have

$$\begin{pmatrix} -cU_N & 1 \\ \kappa & -\tau U_N \end{pmatrix} \begin{pmatrix} A \\ B \end{pmatrix} = \begin{pmatrix} 0 \\ 0 \end{pmatrix} \tag{4.32}$$

Since we require A, B non-zero we need

$$\begin{vmatrix} -cU_N & 1 \\ \kappa & -\tau U_N \end{vmatrix} = 0 \tag{4.33}$$

and so we derive

$$\tau c U_N^2 = \kappa\,. \tag{4.34}$$

Thus, the wavespeed is $U_N = \sqrt{\kappa/\tau c}$.

To calculate the wave amplitude $A(t)$, noting that $B(t)$ then follows from $(4.32)_1$ since,

$$B = cU_N A, \tag{4.35}$$

we differentiate (4.1) with respect to t and take the jump of the result. Thus,

$$\begin{aligned} c[\ddot{\theta}] &= -[\dot{q}_{i,i}]\,, \\ \tau[\ddot{q}_i] + [\dot{q}_i] &= -\kappa'(\theta)[\dot{\theta}\theta_{,i}] - \kappa[\dot{\theta}_{,i}]\,. \end{aligned} \tag{4.36}$$

Since the wave is moving into equilibrium we have U_N =constant and using the product relation (4.13), equations (4.36) become

$$\begin{aligned} c[\ddot{\theta}] &= -[\dot{q}_{i,i}]\,, \\ \tau[\ddot{q}_i] + [\dot{q}_i] &= -\kappa'(\theta)[\dot{\theta}][\theta_{,i}] - \kappa[\dot{\theta}_{,i}]\,. \end{aligned} \tag{4.37}$$

We now wish to derive a Bernoulli equation from (4.37). To this end we use the Hadamard relation (4.27) to note that

$$[\dot{\theta}] = -U_N A \tag{4.38}$$

and so

$$\begin{aligned}\frac{\delta}{\delta t}[\dot{\theta}] &= -\frac{\delta}{\delta t}U_N A = -U_N\frac{\delta A}{\delta t}\\ &=[\ddot{\theta}] + U_N[n_j\dot{\theta}_{,j}]\,.\end{aligned} \tag{4.39}$$

Further,

$$\frac{\delta}{\delta t}[n_i\theta_{,i}] = n_i[\dot{\theta}_{,i}] + U_N n_i[\theta_{,ij}n^j]\,. \tag{4.40}$$

Thus, from the above two equations,

$$\begin{aligned}[\ddot{\theta}] &= -\,U_N\frac{\delta A}{\delta t} - U_N[n_j\dot{\theta}_{,j}]\\ &= -\,2U_N\frac{\delta A}{\delta t} + U_N^2[n^i n^j\theta_{,ij}]\,.\end{aligned} \tag{4.41}$$

In a similar mannner we find

$$[\dot{q}_i] = -U_N B_i\,, \tag{4.42}$$

$$\frac{\delta}{\delta t}[\dot{q}_i] = -U_N\frac{\delta B_i}{\delta t} = [\ddot{q}_i] + U_N[n_j\dot{q}_{i,j}], \tag{4.43}$$

$$\frac{\delta B_i}{\delta t} = \frac{\delta}{\delta t}[q_{i,j}n_j] = [\dot{q}_{i,j}n_j] + U_N[q_{i,jk}n_j n_k]\,, \tag{4.44}$$

$$[\dot{q}_{i,i}] = \frac{\delta B}{\delta t} - U_N[q_{i,jk}n_i n_j n_k]\,, \tag{4.45}$$

where B is given by (4.31), and

$$-U_N\frac{\delta B_i}{\delta t} = [\ddot{q}_i] + U_N\left(\frac{\delta B_i}{\delta t} - U_N[q_{i,jk}n_j n_k]\right), \tag{4.46}$$

so

$$[\ddot{q}_i] = -2U_N\frac{\delta B_i}{\delta t} + U_N^2[q_{i,jk}n_j n_k]\,. \tag{4.47}$$

We now employ (4.38) - (4.47) in (4.37) to obtain

$$-2cU_N\frac{\delta A}{\delta t} + \frac{\delta B}{\delta t} - U_N[q_{i,jk}n_i n_j n_k] + cU_N^2[n_i n_j\theta_{,ij}] = 0\,, \tag{4.48}$$

and

$$\begin{aligned}-2\tau U_N\frac{\delta B}{\delta t} + \kappa\frac{\delta A}{\delta t} - U_N B - U_N\kappa'(\theta)A^2&\\ + \tau U_N^2[q_{i,jk}n_i n_j n_k] - \kappa U_N[n_i n_j\theta_{,ij}] = 0\,.&\end{aligned} \tag{4.49}$$

Recall $B = cU_N A$ and then from (4.48), (4.49), we derive

$$-cU_N \frac{\delta A}{\delta t} - U_N [q_{i,jk} n_i n_j n_k] + cU_N^2 [n_i n_j \theta_{,ij}] = 0\,, \tag{4.50}$$

and

$$\begin{aligned} -\kappa \frac{\delta A}{\delta t} &- U_N B - U_N \kappa'(\theta) A^2 \\ &+ \tau U_N^2 [q_{i,jk} n_i n_j n_k] - \kappa U_N [n_i n_j \theta_{,ij}] = 0\,. \end{aligned} \tag{4.51}$$

Now form (4.50)+λ(4.51) for λ a positive constant. We use the wavespeed relation (4.34) and select $\lambda = 1/\tau U_N$. This removes the $\theta_{,ij}$ and $q_{i,jk}$ terms and leads to the equation

$$\frac{\delta A}{\delta t} + \alpha A + \beta A^2 = 0, \tag{4.52}$$

where

$$\alpha = \frac{U_N}{2}\,, \qquad \beta = \frac{\kappa'(\theta^+)}{2c}\,. \tag{4.53}$$

Equation (4.52) is the same as (4.22) and the solution is, therefore, (4.23). The remarks concerning $A(t)$ after (4.23) apply also to $A(t)$ for (4.52).

4.1.3 More general Maxwell - Cattaneo theory

It has been argued that the basic equations of Maxwell - Cattaneo theory, equations, (4.1), should have coefficients which in general depend on temperature, in order to be compatible with thermodynamics. Arguments along these lines may be seen in the work of (Coleman et al., 1982; Coleman et al., 1986), Dario Graffi in 1984, see (Franchi and Straughan, 1994b), (Morro and Ruggeri, 1987; Morro and Ruggeri, 1988), (Coleman and Newman, 1988). These writers essentially argue that the internal energy ϵ, entropy η, and Helmholtz free energy ψ should depend on $q_i q_i$ where q_i is the heat flux. The development of (Morro and Ruggeri, 1987; Morro and Ruggeri, 1988) employs an appealing use of internal variables, an approach extended to thermoelasticity by (Caviglia et al., 1992) and to porous media by (Fabrizio et al., 2008) and by (Straughan, 2008), pp. 351–353.

In one space dimension the generalized model is effectively given by (Coleman and Newman, 1988) in the form

$$\begin{aligned} \big(c_0(\theta) + a'(\theta) q^2\big)\theta_t + 2a(\theta) q q_t &= -q_x, \\ \tau(\theta) q_t + q &= -\kappa(\theta)\theta_x\,. \end{aligned} \tag{4.54}$$

(Coleman and Newman, 1988) develop a nonlinear wave analysis for system (4.54) and fit the wavespeeds predicted to those found experimentally for NaF and Bi. In this way they are able to make some progress with functional forms for c_0, a, τ and κ.

(Morro and Ruggeri, 1987) perform a complete acceleration wave analysis for their model and also analyse thermal shocks. They also investigate the question of reconciling wavespeeds with thermodynamics. Their model for heat propagation in solids at low temperatures is

$$\begin{aligned} c_0\dot{\theta} &= -q_{i,i}\,, \\ \mathcal{T}\dot{q}_i + (1+\Gamma\dot{\theta})q_i &= -\kappa\theta_{,i}\,, \end{aligned} \tag{4.55}$$

where $c_0, \kappa, \mathcal{T}, \Gamma$ are, in general, functions of θ and for constants $\tilde{A}$ and $\tilde{B}$ they suggest $\mathcal{T}$ and Γ may have forms

$$\begin{aligned} \mathcal{T} &= \kappa(\tilde{A}\theta^{-3} + \tilde{B}\theta^{n-3}), \\ \Gamma &= \kappa\big(5\tilde{A}\theta^{-4} + (5-n)\tilde{B}\theta^{n-4}\big) \end{aligned}$$

for a suitable constant n.

4.1.4 Dual phase lag theory

We have discussed acceleration wave development in detail for the MC theory. Since the approximate models arising from the dual phase lag theory of (Tzou, 1995b; Tzou, 1995a), see section 1.5, are in some sense a generalization of the (Cattaneo, 1948) model we make some comments for acceleration wave propagation in this area. In fact, (Straughan, 2008), pp. 358–360 includes brief comments on acceleration wave propagation in a class of dual phase lag models for a fluid.

If the basic model is based on an energy equation of form

$$c\dot{\theta} = -q_{i,i}\,, \tag{4.56}$$

and then expands a dual phase lag model like

$$q_i(\mathbf{x}, t+\alpha) = -\kappa\big(\theta(\mathbf{x}, t+\tau)\big)\theta_{,i}(\mathbf{x}, t+\tau), \tag{4.57}$$

then we might consider a Taylor expansion of the type

$$\alpha\dot{q}_i + q_i = -\Big(\kappa + \tau\dot{\theta}\kappa' + \frac{\tau^2\dot{\theta}^2}{2}\kappa''(\theta)\Big)\Big(\theta_{,i} + \tau\dot{\theta}_{,i} + \frac{\tau^2}{2}\ddot{\theta}_{,i}\Big)\,. \tag{4.58}$$

Wave motion could be based on equations (4.56) and (4.58), with if need be, an α^2 term on the left of (4.58). While progress is possible with this class of model, an *a priori* knowledge of the exact nature of the expansion, i.e. which terms to retain, would be a distinct advantage.

4.2 Type II rigid heat conductor

A concise and detailed acceleration wave analysis for the (Gurtin and Pipkin, 1968) rigid heat conductor model is provided by (Chen, 1969a). The analysis of (Chen, 1969a) is for a plane wave propagating in one direction. (Chen, 1969b) extends his previous analysis to acceleration waves of arbitrary shape. The Bernoulli equation obtained for the wave amplitude is presented for an arbitrary shaped wave and special attention is paid to acceleration waves of cylindrical and spherical shape. (Lindsay and Straughan, 1976) presented an acceleration wave analysis for a one-dimensional wave in the (Green and Laws, 1972) theory for a rigid body.

The object of this section is to report on work of (Jaisaardsuetrong and Straughan, 2007) who perform an analysis of acceleration wave motion in a (Green and Naghdi, 1991) rigid solid of type II. (Jaisaardsuetrong and Straughan, 2007) retain the equations of (Green and Naghdi, 1991) in their full general nonlinearity, and they completely determine the wave speed and the amplitude of the wave as a function of time. The basic equations of (Green and Naghdi, 1991) type II theory for a rigid solid are described in chapter 1, section 1.10. In the interests of clarity we recap the necessary equations here.

The governing equation is the balance of entropy, namely

$$\rho\dot{\eta} = \rho\xi + \rho s - p_{i,i}, \tag{4.59}$$

where ρ, η, ξ and s are, respectively, density, entropy, internal rate of production of entropy per unit mass, external rate of production of entropy per unit mass, and p_i is the entropy flux vector.

For the thermal displacement variable

$$\alpha(\mathbf{x}, t) = \int_{t_0}^{t} T(\mathbf{x}, s)ds + \alpha_0,$$

there is a temperature function $\theta = \theta(T, \alpha) = \theta(\dot{\alpha}, \alpha)$ such that $\theta > 0$ and $\partial\theta/\partial T > 0$. The functions η, p_i and ξ are expressed in terms of a Helmholtz free energy function $\psi = \psi(\theta, \beta_i)$ by

$$\eta = -\frac{\partial\psi}{\partial T}\Big/\frac{\partial\theta}{\partial T}, \tag{4.60}$$

$$p_i = -\rho\frac{\partial\psi}{\partial\beta_i}\Big/\frac{\partial\theta}{\partial T}, \tag{4.61}$$

and

$$\xi = 2\Lambda\psi_\Lambda\frac{\theta_\alpha}{\theta\theta_T} + \frac{T}{\theta}\left(\frac{\psi_T\theta_\alpha}{\theta_T} - \psi_\alpha\right), \tag{4.62}$$

where β_i and Λ are defined by

$$\beta_i = \alpha_{,i}, \qquad \Lambda = \alpha_{,i}\alpha_{,i} = \beta_i\beta_i. \tag{4.63}$$

To study the simplest acceleration wave we set the external supply of entropy to be zero and so we put $s = 0$. Then, with the aid of equations

(4.60) - (4.62), the governing equation (4.59) may be rewritten as

$$-\frac{\partial \Psi}{\partial t} = \xi + \frac{\partial \mu_i}{\partial x_i}, \tag{4.64}$$

where Ψ and μ_i have been introduced and are

$$\Psi = \frac{\psi_T}{\theta_T}, \qquad \mu_i = \frac{\psi_{\beta_i}}{\theta_T}. \tag{4.65}$$

Equation (4.64) represents the general theory of heat flow in a type II rigid body of (Green and Naghdi, 1991). Special cases follow with the choice of free energy,

$$\psi = c(\theta - \theta \ln \theta) + \frac{k}{2}\beta_i\beta_i \tag{4.66}$$

or, specializing also for the function θ,

$$\psi = c(\theta - \theta \ln \theta) + \frac{k}{2}\beta_i\beta_i, \qquad \theta = a + bT, \tag{4.67}$$

where c, k, a, b are positive constants. When (4.67) holds then $\xi = 0$, see (Green and Naghdi, 1991).

4.2.1 Acceleration waves in type II theory

We define an acceleration wave for a solution to (4.64) to be a two-dimensional surface, $\mathcal{S}$, in $\mathbb{R}^3$, across which $\ddot{\alpha}(\mathbf{x}, \mathbf{t}), \dot{\alpha}_{,\mathbf{i}}(\mathbf{x}, \mathbf{t})$, and $\alpha_{,ij}(\mathbf{x}, \mathbf{t})$ suffer a finite discontinuity, but $\alpha \in C^1(\mathbb{R}^3)$, i.e. in the spatial variables. The jump, $[f]$, of a function f, across $\mathcal{S}$ is defined as in (4.3). The jump is assumed to be even along the wave surface, cf. (Chen, 1973), so that $[f]$ is a function only of t.

We take the jump of equation (4.64) using the forms (4.65) for Ψ and μ_i, noting also the form for ξ from (4.62) and recalling $\theta = \theta(T, \alpha)$, $\psi = \psi(\theta, \beta_i)$. This leads to the following equation for the wave speed V of $\mathcal{S}$

$$\begin{aligned} &V^2\left(\frac{\psi_{TT}}{\theta_T} - \frac{\psi_T\theta_{TT}}{{\theta_T}^2}\right) \\ &- 2\beta_i n_i\left(\frac{2\psi_{T\Lambda}}{\theta_T} - \frac{\theta_{TT}\psi_\Lambda}{{\theta_T}^2}\right)V + 2\frac{\psi_\Lambda}{\theta_T} + 4(\beta_i n_i)^2\frac{\psi_{\Lambda\Lambda}}{\theta_T} = 0. \end{aligned} \tag{4.68}$$

Hence there is a wave moving in the $\pm n_i$ directions, where n_i is the unit outward normal to $\mathcal{S}$, with speed V given by the solutions to (4.68). (Jaisaardsuetrong and Straughan, 2007) note that if ψ is given by (4.66) then equation (4.68) reduces to

$$\frac{-\theta_T c}{\theta}V^2 + k(\beta_i n_i)\frac{\theta_{TT}}{{\theta_T}^2}V + \frac{k}{\theta_T} = 0, \tag{4.69}$$

while if ψ and θ are given by (4.67) then equation (4.68) becomes

$$V^2 = \frac{k\theta}{b^2 c}. \tag{4.70}$$

This implies there are waves moving in opposite directions with speeds $V = \pm b^{-1}\sqrt{k\theta/c}$.

4.2.2 Region with no x variation

To make things more transparent (Jaisaardsuetrong and Straughan, 2007) then restrict attention to the case of a wave moving into a homogeneous region for which

$$\alpha_{,i} \equiv 0. \tag{4.71}$$

In this case the wavespeed equation (4.68) reduces to

$$V^2 = 2\psi_\Lambda \Big/ \left(\frac{\psi_\Lambda \theta_{TT}}{\theta_T} - \psi_{TT} \right). \tag{4.72}$$

Equation (4.69) becomes

$$V^2 = \frac{k\theta}{c{\theta_T}^2}, \tag{4.73}$$

and equation (4.70) remains the same.

We now calculate the amplitudes for a wave entering a homogeneous region.

4.2.3 Amplitude solution

To determine the wave amplitude $A(t) = [\ddot{\alpha}]$ we differentiate (4.64) with respect to t and take the jump of the resulting equation. The result is

$$\begin{aligned} &- \Psi_{TT}[\ddot{\alpha}]^2 - 2\Psi_{T\alpha}T[\ddot{\alpha}] - \Psi_T[\dddot{\alpha}] - \Psi_\alpha[\ddot{\alpha}] - 2\Psi_\Lambda[\dot{\alpha}_{,i}\dot{\alpha}_{,i}] \\ &= \xi_T[\ddot{\alpha}] + 2\left(\frac{\psi_\Lambda}{\theta_T}\right)_T \Big\{ [\dot{\alpha}_{,i}\dot{\alpha}_{,i}] + [\ddot{\alpha}\alpha_{,ii}] \Big\} \\ &\quad + 2T\left(\frac{\psi_\Lambda}{\theta_T}\right)_\alpha [\alpha_{,ii}] + 2\frac{\psi_\Lambda}{\theta_T}[\dot{\alpha}_{,ii}] = 0. \end{aligned} \tag{4.74}$$

The Hadamard relation (4.27) and the wavespeed equation (4.72) are now employed to remove the $[\dddot{\alpha}]$ and $[\dot{\alpha}_{,ii}]$ terms. This leads to the amplitude equation

$$2\frac{\delta A}{\delta t} + \alpha_1 A - \beta A^2 = 0. \tag{4.75}$$

After some calculation one shows that the coefficients α_1 and β have forms

$$\alpha_1 = -\frac{1}{V}\frac{\delta V}{\delta t} - \frac{V^2\theta_T T}{\psi_\Lambda}\left(\frac{\psi_T}{\theta_T}\right)_{T\alpha} - \frac{V^2\theta_T}{2\psi_\Lambda}\left(\frac{\psi_T}{\theta_T}\right)_\alpha$$
$$- \frac{T\theta_T}{\psi_\Lambda}\left(\frac{\psi_\Lambda}{\theta_T}\right)_\alpha - \frac{V^2\theta_T}{2\psi_\Lambda}\left\{\left(\frac{\psi_T\theta_\alpha T}{\theta\theta_T}\right)_T - \left(\frac{\psi_\alpha T}{\theta}\right)_T\right\}, \tag{4.76}$$

and

$$\beta = \frac{V^2\theta_T}{2\psi_\Lambda}\left(\frac{\psi_T}{\theta_T}\right)_{TT} + \frac{\theta_T}{\psi_\Lambda}\left(\frac{\psi_T}{\theta_T}\right)_\Lambda + 2\frac{\theta_T}{\psi_\Lambda}\left(\frac{\psi_\Lambda}{\theta_T}\right)_T. \tag{4.77}$$

While a solution to (4.75) is easily found, (Jaisaardsuetrong and Straughan, 2007) argue that one can understand the physical situation easier when the free energy satisfies (4.66) or (4.67).

In the case ψ satisfies (4.66), α_1 and β are given by

$$\alpha_1 = -\frac{1}{V}\frac{\delta V}{\delta t} + \frac{2T\theta}{\theta_T}(\ln\theta)_{T\alpha} + \frac{\theta}{\theta_T}(\ln\theta)_\alpha$$
$$+ T\frac{\theta_{T\alpha}}{\theta_T} + \frac{2\theta}{\theta_T}\left(\frac{\ln\theta\cdot\theta_\alpha T}{\theta}\right)_T, \tag{4.78}$$

$$\beta = \frac{\theta_T}{\theta} - 3\frac{\theta_{TT}}{\theta_T}. \tag{4.79}$$

(Jaisaardsuetrong and Straughan, 2007) consider the forms for ψ and θ in (4.67), but further assume $T^+ =$ constant. Then one has

$$\alpha_1 = 0, \quad \beta = \frac{b}{(a+bT)}. \tag{4.80}$$

The Bernoulli equation (4.75) reduces to

$$\frac{\delta A}{\delta t} - \frac{\beta}{2}A^2 = 0. \tag{4.81}$$

This equation is solved to see that

$$A(t) = \frac{A(0)}{1 - \dfrac{A(0)\beta t}{2}}. \tag{4.82}$$

Since $\beta > 0$, we see that if $A(0) > 0$ then $A(t)$ blows up in a finite time, $\mathcal{T}$, where

$$\mathcal{T} = \frac{2(a+bT^+)}{ba(0)}.$$

(Jaisaardsuetrong and Straughan, 2007) note that for equation (4.81) one has $\xi = 0$ and it is interesting to note that this may always lead to thermal shock formation.

From (4.82) it is necessary that $A(0) > 0$ for blow-up. This is because in this section the amplitude $A(t)$ is defined as $[\ddot{\alpha}]$. In section 4.1.2 the amplitude was $[\theta_x]$. Since $\dot{\alpha} = T$ the amplitude

$$A(t) = [\ddot{\alpha}] = [\dot{T}] = \dot{T}^-(t)$$

for a wave moving into a region with T^+ constant. From the Hadamard relation (4.5) $[\dot{T}] = -V[T_x]$ and so $A(t) = -VT_x^-(t)$. Thus $A(0) > 0$ yields consequent blow-up of $A(t)$ corresponding to $T_x^-(t) \to -\infty$ which is consistent with a jump in T (thermal shock) forming.

4.3 Acceleration waves with microtemperatures

In this section we commence an acceleration wave analysis for the theory of a rigid body with microtemperatures as described in section 1.12.

An acceleration wave in a rigid solid with microtemperatures is a surface $\mathcal{S}$ such that $\alpha, \beta_i \in C^1(\mathbb{R}^3)$ in the $\mathbf{x}$ variables but the second and higher derivatives possess a finite jump discontinuity across $\mathcal{S}$. It transpires, in fact, that an acceleration wave analysis is rather involved, more so even than the analogous analysis in type II thermoelasticity. Hence, to understand this we begin with the linear equations for an isotropic body. Thus, take the jump of equations (1.139) to find

$$\begin{aligned} &a[\ddot{\alpha}] - K[\alpha_{,ii}] + m[\dot{\beta}_{i,i}] = 0, \\ &b[\ddot{\beta}_i] - d_2[\beta_{i,jj}] - (d_1 + d_3)[\beta_{j,ji}] + m[\dot{\alpha}_{,i}] = 0, \end{aligned} \tag{4.83}$$

for the supply functions s, s_i continuous everywhere. Next, define the wave amplitudes A and B_i by

$$A = [n_i n_j \alpha_{,ij}], \qquad B_i = [n_r n_s \beta_{i,rs}]\,. \tag{4.84}$$

Upon use of the compatibility relations (4.25) and (4.26) together with the Hadamard relation (4.27) one shows from (4.83)

$$\begin{aligned} &aV^2 A - KA = mVB_i n_i, \\ &bV^2 B_i - d_2 B_i - (d_1 + d_3) B_j n_j n_i = mV n_i A. \end{aligned} \tag{4.85}$$

Now write B_i as the sum of its components parallel to the unit normal n_i to $\mathcal{S}$, namely B_{II}, and its tangential components $B_\perp^\alpha$, $\alpha = 1, 2$, i.e.

$$B_i = B_{II} n_i + B_\perp^\alpha x^i_{;\alpha}$$

where $x^i_{;\alpha}$ are tangential vectors to $\mathcal{S}$ in the directions of surface coordinates u_1 and u_2. Thus, (4.85) yield

$$\begin{aligned} &aV^2 A - KA = mVB_{II}\,, \\ &\big[bV^2 - (d_1 + d_2 + d_3)\big] B_{II} = mVA \end{aligned} \tag{4.86}$$

and

$$(bV^2 - d_2)B_\perp^\alpha x^i_{;\alpha} = 0. \tag{4.87}$$

Equation (4.87) immediately shows that the theory with microtemperatures possess a much richer structure than any so far met in this book. Even in the linear, isotropic theory for a rigid solid the thermal structure at the microscopic level is leading to the possibility of a transverse wave with amplitudes $B_\perp^1$ and $B_\perp^2$, and speed V given by

$$V^2 = \frac{d_2}{b}. \tag{4.88}$$

Equations (4.86) are a system in A, B_{II} which lead to the wavespeed equation for a longitudinal wave, namely

$$(aV^2 - K)(bV^2 - D) - m^2V^2 = 0, \tag{4.89}$$

where $D = d_1 + d_2 + d_3$. Thus, if $V_T = \sqrt{K/a}$ denotes the speed of a thermal wave and $V_{MT} = \sqrt{D/b}$ denotes the speed of a "microthermal wave" then (4.89) admits a fast wave with speed V_2 and a slow wave with speed V_1 with

$$V_1^2 < \min\{V_T^2, V_{MT}^2\} < \max\{V_T^2, V_{MT}^2\} < V_2^2.$$

Rather than now proceed immediately to the nonlinear case it is instructive to first develop an acceleration wave analysis for the anisotropic linear equations with microtemperatures, namely, equations (1.137). For the supply functions s, s_i in equations (1.139) continuous we define an acceleration wave as above and take the jumps of equations (1.137) to find

$$\begin{aligned} a[\ddot{\alpha}] - K_{ij}[\alpha_{,ij}] + M_{ij}[\dot{\beta}_{i,j}] &= 0, \\ B_{ij}[\ddot{\beta}_j] - D_{ijrs}[\beta_{r,sj}] + M_{ij}[\dot{\alpha}_{,j}] &= 0. \end{aligned} \tag{4.90}$$

We again employ the compatibility relations (4.25) and (4.26) together with the Hadamard relation (4.27), recalling the wave amplitude definitions (4.84). In this way from (4.90) we obtain

$$\begin{aligned} \big(U_N^2 B_{ij} - Q_{ij}(\mathbf{n})\big)B_j &= U_N M_{ij} n_j A, \\ \big(aU_N^2 - K_{ij}n_i n_j\big)A &= U_N M_{ij} n_j B_i, \end{aligned} \tag{4.91}$$

where

$$Q_{ij}(\mathbf{n}) = D_{iajb} n_a n_b\,, \tag{4.92}$$

and Q_{ij} plays a role of a thermal "acoustic tensor". Equation (4.91) is similar to the jump of the momentum equation in thermoelasticity. However, it is more complicated owing to the B_{ij} term.

In general, we expect a generalized transverse wave from (4.91) and also a generalized longitudinal wave. A generalized longitudinal wave is one where

$B_i = Bn_i^*$ with n_i^* being the unit vector in the direction of $M_{ij}n_j$. We may show that a plane generalized longitudinal wave will exist by appealing to the proof of theorem 2 of (Chadwick and Currie, 1974). Define the matrix $F_{ij}(\mathbf{n})$ by

$$F_{ij}(\mathbf{n}) = Q_{ij}(\mathbf{n}) - U_N^2 B_{ij} \,.$$

Then for any unit vector $\boldsymbol{\omega}$ define the vector field $\mathbf{m}$ by

$$\mathbf{m}(\boldsymbol{\omega}) = \mathbf{M}^{-1}\mathbf{F}(\boldsymbol{\omega})\mathbf{M}\boldsymbol{\omega} - \big[\boldsymbol{\omega}\cdot(\mathbf{M}^{-1}\mathbf{F}(\boldsymbol{\omega})\mathbf{M}\boldsymbol{\omega})\big]\boldsymbol{\omega}\,. \tag{4.93}$$

Note that we require M_{ij} to define a non-singular matrix. The vector $\mathbf{m}$ is a continuous function of $\boldsymbol{\omega}$ and $\mathbf{m}\cdot\boldsymbol{\omega} = 0$. The set of unit vectors forms a sphere S in three-space and so $\mathbf{m}(\boldsymbol{\omega})$ defines a continuous tangent field on S. Thus $\mathbf{m}$ has a zero, see (Chadwick and Currie, 1974), p. 486, and so there is a unit vector $\mathbf{n}$ with $\mathbf{m}(\mathbf{n}) = \mathbf{0}$. Then, from (4.93) $\mathbf{Mn}$ is an eigenvector of $\mathbf{F}(\mathbf{n})$. Thus we have at least one direction $\mathbf{n}$ such that a generalized longitudinal plane wave propagates. The amplitude B_i is in the direction $\mathbf{n}^*$ where $\mathbf{n}^*$ is in the direction $\mathbf{Mn}$. One may thus analyse the wavespeed and amplitude of a plane wave for (4.91).

To complete this section we briefly investigate the propagation of an acceleration wave in a nonlinear rigid body with microtemperatures. To do this we employ equations (1.132) and (1.133) with the constitutive theory (1.136). We suppose for simplicity that the body has a centre of symmetry. Then we define an acceleration wave S as earlier in this section with amplitudes given by (4.84). Expanding (1.132) and (1.133), recalling the centre of symmetry, and taking jumps we find

$$\rho\eta_\theta[\ddot{\alpha}] + \rho\frac{\partial\eta}{\partial\beta_{i,j}}[\dot{\beta}_{i,j}] = \frac{\partial S_k}{\partial T_i}[\dot{\beta}_{i,k}] + \frac{\partial S_k}{\partial\alpha_{,i}}[\alpha_{,ik}]$$

and

$$\rho\frac{\partial\eta_j}{\partial T_j}[\ddot{\beta}_j] + \rho\frac{\partial\eta_i}{\partial\alpha_{,j}}[\dot{\alpha}_{,j}] = \frac{\partial S_{ki}}{\partial\theta}[\dot{\alpha}_{,k}] + \frac{\partial S_{ki}}{\partial\beta_{a,b}}[\beta_{a,bk}]\,.$$

Again employing the compatibility relations (4.25), (4.26) and the Hadamard relation (4.27) we derive

$$\left(\rho\eta_\theta U_N^2 - \frac{\partial S_k}{\partial\alpha_{,i}}n_in_k\right)A = \left(\rho\frac{\partial\eta}{\partial\beta_{i,k}} - \frac{\partial S_k}{\partial T_i}\right)U_Nn_kB_i\,, \tag{4.94}$$

and

$$\left(\rho U_N^2\frac{\partial\eta_i}{\partial T_j} - Q_{ij}\right)B_j = \left(\rho\frac{\partial\eta_i}{\partial\alpha_{,j}} - \frac{\partial S_{ji}}{\partial\theta}\right)U_Nn_jA\,. \tag{4.95}$$

One may define

$$M_{ij} = \rho\frac{\partial\eta_i}{\partial\alpha_{,j}} - \frac{\partial S_{ji}}{\partial\theta}$$

and provided this is invertible one may show there is a generalized plane longitudinal wave in a direction $\mathbf{n}$ with amplitude in the direction $\mathbf{Mn}$. One may show this generalized longitudinal wave gives rise to a fast wave and a slow wave whose wavespeeds U_N^1 and U_N^2 are solutions to the equation

$$\begin{aligned}\Big(\rho\eta_\theta U_N^2 - \frac{\partial S_k}{\partial \alpha_{,p}} n_p n_k\Big)\Big(Q_{ij} - \rho U_N^2 \frac{\partial \eta_i}{\partial T_j}\Big) M_{ja} n_a n_i\\ + U_N^2\Big(\rho\frac{\partial \eta}{\partial \beta_{i,k}} - \frac{\partial S_k}{\partial T_i}\Big) M_{ia} M_{rs} n_k n_a n_r n_s = 0,\end{aligned} \tag{4.96}$$

where $Q_{ij}(\mathbf{n})$ is a (nonlinear) thermal "acoustic tensor" defined by

$$Q_{ia}(\mathbf{n}) = \frac{\partial S_{ki}}{\partial \beta_{a,b}} n_k n_b \,.$$

To the best of my knowledge a theory of acceleration wave propagation in the microtemperatures theory of section (1.12) has not been presented before. Obviously there is much more one may do regarding the calculation of wave amplitudes, studying transverse waves, and other things. It would appear this has potential to be richer even than the comprehensive theory of acceleration waves in thermoelasticity. In the next section we study acceleration waves in a thermoelastic body of type II.

4.4 Type II thermoelasticity

In this section we develop an acceleration wave analysis for the thermoelastic theory of type II introduced by (Green and Naghdi, 1993), described in section 2.3. (Green and Naghdi, 1993) present their equations with a symmetric stress tensor in the current configuration. However, when dealing with acceleration waves we believe it is better to refer everything back to the reference configuration and so we now present the equations with the Piola - Kirchoff stress tensor. While there are many studies of wave motion in the literature which employ the linearized theory of type II thermoelasticity we have not seen an acceleration wave analysis for the fully nonlinear theory as is presented here.

The (Green and Naghdi, 1993) theory essentially starts with an energy equation of form

$$\rho_0\dot\epsilon - \rho_0\theta s + q_{A,A} - S_{Ai}\dot x_{i,A} = 0, \tag{4.97}$$

where $\rho_0, \epsilon, \theta, s, q_A, x_i, S_{Ai}$ are the density, internal energy, temperature, external entropy supply, heat flux, position vector, and the Piola-Kirchoff stress tensor, each referred to the reference configuration. Hence, $_{,A}$ denotes $\partial/\partial X_A$, where X_A is the position in the reference configuration. Introducing the entropy flux vector $p_A = q_A/\theta$ and noting $\epsilon = \psi + \eta\theta$, ψ and η being the Helmholtz free energy function and the entropy, (Green and Naghdi,

1993) deal with a reduced energy equation

$$\rho_0(\dot{\psi} + \eta\dot{\theta}) + \rho_0\theta\xi + p_A\theta_{,A} - S_{Ai}\dot{x}_{i,A} = 0, \tag{4.98}$$

in which ξ is an internal supply of entropy per unit mass. Judicious manipulation of this energy equation allows (Green and Naghdi, 1993) to derive restrictions on the constitutive variables.

To develop an acceleration wave analysis, the key equations of (Green and Naghdi, 1993) are the balance of linear momentum

$$\rho_0\ddot{x}_i = \rho_0 b_i + S_{Ai,A}\,, \tag{4.99}$$

where b_i is a body force, and the balance of entropy

$$\rho_0\dot{\eta} - \rho_0 s - \rho_0\xi + p_{A,A} = 0. \tag{4.100}$$

For type II theory (Green and Naghdi, 1993) show that

$$\begin{aligned} \psi &= \psi(\theta, \alpha_{,B}, F_{jB}),\\ \eta &= \eta(\theta, \alpha_{,B}, F_{jB}),\\ S_{Ai} &= S_{Ai}(\theta, \alpha_{,B}, F_{jB}), \end{aligned} \tag{4.101}$$

where $F_{jB} = \partial x_j/\partial X_B = x_{j,B}$. In particular, (Green and Naghdi, 1993) derive relations equivalent to the following,

$$\eta = -\frac{\partial\psi}{\partial\theta}\,, \qquad S_{Ai} = \rho_0\frac{\partial\psi}{\partial F_{iA}}\,, \qquad p_A = -\rho_0\frac{\partial\psi}{\partial\alpha_{,A}}\,, \qquad \xi = 0. \tag{4.102}$$

In developing a nonlinear acceleration wave analysis we take the body force b_i and external entropy supply s to be zero. Hence, we analyse the equations, where without loss of generality we use $u_i = x_i - X_i$ rather than x_i,

$$\begin{aligned} \rho_0\ddot{u}_i &= S_{Ai,A}\,,\\ \rho_0\dot{\eta} + p_{A,A} &= 0. \end{aligned} \tag{4.103}$$

An acceleration wave in a thermoelastic body of type II is a surface $\mathcal{S}$ in $\mathbb{R}^3$ for which $u_i(\mathbf{X},t)$, $\alpha(\mathbf{X},t) \in C^1(\mathbb{R}^3)$ in their spatial variables, but $\ddot{u}_i$, $\dot{u}_{i,A}$, $u_{i,AB}$, $\ddot{\alpha}$, $\dot{\alpha}_{,A}$, $\alpha_{,AB}$ and their higher derivatives have a jump discontinuity across $\mathcal{S}$.

Using the fact that η, S_{Ai} and p_A are derivatives of ψ, and the constitutive theory in equation (4.101), we expand equations (4.103) and take the jumps recalling the definition of an acceleration wave, to obtain

$$\rho_0[\ddot{u}_i] = \frac{\partial S_{Ai}}{\partial\theta}\,[\theta_{,A}] + \frac{\partial S_{Ai}}{\partial F_{jR}}\,[F_{jR,A}] + \frac{\partial S_{Ai}}{\partial\alpha_{,R}}\,[\alpha_{,RA}]\,, \tag{4.104}$$

and

$$\begin{aligned} \rho_0\bigg(\frac{\partial\eta}{\partial\theta}\,[\dot{\theta}] + \frac{\partial\eta}{\partial F_{iQ}}\,[\dot{F}_{iQ}] + \frac{\partial\eta}{\partial\alpha_{,Q}}\,[\dot{\alpha}_{,Q}]\bigg)\\ = -\frac{\partial p_A}{\partial\theta}\,[\theta_{,A}] - \frac{\partial p_A}{\partial F_{jR}}\,[F_{jR,A}] - \frac{\partial p_A}{\partial\alpha_{,R}}\,[\alpha_{,RA}]\,. \end{aligned} \tag{4.105}$$

Upon use of the Hadamard relation (4.27) in three dimensions and the definition of an acceleration wave we may derive the following relations

$$[\ddot{u}_i] = -U_N[\dot{u}_{i,A}]N^A, \tag{4.106}$$
$$[\dot{u}_{i,A}] = -U_N N^B[u_{i,AB}], \tag{4.107}$$
$$[\ddot{u}_i] = U_N^2 N^A N^B[u_{i,AB}] = U_N^2 A_i, \tag{4.108}$$
$$[\theta_{,A}] = [\dot{\alpha}_{,A}] = -U_N N^B[\alpha_{,AB}], \tag{4.109}$$
$$[\ddot{\alpha}] = [\dot{\theta}] = -U_N[\dot{\alpha}_{,A}]N^A = -U_N[\theta_{,A}]N^A, \tag{4.110}$$
$$= U_N^2 N^A N^B[\alpha_{,AB}] = U_N^2 B, \tag{4.111}$$

where we have introduced the wave amplitudes

$$A_i(t) = N^A N^B[u_{i,AB}], \qquad B(t) = N^A N^B[\alpha_{,AB}], \tag{4.112}$$

and have recalled that $\theta = \dot{\alpha}$.

We employ (4.106) - (4.111) in the jump equations (4.104) and (4.105) to derive the following general amplitude equations,

$$\rho_0 U_N^2 A_i = Q_{ij}A_j + BN_R N_A \frac{\partial S_{Ai}}{\partial \alpha_{,R}} - U_N N_A \frac{\partial S_{Ai}}{\partial \theta} B\,, \tag{4.113}$$

$$\begin{aligned} &\rho_0 \frac{\partial \eta}{\partial \theta} U_N^2 B - B\Big(\rho_0 \frac{\partial \eta}{\partial \alpha_{,A}} N^A U_N + N^A U_N \frac{\partial p_A}{\partial \theta}\Big) \\ &= -N_A N_B \frac{\partial p_A}{\partial \alpha_{,B}} B - A_i\Big(N_A N_B \frac{\partial p_A}{\partial F_{iB}} - \rho_0 U_N N_A \frac{\partial \eta}{\partial F_{iA}}\Big), \end{aligned} \tag{4.114}$$

where Q_{ij} is the acoustic tensor defined by

$$Q_{ij} = N_A N_R \frac{\partial S_{Ai}}{\partial F_{jR}}\,. \tag{4.115}$$

Equations (4.113) and (4.114) are a system of equations in the variables A_1, A_2, A_3 and B and give rise to a polynomial equation for the wavespeeds U_N^2. It is possible to make progress in full generality. However, for the purpose of this section it is likely to be more transparent to consider a wave moving into a particular region and this we now do. Hence, we consider an acceleration wave propagating into a static configuration at uniform deformation and temperature, and we suppose the initial body has a centre of symmetry. In this region ahead of the wave F_{iA} and θ are constant and $\alpha_{,A} \equiv 0$. If we denote quantities evaluated in the region ahead of the wave by a subscript E then since the body has a centre of symmetry, cf. (Truesdell and Noll, 1992), p. 358, or (Spencer, 1980), p. 110, it follows that

$$\frac{\partial S_{Ai}}{\partial \alpha_{,R}}\Big|_E = 0, \qquad \rho_0 \frac{\partial \eta}{\partial \alpha_{,A}}\Big|_E = 0, \qquad \frac{\partial p_A}{\partial \theta}\Big|_E = 0, \qquad \frac{\partial p_A}{\partial F_{iB}}\Big|_E = 0\,.$$

Thus, equations (4.113) and (4.114) now reduce to

$$(Q_{ij} - \rho_0 U_N^2 \delta_{ij})A_j - U_N N_A \frac{\partial S_{Ai}}{\partial \theta} B = 0, \tag{4.116}$$

$$\begin{aligned}\left(\rho_0 \frac{\partial \eta}{\partial \theta} U_N^2 + N_A N_B \frac{\partial p_A}{\partial \alpha_{,B}}\right) B &= \rho_0 U_N N_A \frac{\partial \eta}{\partial F_{iA}} A_i \\ &= -U_N N_A \frac{\partial S_{Ai}}{\partial \theta} A_i \,,\end{aligned} \tag{4.117}$$

where in the last equation (4.102) has been employed.

From (Chen, 1973) equation (4.10) we know that

$$N_A = F_{iA} \frac{|\nabla_{\mathbf{x}} \sigma|}{|\nabla_{\mathbf{X}} \mathcal{S}|} n_i$$

where n_i is the equivalent unit normal to N_A in the current configuration, and σ is the equivalent surface to $\mathcal{S}$ in the current configuration. We define the tensor β_{ij} by

$$\beta_{ij} = \frac{|\nabla_{\mathbf{x}} \sigma|}{|\nabla_{\mathbf{X}} \mathcal{S}|} \left(\frac{\partial S_{Ai}}{\partial \theta} F_{jA} \right) \Big|_E \tag{4.118}$$

and then we may rewrite equation (4.116) as

$$(\tilde{Q}_{ij}(\mathbf{n}) - \rho_0 U_N^2 \delta_{ij})A_j - U_N B \beta_{ij} n_j = 0, \tag{4.119}$$

where $\tilde{Q}_{ij}(\mathbf{n})$ is the tensor $Q_{ij}(\mathbf{N})$ but represented now as a function of $\mathbf{n}$ instead of $\mathbf{N}$.

We now use theorem 2 of (Chadwick and Currie, 1974) to infer there is at least one direction $\mathbf{n}^*$ such that $\boldsymbol{\beta}\mathbf{n}^*$ is an eigenvector of $\mathbf{Q}$. The wave is propagating into an equilibrium region at rest and so $\mathbf{Q}$ is a constant matrix and consequently $\boldsymbol{\beta}\mathbf{n}^*$ is fixed. The matrix $\boldsymbol{\beta}$ is also constant and hence $\mathbf{n}^*$ represents a fixed direction. Hence, we may study the propagation of a plane acceleration wave in the direction of $\mathbf{n}^*$ with its amplitude in the direction $\boldsymbol{\beta}\mathbf{n}^*$. In line with the definition in (Chadwick and Currie, 1974) in classical thermoelasticity we refer to these waves as generalized longitudinal waves.

4.4.1 Wavespeeds

Let now the unit vector in the direction of $\boldsymbol{\beta}\mathbf{n}^*$ be $\mathbf{m}$, so that $\mathbf{A} = A(t)\mathbf{m}$. Then equation (4.119) is

$$(\tilde{Q}_{ij} - \rho_0 U_N^2 \delta_{ij}) m_j A - U_N B N_A \frac{\partial S_{Ai}}{\partial \theta} = 0. \tag{4.120}$$

We multiply this equation by m_i and then the result together with (4.117) give

$$\begin{pmatrix} \tilde{Q}_{ij}m_im_j - \rho_0 U_N^2 & -U_N N_A m_i \dfrac{\partial S_{Ai}}{\partial\theta} \\ U_N N_A m_i \dfrac{\partial S_{Ai}}{\partial\theta} & \rho_0 \dfrac{\partial\eta}{\partial\theta} U_N^2 + N_A N_B \dfrac{\partial p_A}{\partial\alpha_{,B}} \end{pmatrix} \begin{pmatrix} A \\ B \end{pmatrix} = \begin{pmatrix} 0 \\ 0 \end{pmatrix}$$

For a non-zero solution to this we require

$$(U_N^2 - U_M^2)(U_N^2 - U_T^2) + KU_N^2 = 0. \tag{4.121}$$

Here the quantities U_M^2, U_T^2 and K are given by

$$U_M^2 = \frac{\tilde{Q}_{ij}m_im_j}{\rho_0}, \tag{4.122}$$

$$U_T^2 = -\frac{N_A N_B \dfrac{\partial p_A}{\partial\alpha_{,B}}}{\rho_0 \dfrac{\partial\eta}{\partial\theta}}, \tag{4.123}$$

and

$$K = -\frac{N_A \dfrac{\partial S_{Ai}}{\partial\theta} N_B \dfrac{\partial S_{Bi}}{\partial\theta}}{\rho_0^2 \dfrac{\partial\eta}{\partial\theta}}. \tag{4.124}$$

Equation (4.121) admits two solutions under suitable conditions on the wavespeeds and these are attributed to a fast wave (mechanical) and a slow wave (thermal).

Equation (4.121) possesses real solutions if $K \geq (U_M + U_T)^2$ or if $K \leq (U_M - U_T)^2$. The former case is inconsistent with $U_N^2 > 0$ and so we must have

$$K \leq (U_M - U_T)^2.$$

From (4.124) $K < 0$, and then the two wavespeeds $U_N^{(2)}$ and $U_N^{(1)}$ are such that

$$U_N^{(2)2} < \min\{U_M^2, U_T^2\} \leq \max\{U_M^2, U_T^2\} < U_N^{(1)2}. \tag{4.125}$$

The quantities U_M and U_T are the wavespeeds of an acceleration wave in a purely elastic material, or a thermal wave in a rigid solid of type II, respectively. Hence, we have a fast wave travelling faster than either of these two quantities and a slow wave travelling slower. We expect $K < 0$ to hold in practice.

4.4.2 Amplitude behaviour

One may proceed to calculate explicitly the amplitude of an arbitrary shaped wave or even of the plane wave whose speed has been calculated in section 4.4.1, cf. such calculations in (Lindsay and Straughan, 1978; Lindsay and Straughan, 1979). However, the differential geometry involved can obscure the procedure. Hence, we here calculate the amplitude of a one-dimensional acceleration wave for the type II equations (4.103). In order to see clearly which terms disappear due to the body having a centre of symmetry we begin by taking $\partial/\partial X_A$ of each of equations (4.99) and (4.100) and then take the jump of the non-zero terms. This leads to the equations,

$$\begin{aligned} \rho_0[\ddot{u}_{i,A}] =& \frac{\partial S_{Bi}}{\partial\theta}[\dot{\alpha}_{,BA}] + \frac{\partial^2 S_{Bi}}{\partial\theta^2}[\theta_{,A}\theta_{,B}] + \frac{\partial^2 S_{Bi}}{\partial\theta\partial F_{jR}}[\theta_{,B}u_{j,RA}] \\ &+ \frac{\partial S_{Bi}}{\partial F_{jR}}[u_{j,RBA}] + \frac{\partial^2 S_{Bi}}{\partial\theta\partial F_{jR}}[\theta_{,A}u_{j,RB}] \\ &+ \frac{\partial^2 S_{Bi}}{\partial F_{rK}\partial F_{jR}}[u_{r,KA}u_{j,RB}] + \frac{\partial^2 S_{Bi}}{\partial\alpha_{,Q}\partial\alpha_{,R}}[\alpha_{,RB}\alpha_{,QA}], \end{aligned} \tag{4.126}$$

and

$$\begin{aligned} \rho_0\eta_\theta[\ddot{\alpha}_{,A}] &+ \rho_0\eta_{\theta\theta}[\theta_{,A}\dot{\theta}] + \rho_0\frac{\partial^2\eta}{\partial F_{iR}\partial\theta}[\dot{\theta}u_{i,RA}] + \rho_0\frac{\partial\eta}{\partial F_{iQ}}[\dot{u}_{i,QA}] \\ &+ \rho_0\frac{\partial^2\eta}{\partial\theta\partial F_{iQ}}[\dot{F}_{iQ}\theta_{,A}] + \rho_0\frac{\partial^2\eta}{\partial F_{iQ}\partial F_{jR}}[\dot{F}_{iQ}F_{jR,A}] \\ &+ \rho_0\frac{\partial^2\eta}{\partial\alpha_{,Q}\partial\alpha_{,R}}[\dot{\alpha}_{,Q}\alpha_{,RA}] = -\frac{\partial^2 p_B}{\partial\theta\partial\alpha_{,Q}}[\alpha_{,QA}\theta_{,B}] \\ &- \frac{\partial^2 p_B}{\partial F_{iQ}\partial\alpha_{,R}}[\alpha_{,RA}F_{iQ,B}] - \frac{\partial p_B}{\partial\alpha_{,Q}}[\alpha_{,QBA}] \\ &- \frac{\partial^2 p_B}{\partial\theta\partial\alpha_{,Q}}[\alpha_{,QB}\theta_{,A}] - \frac{\partial^2 p_B}{\partial\alpha_{,Q}\partial F_{jR}}[F_{jR,A}\alpha_{,QB}]. \end{aligned} \tag{4.127}$$

We now specialize to the one space dimension case and employ the relation (4.3) for the jump of a product to find from (4.126) and (4.127),

$$\begin{aligned} \rho_0[\ddot{u}_X] =& \frac{\partial S}{\partial\theta}[\dot{\alpha}_{XX}] + \frac{\partial^2 S}{\partial\theta^2}[\theta_X]^2 + 2\frac{\partial^2 S}{\partial\theta\partial F}[\theta_X][u_{XX}] \\ &+ \frac{\partial S}{\partial F}[u_{XXX}] + \frac{\partial^2 S}{\partial F^2}[u_{XX}]^2 + \frac{\partial^2 S}{\partial\alpha_X^2}[\alpha_{XX}]^2, \end{aligned} \tag{4.128}$$

$$\begin{aligned} \rho_0\eta_\theta[\ddot{\alpha}_X] &+ \rho_0\eta_{\theta\theta}[\dot{\alpha}_X][\ddot{\alpha}] + \rho_0\eta_{\theta F}[\ddot{\alpha}][u_{XX}] + \rho_0\eta_F[\dot{u}_{XX}] \\ &+ \rho_0\eta_{F\theta}[\dot{u}_X][\dot{\alpha}_X] + \rho_0\eta_{FF}[\dot{u}_X][u_{XX}] + \rho_0\eta_{\alpha_X\alpha_X}[\dot{\alpha}_X][\alpha_{XX}] \\ &= -2p_{\theta\alpha_X}[\alpha_{XX}][\dot{\alpha}_X] - 2p_{F\alpha_X}[\alpha_{XX}][u_{XX}] - p_{\alpha_X}[\alpha_{XXX}], \end{aligned} \tag{4.129}$$

where u, S, F denote the one-dimensional counterparts of u_i, S_{Ai} and F_{iA}.

We now employ the Hadamard relation (4.5) to derive the following expressions,

$$
\begin{aligned}
[u_{Xt}] &= -U_N[u_{XX}], \\
[\ddot{u}_X] &= -2U_N\frac{\delta A}{\delta t} + U_N^2[u_{XXX}], \\
[\dot{u}_{XX}] &= \frac{\delta A}{\delta t} - U_N[u_{XXX}].
\end{aligned}
\tag{4.130}
$$

The one-dimensional equivalent of equations (4.116) and (4.117) are the equations

$$
\begin{aligned}
(S_F - \rho_0 U_N^2)A &= U_N\frac{\partial S}{\partial \theta}B, \\
(\rho_0\eta_\theta U_N^2 + p_{\alpha_X})B &= A\rho_0 U_N \eta_F.
\end{aligned}
\tag{4.131}
$$

One now employs (4.130) in (4.128) and (4.129) to derive

$$
\begin{aligned}
-2\rho_0 U_N\frac{\delta A}{\delta t} - \frac{\partial S}{\partial \theta}\frac{\delta B}{\delta t} + (\rho_0 U_N^2 - S_F)[u_{XXX}] + \frac{\partial S}{\partial \theta}U_N[\alpha_{XXX}] \\
- S_{\theta\theta}U_N^2B^2 + 2S_{\theta F}U_N BA - S_{FF}A^2 - S_{\alpha_X\alpha_X}B^2 = 0,
\end{aligned}
\tag{4.132}
$$

$$
\begin{aligned}
-2\rho_0\eta_\theta U_N\frac{\delta B}{\delta t} + \rho_0\eta_F\frac{\delta A}{\delta t} + (\rho_0\eta_\theta U_N^2 + p_{\alpha_X})[\alpha_{XXX}] \\
- \rho_0\eta_F U_N[u_{XXX}] - (\rho_0\eta_{\theta\theta}U_N^3 + \rho_0\eta_{\alpha_X\alpha_X}U_N + 2p_{\theta\alpha_X}U_N)B^2 \\
+ (2\rho_0\eta_{\theta F}U_N^2 + 2p_{F\alpha_X})BA - \rho_0\eta_{FF}U_N A^2 = 0.
\end{aligned}
\tag{4.133}
$$

Now form the combination (4.132)+(B/A)(4.133). Use (4.131) to eliminate B from the result. After further use of (4.131) and use of the constitutive relations (4.102) one may show $A(t)$ satisfies the equation

$$
2\rho_0 U_N\left[1 + \frac{\eta_\theta(\rho_0 U_N^2 - S_F)}{(\rho_0\eta_\theta U_N^2 + p_{\alpha_X})}\right]\frac{\delta A}{\delta t} + \zeta A^2 = 0,
\tag{4.134}
$$

where

$$
\begin{aligned}
\zeta =& \rho_0\psi_{FFF} + \rho_0^3 U_N^2(\psi_{F\theta})^2\{3U_N^2\psi_{F\theta\theta} + 3\psi_{F\alpha_X\alpha_X}\} \\
&+ \frac{3\rho_0\psi_{\theta FF}}{\psi_{F\theta}}(U_N^2 - \psi_{FF}) \\
&+ \rho_0^3 U_N^2\psi_{F\theta}(U_N^2 - \psi_{FF})\{\psi_{\theta\theta\theta}U_N^2 + 3\psi_{\theta\alpha_X\alpha_X}\}.
\end{aligned}
\tag{4.135}
$$

Let us denote by ζ_1 the coefficient

$$
\zeta_1 = \frac{\zeta}{2\rho_0 U_N[1 + \eta_\theta(\rho_0 U_N^2 - S_F)/(\rho_0\eta_\theta U_N^2 + p_{\alpha_X})]}.
$$

Then we solve equation (4.134) to find

$$
A(t) = \frac{A(0)}{1 + \zeta_1 t A(0)}.
$$

The amplitude behaviour depends on sgn ζ_1. If $\zeta_1 > 0$ then $A(0) > 0$ results in $A(t)$ decaying to zero. Under the same circumstances $A(0) < 0$ leads to $|A(t)| \to \infty$ as $t \to -1/\zeta_1 A(0)$. This is believed to indicate the beginning of a thermal shock wave, cf. the calculations of (Fu and Scott, 1991), the numerical work of (Christov et al., 2006; Christov et al., 2007), and chapter 5, section 5.1, of this book.

4.5 Type III thermoelasticity

As we have seen in chapter 2 (Green and Naghdi, 1992) also develop another thermoelasticity theory based on their 1991 work which they call type III thermoelasticity. This theory would appear to have the potential for heat transport at finite speed, but there *is* dissipation in this theory. The linearized isotropic equations are derived in (Green and Naghdi, 1992), equations (3.19), (3.20) and these writers study the behaviour of one-dimensional waves in the framework of their linearized theory.

(Quintanilla and Straughan, 2004) observe that most of the work prior to 2004 dealing with type II or type III thermoelasticity is analysing the linearized theory. They tackled the nonlinear theory directly and established a fundamental difference between Green-Naghdi thermoelasticity of type II and that of type III. We have seen in section 4.4 that in the theory of type II (Green and Naghdi, 1993), a nonlinear acceleration wave analysis allows both a mechanical and a thermal wave to propagate. However, (Quintanilla and Straughan, 2004) show that in the theory of type III this is no longer true. They demonstrate that there is only one wave and they reconcile this to the fact that it in some ways resembles the situation in classical thermoelasticity, cf. (Chen, 1973), (Chadwick and Currie, 1974; Chadwick and Currie, 1975), (Coleman and Gurtin, 1965), (Iesan and Scalia, 1996), (McCarthy, 1972). As (Quintanilla and Straughan, 2004) remark, they believe that this is a highly relevant result in placing the Green-Naghdi type II and type III theories in the context of "hyperbolic thermoelasticity".

We now describe the nonlinear acceleration wave analysis of (Quintanilla and Straughan, 2004) for type III thermoelasticity.

The governing equations of type III thermoelasticity are

$$\rho_0 \ddot{u}_i = \rho_0 F_i + S_{Ai,A} \,, \tag{4.136}$$

$$\rho_0 \dot{\eta} = \rho_0 s + \rho_0 \xi - p_{A,A} \,. \tag{4.137}$$

Here $u_i, \rho_0, F_i, \eta, s, \xi, p_A$ and S_{Ai} are the displacement, density, body force, entropy, external rate of supply of entropy per unit mass, internal rate of production of entropy per unit mass, the entropy flux vector, and the Piola - Kirchoff stress tensor, respectively.

In terms of the absolute temperature θ and with the heat flux given by $q_A = \theta p_A$, the constitutive equations are derived by (Green and Naghdi,

1992) in terms of a Helmholtz free energy

$$\psi = \psi(\theta, F_{iA}, \alpha_{,A}) . \tag{4.138}$$

Here F_{iA} is the deformation gradient, i.e. $F_{iA} = \partial x_i / \partial X_A$. Thermodynamics requires, cf. chapter 2,

$$\eta = -\frac{\partial \psi}{\partial \theta}, \qquad S_{Ai} = \rho_0 \frac{\partial \psi}{\partial F_{iA}} \tag{4.139}$$

whence using (4.138)

$$\eta = \eta(\theta, F_{jR}, \alpha_{,R}), \qquad S_{Ai} = S_{Ai}(\theta, F_{jR}, \alpha_{,R}). \tag{4.140}$$

It is interesting to note that these relations hold for both type II and type III thermoelasticity. For type II theory we saw in section 4.4 that $\xi = 0$. In type III theory this is not true and we must have

$$\xi = \xi(\theta, F_{jR}, \alpha_{,R}, \theta_{,R}) . \tag{4.141}$$

As observed by (Quintanilla and Straughan, 2004) a fundamental difference between type II and type III thermoelasticity is due to the forms for the entropy flux vector. We saw in section 4.4 that for type II theory we may write p_A as

$$p_A = -\rho_0 \frac{\partial \psi}{\partial \alpha_{,B}} = p_A(\theta, F_{jR}, \alpha_{,R}) . \tag{4.142}$$

No such relation is available for type III thermoelasticity. All we may assert is that p_A has the functional form

$$p_A = p_A(\theta, F_{jR}, \alpha_{,R}, \theta_{,R}) . \tag{4.143}$$

Note, unlike (4.142), equation (4.143) has p_A depending explicitly on $\theta_{,R} = \dot{\alpha}_{,R}$.

4.5.1 Fundamental jump relations

To develop an acceleration wave analysis for a type III thermoelastic body we follow (Quintanilla and Straughan, 2004) and employ the global entropy balance law of (Green and Naghdi, 1995b), namely

$$\frac{d}{dt} \int_P \rho_0 \eta \, dv = \int_P \rho_0 (s + \xi) dv - \int_{\partial P} k \, da, \tag{4.144}$$

where $k = p_A N_A$, with P being an arbitrary volume in the thermoelastic body with boundary ∂P, and N_A is the unit outward normal to ∂P.

We define an acceleration wave in a thermoelastic body of type III to be a singular surface $\mathcal{S}$ across which the displacement u_i, the thermal displacement α, and their first derivatives are continuous while the second and higher derivatives, in general, possess finite discontinuities. The amplitudes

$A_i(t)$ and $B(t)$ of the acceleration wave are defined as

$$A_i(t) = [\ddot{u}_i], \qquad B(t) = [\ddot{\alpha}]. \tag{4.145}$$

We require the integrated form of the entropy balance law (4.144) evaluated across $\mathcal{S}$, and this is, cf. (Iesan and Scalia, 1996) p. 30,

$$[\rho_0 \eta] U_N + N_A [p_A] = 0 \qquad \text{on } \mathcal{S}, \tag{4.146}$$

where U_N is the wavespeed. From equation (4.140) we know $\eta = \eta(\theta, F_{iA}, \alpha_{,A})$ and due to our definition of an acceleration wave η is continuous across $\mathcal{S}$, therefore $[\rho_0 \eta] = 0$. Thus equation (4.146) reduces to

$$N_A [p_A] = 0. \tag{4.147}$$

Unlike type II thermoelasticity where $[p_A] = 0$ holds automatically due to the form of p_A, it is not obviously true for type III theory because of the representation (4.143). However, θ is continuous across $\mathcal{S}$, and since $q_A = \theta p_A$, equation (4.147) now leads to

$$N_A [q_A] = 0. \tag{4.148}$$

The presence of $\theta_{,R}$ in the constitutive form for $q_A = q_A(\theta, F_{iA}, \alpha_{,A}, \theta_{,A})$ means that equation (4.148) is not automatically satisfied.

At this point (Quintanilla and Straughan, 2004) follow (Coleman and Gurtin, 1965) and restrict attention to one-dimensional waves, and they generalize the argument of (Coleman and Gurtin, 1965) relating to homothermal waves. In one dimension equation (4.148) is $[q] = 0$. A definite heat conductor is defined, generalizing the definition in (Coleman and Gurtin, 1965), to be one for which $q(\theta, F, \alpha_X, \theta_X)$ is a strictly monotone function of θ_X, for fixed θ, F. Then analogous to equation (4.19) of (Coleman and Gurtin, 1965) one sees that

$$[q] = q(\theta^-, F^-, \alpha_X^-, \theta_X^-) - q(\theta^+, F^+, \alpha_X^+, \theta_X^+).$$

Across an acceleration wave $\mathcal{S}$, by definition θ, F and α_X are continuous. Therefore, since q is a strictly monotone function of θ_X, if $[\theta_X] (= [\dot{\alpha}_X]) \neq 0$ then $[q] \neq 0$. This contradicts the fact that $[q] = 0$. Hence we must have $[\theta_X] = 0$. Therefore, for type III thermoelasticity, an acceleration wave in a definite conductor is homothermal, i.e. $[\ddot{\alpha}] = 0$. Even though $[\ddot{\alpha}] = 0$, the higher derivatives need not have zero jumps, cf. the arguments of (McCarthy, 1972) in classical thermoelasticity. Hence, $[\alpha_{ttt}] = [\ddot{\theta}] \neq 0$, subscript t denoting partial differentiation with respect to time at fixed $\mathbf{X}$.

The amplitudes in one-dimension are

$$A(t) = [u_{XX}], \qquad B(t) = [\theta_{XX}]. \tag{4.149}$$

The equations of motion for zero body force and external entropy supply are,

$$\rho_0 \ddot{u} = S_X, \tag{4.150}$$

$$\rho_0 \dot{\eta} = \rho_0 \xi - p_X, \tag{4.151}$$

where S and p are the stress tensor and entropy flux. By taking the jumps of (4.150) and (4.151) one derives the wavespeed equation as

$$U_N^2 = \frac{1}{\rho_0} \frac{\partial S}{\partial F}. \tag{4.152}$$

A direct relation between A and B is found from the jump of equation (4.151) as

$$A\Big(\frac{\partial p}{\partial F} - \rho_0 U_N \frac{\partial \eta}{\partial F}\Big) = -\frac{\partial p}{\partial \theta_X} B. \tag{4.153}$$

The amplitude equation is found from equation (4.150) and (Quintanilla and Straughan, 2004) present this as

$$2\rho_0 \frac{\delta A}{\delta t} = \zeta_1 A + \zeta_2 A^2. \tag{4.154}$$

The coefficients ζ_1 and ζ_2 are given by

$$\begin{aligned}\zeta_1 =& \frac{-3\rho_0}{U_N}\frac{\delta U_N}{\delta t} - \frac{S_\theta}{U_N}\frac{(\rho_0 U_N \eta_F - p_F)}{p_{\theta_X}} - \frac{\theta_X^+ S_{\theta F}}{U_N} + \frac{\dot{\theta}^+ S_{F\theta}}{U_N^2} + \frac{\theta_X^+ S_{F\alpha_X}}{U_N^2}\\ &+ \frac{S_{FF}}{U_N}\Big(\dot{u}_X^+ - \frac{u_{XX}^+}{U_N}\Big) + \frac{S_{\alpha_X}}{U_N^2}\frac{(\rho_0 U_N \eta_F - p_F)}{p_{\theta_X}} - \frac{\alpha_{XX}^+}{U_N} S_{\alpha_X F},\end{aligned}$$

and

$$\zeta_2 = \frac{S_{FF}}{U_N}. \tag{4.155}$$

(Quintanilla and Straughan, 2004) note that the amplitude $A(t)$ follows from (4.154) and the development of the acceleration wave into possible shock formation may be studied, cf. (Fu and Scott, 1991) and section 5.1 of this monograph. Once $A(t)$ is known, the thermal amplitude B follows from (4.153). Thus, the mechanical wave determines the thermal wave behaviour. Therefore, acceleration waves in type III thermoelasticity are very different from those in type II where separate mechanical and thermal waves may propagate.

4.6 Acceleration waves in a type II fluid

The basic equations for a type II fluid are described in section 3.3. They consist of the equations of continuity of mass, balance of linear momentum, and balance of entropy and are, for zero body force b_i and zero entropy supply s, respectively,

$$\dot{\rho} + \rho \frac{\partial v_i}{\partial x_i} = 0, \qquad (4.156)$$

$$\rho \dot{v}_i = \frac{\partial T_{ji}}{\partial x_j}, \qquad (4.157)$$

$$\rho \dot{\eta} = -\frac{\partial p_i}{\partial x_i}, \qquad (4.158)$$

where ρ and v_i are density and velocity, and T_{ji}, η and p_i are the Cauchy stress tensor, entropy and entropy flux vector, with a superposed dot denoting the material time derivative. Upon writing the stress, entropy and entropy flux vector in terms of the Helmholtz free energy function ψ, namely

$$T_{ij} = -p\delta_{ij} - \frac{\rho}{2}\left(\frac{\partial \psi}{\partial \alpha_{,i}} \alpha_{,j} + \frac{\partial \psi}{\partial \alpha_{,j}} \alpha_{,i} \right),$$

$$\eta = -\frac{\partial \psi}{\partial \theta}, \qquad p_i = -\frac{\partial \psi}{\partial \alpha_{,i}},$$

$$\psi = \psi(\rho, \theta, \alpha_{,i}),$$

we may expand equations (4.156) - (4.158) to see that the governing equations for a type II fluid become

$$\frac{\partial \rho}{\partial t} + \frac{\partial \rho}{\partial x_i} v_i + \rho \frac{\partial v_i}{\partial x_i} = 0, \qquad (4.159)$$

$$\rho\left(\frac{\partial v_i}{\partial t} + v_j \frac{\partial v_i}{\partial x_j} \right) = -\frac{\partial p}{\partial x_i} - \frac{1}{2} \frac{\partial}{\partial x_j} \left[\rho \left(\frac{\partial \psi}{\partial \alpha_{,j}} \alpha_{,i} + \frac{\partial \psi}{\partial \alpha_{,i}} \alpha_{,j} \right) \right], \qquad (4.160)$$

and

$$-\rho \left\{ \frac{\partial^2 \psi}{\partial t \partial \theta} + v_i \left(\frac{\partial^2 \psi}{\partial x_i \partial \theta} \right) \right\} = \frac{\partial}{\partial x_i} \left(\rho \frac{\partial \psi}{\partial \alpha_{,i}} \right). \qquad (4.161)$$

These five equations represent a hyperbolic system for the density ρ, velocity v_i and thermal displacement α.

An acceleration wave for a type II fluid is defined to be a two-dimensional surface $\mathcal{S}$ in $\mathbb{R}^3$ such that $v_i, \rho, \alpha, \dot{\alpha}$, and $\alpha_{,i}$ are continuous throughout $\mathbb{R}^3$, but their derivatives $\dot{v}_i, v_{i,j}, \dot{\rho}, \rho_{,i}, \ddot{\alpha}, \dot{\alpha}_{,i}$ and $\alpha_{,ij}$, along with higher derivatives, suffer a finite discontinuity (jump) across $\mathcal{S}$.

We now follow (Quintanilla and Straughan, 2008) and consider an acceleration wave moving into an equilibrium region for which $v_i^+ \equiv 0$, $\rho^+ \equiv$ constant, $\theta^+ \equiv$ constant, and $\alpha_{,i}^+ \equiv 0$. In addition, we shall suppose the body possesses a centre of symmetry.

One begins by taking the jumps of equations (4.159) - (4.161) to find

$$[\rho_t] + \rho[v_{i,i}] = 0, \qquad (4.162)$$

$$\rho[v_{i,t}] = -p_\rho[\rho_{,i}] - p_\theta[\theta_{,i}], \tag{4.163}$$

and

$$-\rho\psi_{\theta\theta}[\dot\theta] - \rho\psi_{\theta\rho}[\dot\rho] = \psi_{\alpha_{,i}\alpha_{,j}}[\alpha_{,ji}]\,. \tag{4.164}$$

The amplitudes A^i, B and C are defined by

$$A^i = [v^i_{,j}n^j], \qquad B = [n^i\rho_{,i}], \qquad C = [\alpha_{,ij}n^i n^j], \tag{4.165}$$

where n^i is the unit normal to $\mathcal{S}$ in the + direction. Noting that from the compatibility equations (4.25), (4.26),

$$A^i n_j = [v^i_{,j}], \qquad Bn_i = [\rho_{,i}], \qquad Cn_in_j = [\alpha_{,ij}],$$

we see that employing the Hadamard relation (4.27) in (4.162) - (4.164) yields the equations

$$-VB + \rho A^i n_i = 0, \tag{4.166}$$
$$-\rho V A_i = -p_\rho B n_i + p_\theta V n_i C, \tag{4.167}$$
$$-\rho\psi_{\theta\theta}V^2 C + \rho\psi_{\rho\theta}VB = \rho\psi_{\alpha_{,i}\alpha_{,j}}n^i n^j C, \tag{4.168}$$

where V is the wavespeed at $\mathcal{S}$.

It follows from equation (4.167) that we must have $A_i = An_i$ with $A = [n^i n_j v^j_{,i}]$ so that the acceleration wave $\mathcal{S}$ must be a longitudinal wave. Then (4.166) - (4.168) may be written as (taking the inner product of (4.167) with n_i)

$$\begin{pmatrix} \rho & -V & 0 \\ -\rho V & p_\rho & -p_\theta V \\ 0 & \rho\psi_{\rho\theta}V & -\rho\psi_{\theta\theta}V^2 - \rho\psi_{\alpha_{,i}\alpha_{,j}}n^i n^j \end{pmatrix} \begin{pmatrix} A \\ B \\ C \end{pmatrix} = \begin{pmatrix} 0 \\ 0 \\ 0 \end{pmatrix}$$

A non-zero solution of this system requires that

$$\begin{vmatrix} \rho & -V & 0 \\ -\rho V & p_\rho & -p_\theta V \\ 0 & \rho\psi_{\rho\theta}V & -\rho\psi_{\theta\theta}V^2 - \rho\psi_{\alpha_{,i}\alpha_{,j}}n^i n^j \end{vmatrix} = 0$$

Expansion of this determinant leads to the wavespeed equation

$$(V^2 - p_\rho)(\psi_{\theta\theta}V^2 + \psi_{\alpha_{,i}\alpha_{,j}}n^i n^j) + \psi_{\rho\theta}p_\theta V^2 = 0. \tag{4.169}$$

Let us observe that $V_2 = \sqrt{p_\rho}$ is the wavespeed of an acoustic wave in a classical theory whereas $V_1 = \sqrt{-\psi_{\alpha_{,i}\alpha_{,j}}n^i n^j/\psi_{\theta\theta}}$ is the wavespeed of a thermal wave in the current version of the Green - Naghdi type II theory, see section 4.2. Precise forms for V^2 (and hence V) follow from the quadratic equation (4.169) provided we specify a form for the Helmholtz free energy function ψ. One may rearrange the last coefficient in equation (4.169) as

$$K = \psi_{\rho\theta}p_\theta = \rho^2(\psi_{\rho\theta})^2. \tag{4.170}$$

From this relation (Quintanilla and Straughan, 2008) deduce that $\psi_{\rho\theta} = 0$ leads to two distinct waves, a pressure wave with speed V_2 and a thermal wave with speed V_1. In general, one expects $\psi_{\rho\theta} \neq 0$, and so (4.169) demonstrates that we have two connected waves. To clarify this, we rewrite equation (4.169) in the form

$$(V^2 - V_1^2)(V^2 - V_2^2) = -\frac{K}{\psi_{\theta\theta}} V^2. \tag{4.171}$$

It is not unreasonable to expect ψ to be such that $\psi_{\theta\theta} < 0$ wherein from equation (4.171) we may deduce that there are two waves which move with speeds V_L and V_U such that

$$0 < V_L^2 < \min\{V_1^2, V_2^2\} < \max\{V_1^2, V_2^2\} < V_U^2.$$

We interpret these as a fast wave, speed V_U, and a slow wave, speed V_L. We know that in low temperature solids the fast wave is a mechanical wave and the slow one is associated with the thermal field, cf. section 1.1. It is to be anticipated that V_U, V_L will have a similar interpretation here.

(Quintanilla and Straughan, 2008) show how one may calculate an amplitude equation for a one-dimensional acceleration wave. Let now $\mathcal{S}$ be a one-dimensional acceleration wave moving along the $x-$axis. Put $\mathbf{v} = (u(x,t),0,0)$ with $\rho(x,t)$, $\alpha(x,t)$, and define the one-dimensional amplitudes by

$$A(t) = [u_x] = u_x^- - u_x^+, \qquad B(t) = [\rho_x], \qquad C(t) = [\alpha_{xx}]. \tag{4.172}$$

The governing equations in one-dimension are

$$\begin{aligned} &\rho_t + u\rho_x + \rho u_x = 0, \\ &\rho(u_t + uu_x) = -p_x - (\rho\psi_{\alpha_x}\alpha_x)_x\,, \\ &-\rho\big(\psi_{\theta t} + u(\psi_\theta)_x\big) = (\rho\psi_{\alpha_x})_x\,. \end{aligned} \tag{4.173}$$

To determine the amplitudes we differentiate equations (4.173) with respect to x and take the jumps of the resulting three equations. Additionally one uses the one-dimensional version of equations (4.166) and (4.167) in the forms

$$\rho A = VB, \qquad \text{and} \qquad p_\theta V^2 C = \rho(p_\rho - V^2)A, \tag{4.174}$$

and then one eliminates B and C to derive a Bernoulli equation for $A(t)$.

In fact, one may show

$$a\,\frac{\delta A}{\delta t} + bA^2 = 0, \tag{4.175}$$

where the coefficients a and b are given by

$$a = 2\left(\frac{U_M^2}{V^2} - \frac{(V^2 - U_M^2)^2}{\kappa V^2}\right), \tag{4.176}$$

and

$$\begin{aligned} b =& 3 + \frac{P_\rho \rho^2}{V^2} + 3(V^2 - U_M^2)\Big(\frac{\rho\psi_{\rho\rho\theta}}{V^2\psi_{\rho\theta}} - \frac{\psi_{\theta\theta}}{\rho^2\psi_{\rho\theta}}\Big) \\ &+ \frac{3(V^2 - U_M^2)^2}{V^4 p_\theta \psi_{\rho\theta}}(\rho V^2\psi_{\rho\theta\theta} + \rho\psi_{\rho\alpha_x\alpha_x} + \psi_{\alpha_x\alpha_x} - V^2\psi_{\theta\theta}) \\ &+ \frac{(V^2 - U_M^2)^3}{V^4 p_\theta^2 \psi_{\rho\theta}}(V^2\psi_{\theta\theta\theta} + 3\psi_{\theta\alpha_x\alpha_x}) . \end{aligned} \tag{4.177}$$

The solution to equation (4.175) is

$$A(t) = \frac{A(0)}{1 - (b/a)tA(0)} . \tag{4.178}$$

If $(b/a)A(0) > 0$ there is always blow-up of $A(t)$ in a finite time. Once $A(t)$ is known the other amplitudes $B(t)$ and $C(t)$ then follow from equation (4.174).

It is of interest to note that as the thermal effects disappear then $V^2 \to U_M^2$ and a and b reduce to forms consistent with that for a classical perfect fluid, namely,

$$\frac{\delta A}{\delta t} = A^2\Big(\frac{3}{2} + \frac{\rho^2(d/d\rho)(p_\rho/\rho)}{2V^2}\Big) .$$

4.7 Acceleration waves in a type III fluid

In this section we develop an acceleration wave analysis for the inviscid theory of a type III fluid presented in section 3.4.2.

The basic equations are those of balance of mass, momentum, and entropy as given in (3.49) - (3.51) which for zero body force b_i and zero external entropy supply s may be written

$$\dot\rho + \rho v_{i,i} = 0, \tag{4.179}$$

$$\rho\dot v_i = T_{ji,j} , \tag{4.180}$$

$$\rho\dot\eta = -p_{i,i} + \rho\xi . \tag{4.181}$$

To study acceleration waves in this theory we find it necessary to begin with the integrated form of (4.181) rather than the local form as given. If $\mathcal{P}$ denotes a volume in the fluid with boundary $\partial\mathcal{P}$ then the integrated form is, see (Green and Naghdi, 1991), equation (7.19), see also (Green and Naghdi, 1977),

$$\frac{d}{dt}\int_{\mathcal{P}} \rho\eta \, dV = \int_{\mathcal{P}} \rho\xi \, dV - \oint_{\partial\mathcal{P}} k \, dA \tag{4.182}$$

where $k = p_i n_i$, n_i being the unit outward normal to $\partial\mathcal{P}$. The quantities dV and dA denote the volume and surface area integral elements, respectively.

At the outset we define an acceleration wave in an inviscid fluid of type III to be surface $\mathcal{S}$ in $\mathbb{R}^3$ across which v_i, ρ, α_t and $\alpha_{,i}$ are continuous but the functions $v_{i,t}, v_{i,j}, \rho_t, \rho_{,i}, \alpha_{tt}, \alpha_{,ti}, \alpha_{,ij}$ and their higher derivatives possess a finite discontinuity. As usual, + and − denote the limits on $\mathcal{S}$ as approached from the right and left, and $[f] = f^- - f^+$. We use a "pillbox" argument on equation (4.182) to see that this equation evaluated across $\mathcal{S}$ yields, cf. the procedure in (Iesan and Scalia, 1996), p. 30,

$$[\rho\eta]V + n_i[p_i] = 0. \tag{4.183}$$

Since $\eta = -\partial\psi/\partial\theta$ (see equation (3.57)) and $\psi = \psi(\rho, \theta, \alpha_{,i})$ (see equation (3.56)) $[\rho\eta] = 0$ and so (4.183) becomes

$$n_i[p_i] = 0.$$

Further, since $q_i = \theta p_i$ and $\theta = \dot{\alpha}$, we infer from this that

$$n_i[q_i] = 0. \tag{4.184}$$

We now follow the article of (Coleman and Gurtin, 1965) in viscoelasticity. Suppose we consider a one-dimensional wave and the type III fluid is a definite heat conductor. This means that q (the one-dimensional component of q_i) is a *strictly monotone function* of $\theta_x = \dot{\alpha}_x$ for ρ, θ, α_x fixed. Since the one-dimensional component p of p_i is such that $p = p(\rho, \theta, \alpha_x, \theta_x)$ it follows that $q = q(\rho, \theta, \alpha_x, \theta_x)$. Of course, ρ, θ and α_x are continuous across $\mathcal{S}$. Then, from (4.184) $[q] = 0$. But

$$[q] = q(\rho^-, \theta^-, \alpha_x^-, \theta_x^-) - q(\rho^+, \theta^+, \alpha_x^+, \theta_x^+). \tag{4.185}$$

Since q is continuous in its arguments ρ, θ, α_x and is a strictly monotone function of θ_x which does have a discontinuity across $\mathcal{S}$ it then follows that since $\theta_x^- \neq \theta_x^+$, $[q] \neq 0$. This contradicts (4.184). Therefore, we conclude that an acceleration wave in a type III inviscid fluid which is a definite heat conductor must be such that $[\theta_x] = 0$, and so $[\alpha_{tt}] = 0$, i.e. the wave is homothermal (in a sense analogous to the definition in (Coleman and Gurtin, 1965)). However, even though $[\alpha_{tt}] = 0$ it is still true that $[\alpha_{ttt}] \neq 0$, with non-zero jumps also for other third and higher derivatives.

The above analysis shows that an acceleration wave for a type III fluid behaves very differently from one in a type II fluid where we have already seen both mechanical and thermal waves propagate. The situation is, therefore, analogous to that for propagation of an acceleration wave in a type II or type III thermoelastic body as shown earlier in this chapter.

To continue with an acceleration wave analysis we now have a surface $\mathcal{S}$ across which $\rho_t, \rho_{,i}, v_{i,t}, v_{i,j}, \alpha_{ttt}, \alpha_{,itt}, \alpha_{,ijt}$ and $\alpha_{,ijk}$ and their higher derivatives suffer a finite discontinuity but their lower derivatives are continuous.

We now expand equations (4.179) and (4.180), recalling from (3.65) that

$$T_{ij} = -p\delta_{ij} - \rho\alpha_{,i}\frac{\partial\psi}{\partial\alpha_{,j}}, \qquad p = \rho^2\frac{\partial\psi}{\partial\rho},$$

with $\psi = \psi(\rho, \theta, \alpha_{,i})$. Then, remembering the differentiability properties of ρ, v_i and α across $\mathcal{S}$ we see that

$$\begin{aligned}
&[\rho_t] + v_i[\rho_{,i}] + \rho[v_{i,i}] = 0,\\
&\rho[v_{i,t}] + \rho v_j[v_{i,j}] = -\frac{\partial p}{\partial \rho}[\rho_{,i}] - [\rho_{,j}]\alpha_{,i}\frac{\partial \psi}{\partial \alpha_{,j}} - \rho\alpha_{,i}\frac{\partial^2 \psi}{\partial \rho \partial \alpha_{,j}}[\rho_{,j}]\,.
\end{aligned} \tag{4.186}$$

If we denote the wavespeed at $\mathcal{S}$ by u_n then the Hadamard relation (4.27) may be used to show

$$[\dot{\psi}] = \frac{\delta}{\delta t}[\psi] - (u_n n^k - v^k)[\psi_{,k}]\,.$$

We put $U = u_n - v^k n_k$ at $\mathcal{S}$ and define the wave amplitudes B and C_i by

$$B = [n^i \rho_{,i}], \qquad C^i = [v^i_{,j} n^j]\,. \tag{4.187}$$

Then using the compatibility relations we have $[\dot{v}_i] = -UC^i$, $[\dot{\rho}] = -UB$, $[\rho_{,i}] = Bn_i$ and $[v^i_{,j}] = C^i n_j$ so that (4.186) may be rewritten as

$$\begin{aligned}
&-UB + \rho C^i n_i = 0,\\
&-\rho U C^i + \frac{\partial p}{\partial \rho} B n^i + \alpha_{,i}\frac{\partial \psi}{\partial \alpha_{,j}} B n^j + \rho \alpha_{,i}\frac{\partial^2 \psi}{\partial \rho \partial \alpha_{,j}} B n^j = 0.
\end{aligned} \tag{4.188}$$

Unlike the situation for a classical fluid, or one of Green-Laws type, cf. (Lindsay and Straughan, 1978), we are not immediately able to deduce from (4.188) that $\mathcal{S}$ is a longitudinal wave. The type III thermal effects are playing a strong role.

To calculate the wavespeeds from (4.188) we must write it as a system in $(B, C^1, C^2, C^3)^T$, i.e.

$$\begin{pmatrix}
-U & \rho n_1 & \rho n_2 & \rho n_3\\
\frac{\partial p}{\partial \rho} n^1 + \alpha_{,1}\psi_{\alpha_{,j}} n^j + \rho\alpha_{,1}\psi_{\rho\alpha_{,j}} n^j & -\rho U & 0 & 0\\
\frac{\partial p}{\partial \rho} n^2 + \alpha_{,2}\psi_{\alpha_{,j}} n^j + \rho\alpha_{,2}\psi_{\rho\alpha_{,j}} n^j & 0 & -\rho U & 0\\
\frac{\partial p}{\partial \rho} n^3 + \alpha_{,3}\psi_{\alpha_{,j}} n^j + \rho\alpha_{,3}\psi_{\rho\alpha_{,j}} n^j & 0 & 0 & -\rho U
\end{pmatrix}
\begin{pmatrix} B\\ C^1\\ C^2\\ C^3 \end{pmatrix} = \begin{pmatrix} 0\\ 0\\ 0\\ 0 \end{pmatrix}$$

Upon requiring $(B, C^1, C^2, C^3) \neq (0,0,0,0)$ and evaluating the 4×4 determinant we are able to show that

$$\begin{aligned}
\rho^3 U^4 &- \rho^3 U^2\Big(\frac{\partial p}{\partial \rho} n_2^2 + n_2\alpha_{,2}\psi_{\alpha_{,j}} n^j + \rho n_2 \alpha_{,2}\psi_{\rho\alpha_{,j}} n^j\Big)\\
&- \rho^3 U^2\Big(\frac{\partial p}{\partial \rho} n_1^2 + n_1\alpha_{,1}\psi_{\alpha_{,j}} n^j + \rho n_1 \alpha_{,1}\psi_{\rho\alpha_{,j}} n^j\Big)\\
&- \rho^3 U^2\Big(\frac{\partial p}{\partial \rho} n_3^2 + n_3\alpha_{,3}\psi_{\alpha_{,j}} n^j + \rho n_3 \alpha_{,3}\psi_{\rho\alpha_{,j}} n^j\Big) = 0.
\end{aligned} \tag{4.189}$$

From this equation we deduce that either

$$U^2 = 0$$

which is a standing wave, or we have a propagating wave with wavespeed U given by

$$
\begin{aligned}
U^2 = & \frac{\partial p}{\partial \rho} + \frac{1}{2}(\alpha_{,i}\psi_{\alpha_{,j}} + \alpha_{,j}\psi_{\alpha_{,i}})n_i n_j \\
& + \frac{\rho}{2}(\alpha_{,i}\psi_{\alpha_{,j}} + \alpha_{,j}\psi_{\alpha_{,i}})n_i n_j \,.
\end{aligned}
\tag{4.190}
$$

Of course, equation (4.181) still plays an important role. Recalling the definitions and differentiability properties at $\mathcal{S}$ we see that expanding and taking jumps (4.181) yields

$$
\rho\eta_\rho[\dot\rho] = -\frac{\partial p_i}{\partial \rho}[\rho_{,i}] - \frac{\partial p_i}{\partial \gamma_j}[\dot\alpha_{,ji}] \,. \tag{4.191}
$$

In deriving (4.191) we encounter the terms

$$
\rho\eta_\theta\dot\theta, \;\; \rho\eta_{\alpha_{,i}}\dot{\overline{\alpha_{,i}}}\,, \;\; \kappa\gamma_i\gamma_i/\theta^2, \;\; p_{i\theta}\theta_{,i}\,, \;\; p_{i\alpha_{,j}}\alpha_{,ji}\,,
$$

but these are continuous across $\mathcal{S}$. We next use the expressions for $[\dot\rho]$ and $[\rho_{,i}]$ and then note that if we define

$$
A(t) = [n^r n_a n_b \alpha_{,rab}],
$$

then one can show $[\dot\alpha_{,ji}] = -U n_i n_j A$. Thus, (4.191) is a relation between A and B, namely

$$
B\Big(\frac{\partial p_i}{\partial \rho} n_i - \rho\eta_\rho U\Big) = U n_i n_j \frac{\partial p_i}{\partial \gamma_j} A \,. \tag{4.192}
$$

Once we determine the amplitudes B and C_i, then (4.192) yields the thermal amplitude A.

To determine the amplitudes B and C_i we must differentiate equations (4.179) and (4.180) with respect to t or x_i and then use the wavespeed relation (4.190) together with (4.192) to derive a Bernoulli equation for B or C_i. Once either of these is known the other follows from (4.188) and then the solution is completed by determining A from (4.192).

Although the ideas of acceleration waves have been under constant development for over forty years, they are still being employed with much effect in the current literature. In fact, the use of acceleration waves and related analyses have proved extremely useful in recent investigations of wave motion in various continuous and random media, and in a variety of thermodynamic states, see e.g. (Chen, 1969a; Chen, 1969b), (Christov et al., 2006; Christov et al., 2007), (Christov and Jordan, 2008; Christov and Jordan, 2009), (Ciarletta and Iesan, 1993), (Ciarletta and Straughan, 2006; Ciarletta and Straughan, 2007b; Ciarletta and Straughan, 2007a), (Ciarletta et al., 2007), (Curro et al., 2009), (Eremeyev, 2005), (Fabrizio, 1994), (Fabrizio and Morro, 2003), (Franchi, 1985), (Fu and Scott, 1988; Fu and Scott, 1990; Fu and Scott, 1991), (Gultop, 2006), (Iesan and Scalia, 2006), (Jordan, 2004; Jordan, 2005a; Jordan, 2005b; Jordan,

2006; Jordan, 2007; Jordan, 2008b; Jordan, 2008a), (Jordan and Christov, 2005), (Jordan and Feuillade, 2004), (Jordan and Puri, 1999; Jordan and Puri, 2005), (Jordan and Straughan, 2006), (Kameyama and Sugiyama, 1996), (Lin and Szeri, 2001), (Mariano and Sabatini, 2000), (Marasco, 2009a; Marasco, 2009b), (Marasco and Romano, 2009), (Mentrelli et al., 2008), (Morro, 1978; Morro, 2006), (Ostoja-Starzewski and Trebicki, 1999; Ostoja-Starzewski and Trebicki, 2006), (Rai, 2003), (Truesdell and Rajagopal, 1999), (Ruggeri and Sugiyama, 2005), (Sabatini and Augusti, 2001), (Straughan, 1986; Straughan, 2008; Straughan, 2009a), (Sugiyama, 1994), (Valenti et al., 2004), (Weingartner et al., 2006; Weingartner et al., 2008), (Whitham, 1974).

4.8 Exercises

Exercise 4.8.1 *Define an acceleration wave for the equation (in 3-D)*

$$\frac{\partial u}{\partial t} + u\frac{\partial u}{\partial x} = \lambda u^2,$$

λ being a constant. Find the wavespeed V. Derive the amplitude equation for $a(t) = [u_t]$. Solve this equation for a wave moving into a region where $u^+ = \alpha$ is constant, and $\lambda = 0$. If $a(0)/u^+ > 0$ what happens?

Exercise 4.8.2 *Determine the wavespeeds of an acceleration wave for the system*

$$\begin{aligned} \frac{\partial u}{\partial t} + u\frac{\partial u}{\partial x} + v\frac{\partial v}{\partial x} &= u^3 - v^3, \\ \frac{\partial v}{\partial t} + v\frac{\partial v}{\partial x} + u\frac{\partial u}{\partial x} &= v^3 - u^3. \end{aligned} \tag{4.193}$$

Find and solve the amplitude equation for an acceleration wave to (4.193) with $u^+ = v^+ =$ constant.

Exercise 4.8.3 *With the wave amplitude $[u_x] = a(t)$ derive an acceleration wave analysis for the equation, see (Jordan and Puri, 2005),*

$$u_{tt} - c^2 u_{xx} = 3\beta(u^2 u_x)_x \,. \tag{4.194}$$

Hint. Write equation (4.194) as a special case of Shablovskii's equation

$$u_{tt} - \big(\kappa(u)u_x\big)_x = 0.$$

Use the decomposition $u_t = -q_x, q_t = -\kappa(u)u_x$, and define an acceleration wave to be a surface $\mathcal{S}$ across which u, q are continuous but their first and higher derivatives may have a finite discontinuity. Calculate the wavespeed, V, and find and solve the amplitude equation for an acceleration wave moving into a region for which $u =$ constant, so that $u^+ =$ constant.

Exercise 4.8.4 *With the wave amplitude* $[u_x] = a(t)$ *derive an acceleration wave analysis for the equation, see (Jordan, 2008b),*

$$u_{tt} + k_1 u_t - c^2 u_{xx} = 3\beta(u^2 u_x)_x \,. \tag{4.195}$$

Hint. See the hint in exercise 4.8.3.

Exercise 4.8.5 *Consider the partial differential equation, see (Jordan, 2006),*

$$u_{tt} + \delta u_t - u_{xx} = -(\epsilon_1 u_x^2 + \epsilon_2 u_t^2)_t \,. \tag{4.196}$$

Define an acceleration wave for equation (4.196) to be a surface $\mathcal{S}$ *across which* u, u_t, u_x *are continuous (and continuous everywhere), but higher derivatives have a finite discontinuity across* $\mathcal{S}$. *Define* $a(t) = [u_{xx}]$ *to be the amplitude of the acceleration wave. Find the wavespeed* V *and the amplitude equation. Solve the amplitude equation. What can you deduce from its solution?*

Exercise 4.8.6 *As shown in section 1.8 another alternative to allow a temperature wave with finite speed was suggested by (Green and Laws, 1972). If we specialize the theory of section 1.8 to one space dimension then (Green and Laws, 1972) proposed the conservation law for temperature be replaced by*

$$\frac{\partial}{\partial t}\,\phi(\theta,\dot{\theta}) = -\frac{\partial q}{\partial x}\,, \tag{4.197}$$

where $q = -\kappa(\theta)\theta_x$, *and* $\phi = \phi(\theta,\dot{\theta})$ *is a "generalized temperature".*

Find the wavespeed of an acceleration wave to (4.197) and calculate the solution to the amplitude equation for the amplitude $a(t) = [\theta_{xx}]$, *cf. (Lindsay and Straughan, 1976).*

Exercise 4.8.7 *(See (Straughan, 2010a).) The Cattaneo - Christov equations for a compressible fluid, see section 3.1.2, may be written*

$$\rho c_p\Big(\frac{\partial T}{\partial t} + v_i\frac{\partial T}{\partial x_i}\Big) = -\frac{\partial q_i}{\partial x_i}\,, \tag{4.198}$$

$$\tau\Big(\frac{\partial q_i}{\partial t} + v_j\frac{\partial q_i}{\partial x_j} - q_j\frac{\partial v_i}{\partial x_i} + q_i\frac{\partial v_r}{\partial x_r}\Big) + q_i = -\kappa T_{,i}\,. \tag{4.199}$$

Replace equation (4.198) by the energy balance law, cf. equation $(3.2)_3$, *with the heat supply* $r = 0$,

$$\rho\dot{\epsilon} + \frac{\partial q_i}{\partial x_i} - t_{ij}d_{ij} = 0,$$

where the internal energy ϵ *now depends on density,* ρ, *and temperature,* T. *For an inviscid fluid* $t_{ij} = -p\delta_{ij}$ *where the pressure* p *will have the form of* $p = \rho^2\partial\psi/\partial\rho$, ψ *being the Helmholtz free energy function. Show that one then derives the equations for nonlinear behaviour in a Cattaneo-Christov gas*

$$\begin{aligned}
&\rho\epsilon_T\dot{T} + \rho^2 T\psi_{T\rho}d_{ii} = -q_{i,i}\,,\\
&\tau(q_{i,t} + v_j q_{i,j} - q_j v_{i,j} + v_{m,m}q_i) + q_i = -\kappa T_{,i}\,,\\
&\dot{\rho} + \rho v_{i,i} = 0,\\
&\rho\dot{v}_i = -p_{,i}\,.
\end{aligned} \tag{4.200}$$

Develop an acceleration wave analysis for a solution to system (4.200) *and show the wavespeed* U *satisfies the quadratic equation*

$$(U^2 - U_T^2)(U^2 - U_M^2) + U^2\kappa_1 = 0 \tag{4.201}$$

where

$$\kappa_1 = \frac{\rho^2(\psi_{\rho T})^2}{\psi_{TT}}$$

with U_T *and* U_M *being the speed of a thermal and a mechanical wave, respectively, and*

$$U_T^2 = \frac{\kappa}{\tau\rho(\partial\epsilon/\partial T)}\,, \qquad U_M^2 = \frac{\partial p}{\partial\rho}\,.$$

Suppose $\kappa_1 < 0$. *How do you interpret equation* (4.201)*?*

5

Shock waves and numerical solutions

In chapter 4 we have studied the evolutionary behaviour of an acceleration wave in various continuum theories where the temperature field could propagate as a wave. For example, in the Maxwell-Cattaneo theory of section 4.1 we saw that an acceleration wave was a singular surface $\mathcal{S}$ such that the temperature, θ, and the heat flux, q_i were continuous across $\mathcal{S}$, but $\theta_{,t}, \theta_{,i}, q_{i,t}, q_{i,j}$ and higher derivatives possessed a discontinuity across $\mathcal{S}$. When the wave amplitude of an acceleration wave becomes infinite a shock wave can form, cf. section 5.1. For the Maxwell-Cattaneo equations, equations (4.1), a shock wave is a singular surface across which θ and q_i themselves have a finite discontinuity.

The study of thermal shock waves is of much importance in its own right. There are many applications of such waves. For example, (Yang, 1993) notes that thermal shock waves are observed in all organisms at the cellular level. This will in turn result in an accumulation of heat shock proteins in cells. Also, extreme heat waves have been observed on planetary bodies. (Bryner, 2009) observes that the planet HD80606b, which has four times the mass of Jupiter and is some 200 light years from Earth, has temperature variations of over 555°C in only a six hour period. This leads to large shock wave storms which travel faster than the speed of sound generating increasing heat and high speed winds.

Thermal shock waves have been studied theoretically for some time. For example, (Atkin and Fox, 1984) used a discontinuity analysis to study thermal shock evolution in a model for liquid helium II. They allowed a discontinuity in the temperature field and in the superfluid velocity. The same writers in (Atkin and Fox, 1985) studied thermal shocks in a model for liquid helium II when the waves are spherically symmetric. (Shablovskii, 1984; Shablovskii, 1985; Shablovskii, 1987) also analyses thermal shock waves but employs equations of Maxwell-Cattaneo type. For example, (Shablovskii, 1984) employs the equations

$$\begin{aligned} cT_t &= -q_x \,, \\ \gamma q_t + q &= -\lambda(T) T_x \end{aligned}$$

in one space dimension, where T, q are temperature and heat flux, for a temperature dependent thermal conductivity $\lambda(T)$. (Shablovskii, 1984) argues that for high - intensity heat transfer, or when the properties of the body are such that one may neglect the dissipation term, then one may study the temperature field equation

$$T_{tt} = \big(\kappa(T) T_x\big)_x \tag{5.1}$$

where $\kappa(T) = \lambda(T)/c\gamma$, c, γ, constants. A similar deduction is made in (Shablovskii, 1987) although there c, γ and λ are allowed to be functions of T. (Shablovskii, 1984) studies a characteristic solution (simple wave) for equation (5.1) and shows that a temperature shock may form. Thermal shocks in the (Morro and Ruggeri, 1988) generalization of the Cattaneo equations, equations (1.48), are investigated by (Ruggeri et al., 1990), (Tarkenton and Cramer, 1994), and (Ruggeri et al., 1996). (Ruggeri et al., 1996) use a characteristic solution (simple wave) to show how a solution may steepen in a finite time (or finite distance) to form a discontinuity which essentially corresponds to a thermal shock.

Interactions of acceleration waves and shock waves are analysed by (Morro, 1978) and by (Mentrelli et al., 2008), where further appropriate references may be found.

Numerical analysis of thermal shock evolution or of the development of a thermal shock is a "hot" topic in the current research literature. Many numerical methods have been used, mostly on Cattaneo - like systems. For example, (Glass et al., 1986) employ a MacCormack predictor-corrector method, (Cramer et al., 2001) uses cellular automata, an angled derivative method is is used by (McCartin and Causley, 2006), (Reverberi et al., 2008) use a Hartree hybrid method, (Roy et al., 2009) uses a multiple scales technique, and (Christov and Jordan, 2010) employ a Godunov argument. These articles all analyse some form of Cattaneo system. The Godunov method was also used by (Christov et al., 2006), while (Jordan and Christov, 2005) and (Jordan, 2007) employed an accurate finite difference method. (Shen and Zhang, 2003) used a high order characteristics based TVD scheme on a dual phase lag model, while (Bargmann and Steinmann, 2006; Bargmann and Steinmann, 2008) and (Bargmann et al., 2008a) employs a finite element technique on Green-Naghdi type II and type III models, see section 5.2. Other numerical schemes have been employed to obtain approximate solutions to hyperbolic models and some of these are discussed in chapter 9. The area of producing numerical solutions for thermal wave problems is one which is gaining impetus and with increasing applications of heat wave theories and the methods used to develop hyperbolic - like models, there is no doubt there will be much further development and analysis of numerical schemes to solve problems involving finite speed of propagation heat transfer.

5.1 Shock development

A general theory of shock development starting with the equation

$$\frac{\partial u}{\partial t} + c(u)\frac{\partial u}{\partial x} = 0, \qquad x \in \mathbb{R},\ t > 0, \tag{5.2}$$

with the initial data

$$u(x,0) = f(x), \tag{5.3}$$

is given in detail in (Whitham, 1974). Indeed, a shock wave (where u is discontinuous) may form for a solution to equation (5.2) even if the initial data function is a C^∞ function (see exercise 5.4.3). The book of (Whitham, 1974) lucidly describes shock formation and related topics in chapters 2 to 4. He then gives a beautiful account of similar work in systems of partial differential equations, including, in particular, the theory of Riemann invariants in sections 5.3, 5.4, 6.7 and 6.8.

We are primarily interested in thermal shock waves in this chapter and in the development of an acceleration thermal wave into a thermal shock. (Shablovskii, 1984) commences with the system of equations

$$\begin{aligned} c\frac{\partial \theta}{\partial t} &= -\frac{\partial q}{\partial x}, \\ \tau\frac{\partial q}{\partial t} + q &= -k(\theta)\frac{\partial \theta}{\partial x} \end{aligned} \tag{5.4}$$

where c, τ are constants, although the thermal conductivity k is a function of the temperature θ. Here q is the heat flux and the system is considered in one space dimension. (Shablovskii, 1984) essentially argues that one may discard the q term and work with the reduced system

$$\begin{aligned} c\frac{\partial \theta}{\partial t} &= -\frac{\partial q}{\partial x}, \\ \tau\frac{\partial q}{\partial t} &= -k(\theta)\frac{\partial \theta}{\partial x}. \end{aligned} \tag{5.5}$$

If we put $\kappa = k/c\tau$, $\kappa = \kappa(\theta)$, then one may eliminate q and show θ satisfies what we call Shablovskii's equation

$$\frac{\partial^2 \theta}{\partial t^2} = \frac{\partial}{\partial x}\left(\kappa(\theta)\frac{\partial \theta}{\partial x}\right). \tag{5.6}$$

Interestingly, (Shablovskii, 1987) continues his investigation by dropping q from $(5.4)_2$, but in that work he allows c, τ and k to depend on temperature θ. (Then, or course, one obtains a different equation to (5.6).)

(Shablovskii, 1984) defines the variables $R(x,t)$ and $L(x,t)$ with $L = \int_{\theta_0}^{\theta} \kappa(\xi)d\xi$. Then equation (5.6) is written as an equivalent system

$$\begin{aligned} \frac{\partial \theta}{\partial t} &= \frac{\partial R}{\partial x}, \\ \frac{\partial R}{\partial t} &= \frac{\partial L}{\partial x}. \end{aligned} \tag{5.7}$$

(Shablovskii, 1984) shows how one relates the function R to the heat flux q. Writing (5.7) as a hyperbolic system of form

$$\frac{\partial u_i}{\partial t} + a_{ij}\frac{\partial u_j}{\partial x} = 0$$

we see that the matrix $\mathbf{a}$ with components a_{ij} is given by

$$\mathbf{a} = \begin{pmatrix} 0 & 1 \\ \kappa & 0 \end{pmatrix}$$

where $\mathbf{u} = (\theta, R)^T$. The eigenvalues of $\mathbf{a}$ are found to be $\pm\kappa$ and the corresponding left eigenvectors are $(1, \pm\sqrt{\kappa})^T$. Then, as (Shablovskii, 1984) points out one finds the Riemann invariants r, s to be

$$r = R + \int_{\theta_0}^{\theta} \sqrt{\kappa'(\xi)}d\xi, \qquad s = R - \int_{\theta_0}^{\theta} \sqrt{\kappa'(\xi)}d\xi, \tag{5.8}$$

where θ_0 is a starting temperature, and $\kappa' = d\kappa/d\theta$. The Riemann invariants satisfy the characteristic equations

$$\begin{aligned} \frac{dr}{dt} &= 0 \qquad \text{on} \qquad \frac{dx}{dt} = \sqrt{\kappa'(\theta)} \\ \frac{ds}{dt} &= 0 \qquad \text{on} \qquad \frac{dx}{dt} = -\sqrt{\kappa'(\theta)}\,. \end{aligned}$$

(Shablovskii, 1984) studies a simple wave for this theory in which $r = r_0$, where r_0 is a constant throughout. It is shown in (Shablovskii, 1984) that the solution to equation (5.6) may break down in a finite time using simple wave theory. The breakdown will be in θ_t, θ_x and (Shablovskii, 1984) further studies the thermal shock.

(Ruggeri et al., 1996) begin with the nonlinear equations for heat transport in a rigid heat conductor

$$\begin{aligned} \rho\epsilon_t + q_{i,i} &= 0, \\ (\alpha q_i)_t + \nu_{,i} &= -\frac{\nu'}{\kappa}\,q_i\,, \end{aligned} \tag{5.9}$$

where ϵ is the internal energy, and $\epsilon, \alpha, \nu, \kappa$ are functions of temperature θ, $\nu' = d\nu/d\theta$. This is a nonlinear Cattaneo-like system of equations similar to that derived by (Morro and Ruggeri, 1988), cf. page 14. (Ruggeri et al., 1996) argue that the coefficient of the q_i term in equation (5.9) may be very small and they neglect this quantity. This leads them to study the system

$$\begin{aligned}\frac{\partial \theta}{\partial t} &= -\frac{1}{\rho c_v}\frac{\partial q}{\partial x}\,,\\ \frac{\partial q}{\partial t} &= -\frac{\nu'}{\alpha}\frac{\partial \theta}{\partial x} + \frac{\alpha'}{\rho c_v \alpha} q \frac{\partial q}{\partial x}\end{aligned} \tag{5.10}$$

where c_v is the specific heat of the rigid body. (Ruggeri et al., 1996) obtain a solution to equations (5.10) which is a simple wave and they show that a finite time discontinuity may form. They calculate the critical time of formation. (Ruggeri et al., 1996) apply their results specifically to discontinuity in temperature formation in a crystal of sodium fluoride.

Blow-up of gradients in thermoelasticity is treated by (Dafermos, 1985) and by (Dafermos and Hsiao, 1986) and while their analysis is for a model corresponding to classical nonlinear thermoelasticity, the arguments are very interesting. (Dafermos, 2006) is another very interesting article dealing with continuous solutions to the conservation law

$$\frac{\partial u}{\partial t} + \frac{\partial}{\partial x} f(u) = 0$$

and the equation

$$\frac{\partial u}{\partial t} + \frac{\partial}{\partial x} f(u) = g.$$

(Dafermos and Hsiao, 1986) derive the Bernoulli equation for the amplitude of an acceleration wave in thermoelasticity. However, they are concerned with showing that solutions to the equations of thermoelasticity with large initial data generally blow up in a finite time. They specifically note that the situation is different from isothermal nonlinear elasticity. They use the Riemann invariants but observe that the "*equations now contain coupling terms that depend on T*" (temperature) "*and its derivatives. Therefore, it is no longer possible to establish boundedness or blow-up by restricting attention on a fixed distinct characteristic*".

(Dafermos, 1985) deals with singularity formation in thermoelasticity. However, his illuminating article contains analysis of several models which bear resemblance to some of the heat transfer theories discussed in chapter 1.

(Dafermos, 1985) shows that the Cauchy problem for the equation

$$u_t + u u_x = \mu u_{xx}\,, \qquad x \in \mathbb{R},\, t > 0,$$

with the initial data

$$u(x,0) = u_0(x), \tag{5.11}$$

has a unique C^∞ solution for u_0 bounded and measureable. He then shows that the equation

$$u_t + uu_x = 0\,, \quad x \in \mathbb{R},\, t > 0, \tag{5.12}$$

together with initial data (5.11) behaves very differently. When $du_0/dx \geq 0$ he shows there is a C^1 solution, but if $du_0/dx < 0$ there is a local solution which breaks down at $t = -[\inf_x du_0/dx]^{-1}$. (Dafermos, 1985) then considers the equation

$$u_t + uu_x + \mu u = 0\,, \quad x \in \mathbb{R},\, t > 0, \tag{5.13}$$

together with initial data (5.11). In this case he shows that there is a global C^1 solution provided $du_0/dx \geq -\mu$, with exponential decay in time of solutions if $du_0/dx > -\mu$. When $du_0/dx < -\mu$ he shows there is a local C^1 solution which breaks down at $t = \mu^{-1} \log\left[a/(a+\mu)\right]$ where $a = \inf_x du_0/dx$.

(Dafermos, 1985) observes that, ... "*the advantage of the method of characteristics lies in that it yields explicitly the threshold amplitude beyond which waves break as well as the critical time the first wave breaks. On the other hand, the method is very special and it may be expected to work only when the equations are relatively simple*".

(Dafermos, 1985) also shows that if du_0/dx and d^2u_0/dx^2 are in $L^2(-\infty,\infty)$ and $\|u_{0x}\|\,\|u_{0xx}\| < 2\mu^2/25$, ($\|\cdot\|$ being the norm on $L^2(-\infty,\infty)$), then (5.13) and (5.11) possesses a C^1 solution such that u_x and u_{xx} are in $L^2(-\infty,\infty)$ for any $t \geq 0$ and they decay to zero in $L^2(-\infty,\infty)$ as $t \to \infty$.

(Dafermos, 1985) also considers the "memory" equation

$$u_t + uu_x + \int_0^t a'(t-s)uu_x ds = 0\,, \quad x \in \mathbb{R},\, t > 0, \tag{5.14}$$

and the partial differential equation

$$u_{tt} + u_t u_{tx} = \mu u_{txx}\,. \tag{5.15}$$

The article of (Dafermos, 1985) also considers thermoviscoelasticity, fading memory in thermoelasticity, development of singularities in thermoelasticity, the entropy admissibility criterion to address the uniqueness issue, the Lax admissibility criterion for shock waves, and further criteria known as the viscosity criterion and the entropy rate admissibility criterion. The subject of conservation laws in continuum mechanics, hyperbolic equations, and hyperbolic systems is covered comprehensively in the masterpiece of (Dafermos, 2010).

(Fu and Scott, 1991) is a highly relevant article for the discussion in this section. They consider the equations of nonlinear isothermal elasticity in one space dimension. They assume that the body is an elastic half - space which is prestrained but is also quiescent. They allow a disturbance on the boundary which starts with a discontinuity in acceleration, rises smoothly and then decays to zero again, touching the boundary once more with a discontinuity in the acceleration. They solve the equations using Riemann invariants and find the time (or the point on the boundary) exactly for solution blow - up in acceleration, for a simple wave problem. However, they also employ an acceleration wave discontinuity analysis, like that described in chapter 4. Thus, they determine the critical time (or place) where the acceleration wave amplitude becomes infinite. Their striking result is that the critical time for acceleration wave blow-up is *exactly the same* as the critical time where the Riemann invariant solution becomes discontinuous. They conclude that, ... "*the acceleration wave amplitude predicted by singular surface theory becomes infinite at precisely the place where simple wave theory predicts that a shock will first begin to form*".

P. M. Jordan and his co-workers have presented a very interesting series of papers which employ numerical and analytical methods to study, *mutatis mutandis*, the development of a weak discontinuity (an acceleration wave) into a strong discontinuity (a shock wave). While several of these analyses are not directly applicable to heat wave propagation, typically arising in acoustic wave propagation in a gas or in a saturated porous medium, they are so pertinent to the work of the present section that they are highly relevant. The paper of (Christov and Jordan, 2010) is dealing with a temperature dependent thermal conductivity Cattaneo model and is specifically mentioned in section 5.3. The article of (Bargmann et al., 2008a) deals with a model for heat propagation in Green-Naghdi type II and type III materials and this is discussed in section 5.2.

We list below the mathematical equations which arise, with very brief details of the methods employed. We stress that most of these articles cited below present very interesting comparisons of how a solution with a disturbance present in the boundary does evolve into a shock wave, and the critical shock time or critical distance which is obtained from acceleration wave theory. Of course, the articles deal with other things, especially with detailed discussion of what the mathematical results mean to the physical questions being posed.

(Jordan et al., 2000) deals with the equations

$$\frac{\partial^2 u}{\partial t^2} - c^2\frac{\partial^2 u}{\partial x^2} + r\frac{\partial u}{\partial t} = 0 \tag{5.16}$$

and

$$\frac{\partial^2 u}{\partial t^2} - c^2\frac{\partial^2 u}{\partial x^2} - a^2\frac{\partial^3 u}{\partial x^2 \partial t} = 0 \tag{5.17}$$

and they develop analytical solutions using Laplace transforms. (Jordan and Puri, 2005) study the equation

$$\frac{\partial^2 u}{\partial t^2} - c^2\frac{\partial^2 u}{\partial x^2} = 3\beta\frac{\partial}{\partial x}\left(u^2\frac{\partial u}{\partial x}\right) \tag{5.18}$$

using acceleration wave methods and finite differences. It is interesting that this is an example of Shablovskii's equation (5.1). (Jordan and Christov, 2005) analyse the equation

$$\frac{\partial^2 u}{\partial t^2} - c^2 \frac{\partial^2 u}{\partial x^2} = \beta \frac{\partial^2 (u^2)}{\partial t^2} \tag{5.19}$$

using acceleration wave theory and finite differences. (Jordan, 2004) investigates the equation

$$\frac{\partial^2 u}{\partial t^2} - c^2 \frac{\partial^2 u}{\partial x^2} - k_1 \frac{\partial^3 u}{\partial x^2 \partial t} + \frac{\partial}{\partial t} \left[\left(\frac{\partial u}{\partial x} \right)^2 + k_2 \left(\frac{\partial u}{\partial t} \right)^2 \right] = 0 \tag{5.20}$$

while (Jordan, 2006) deals with the equation

$$\frac{\partial^2 u}{\partial t^2} - c^2 \frac{\partial^2 u}{\partial x^2} + k_1 \frac{\partial u}{\partial t} + \frac{\partial}{\partial t} \left[k_2 \left(\frac{\partial u}{\partial x} \right)^2 + k_3 \left(\frac{\partial u}{\partial t} \right)^2 \right] = 0\,. \tag{5.21}$$

Both of these articles use acceleration wave techniques and travelling wave analysis. (Christov et al., 2006) analyse the partial differential equations

$$\frac{\partial^2 u}{\partial t^2} - c^2 \frac{\partial^2 u}{\partial x^2} + 2 \frac{\partial u}{\partial x} \frac{\partial^2 u}{\partial x \partial t} + k_1 \frac{\partial u}{\partial t} \frac{\partial^2 u}{\partial x^2} + k_2 \left(\frac{\partial u}{\partial x} \right)^2 \frac{\partial^2 u}{\partial x^2} = 0 \tag{5.22}$$

and

$$\frac{\partial^2 u}{\partial t^2} + k_1 \frac{\partial u}{\partial t} \frac{\partial^2 u}{\partial t^2} - \frac{\partial^2 u}{\partial x^2} = 0\,. \tag{5.23}$$

They employ acceleration wave methods but notably, use a MUSCL-Hancock numerical scheme which is based on a high resolution Godunov method. (Jordan, 2008b) studies the equation

$$\frac{\partial^2 u}{\partial t^2} + k_1 \frac{\partial u}{\partial t} - \frac{\partial^2 u}{\partial x^2} = k_2 \frac{\partial}{\partial x} \left(u^2 \frac{\partial u}{\partial x} \right) \tag{5.24}$$

with finite difference and acceleration wave methods. (Christov and Jordan, 2008) analyse the equation

$$\tau \frac{\partial^2 u}{\partial t^2} + \frac{\partial u}{\partial t} - k_1 \frac{\partial^2 u}{\partial x^2} + k_2 \left(1 - \frac{u}{k_3} \right) \frac{\partial u}{\partial x} = 0\,, \tag{5.25}$$

using a Godunov numerical method and also acceleration wave techniques. The same equation is studied by means of a Cole-Hopf transformation by (Jordan, 2010a) who also analyses the same equation without the $\partial u/\partial t$ term. (Christov and Jordan, 2009) investigate the equation

$$\frac{\partial^2 u}{\partial t^2} = c^2 \frac{\partial^2 u}{\partial x^2} + c^2 \frac{\partial}{\partial x} \left[\beta(u_x^2) \frac{\partial u}{\partial x} \right] - \delta \frac{\partial u}{\partial t} \tag{5.26}$$

with a Godunov numerical technique, but also employ travelling wave and acceleration wave methods. (Jordan, 2007) analyses a traffic flow equation

$$\frac{\partial^2 u}{\partial t^2} - c^2 \frac{\partial^2 u}{\partial x^2} = k_1 \frac{\partial u}{\partial t} \left(1 - \frac{u}{k_2} \right) \tag{5.27}$$

studying acceleration waves, shocks, and using a finite difference numerical technique. This work is further described in section 5.2. The work of (Christov and Jordan, 2010) concentrates on the equation

$$\tau\frac{\partial^2 u}{\partial t^2}+\frac{\partial u}{\partial t}=k_1\frac{\partial}{\partial x}\Big[(1+k_2u)\frac{\partial u}{\partial x}\Big] \tag{5.28}$$

and they analyse acceleration waves, shock waves, and use a Godunov numerical technique. (Bargmann et al., 2008b) study the equation

$$(1+\lambda u)\frac{\partial^2 u}{\partial t^2}-(1+\lambda u)^2\frac{\partial^2 u}{\partial x^2}=\lambda\Big(\frac{\partial u}{\partial t}\Big)^2. \tag{5.29}$$

Their work uses travelling waves, acceleration waves, and finite differences and arises from Green-Naghdi type II and type III theory.

To complete this section we consider the development of an acceleration wave into a thermal shock wave for two specific examples. One is for Shablovskii's equation, (5.6), the other involves the temperature - dependent thermal conductivity version of the Maxwell - Cattaneo model, cf. section 5.3.

We write Shablovskii's equation, (5.6), in the form

$$\begin{aligned}\frac{\partial^2\theta}{\partial t^2}=&\frac{\partial}{\partial x}\Big(\kappa(\theta)\frac{\partial\theta}{\partial x}\Big)\\ =&\kappa(\theta)\frac{\partial^2\theta}{\partial x^2}+\kappa'(\theta)\Big(\frac{\partial\theta}{\partial x}\Big)^2.\end{aligned} \tag{5.30}$$

Equation 5.30 is here defined for $x\in(0,1)$ and $t>0$. On the boundary $x=1$ we assume $\theta=0$, while at $x=0$, θ satisfies

$$\theta(0,t)=\big[H(t)-H(t-t_w)\big]\,\sin\Big(\frac{\pi t}{t_w}\Big), \tag{5.31}$$

where H is the Heaviside function and $t_w>0$ is a constant. The initial data we employ are

$$\theta(x,0)=0,\qquad\qquad \frac{\partial\theta}{\partial t}(x,0)=0.$$

Thus, the boundary initial value problem under consideration allows us to study the effect of a disturbance starting at $x=0$, $t=0$ which propagates into the spatial domain as t increases. Due to the boundary condition (5.31) θ is everywhere continuous but its derivatives θ_x and θ_t are not. In fact they suffer a jump at one point in x (starting at $x=0$) and the discontinuity moves along a curve Σ in the (x,t) plane. This is thus an acceleration wave. (Such problems have been investigated at length in a variety of contexts by (Jordan and Christov, 2005), (Jordan and Puri, 2005), (Jordan, 2008b; Jordan, 2010b), (Christov and Jordan, 2010).)

If we let $a(t) = [\theta_x]$ be the wave amplitude then writing equation (5.30) as the system

$$\frac{\partial\theta}{\partial t} = -\frac{\partial q}{\partial x}$$
$$\frac{\partial q}{\partial t} = -\kappa\frac{\partial\theta}{\partial x}$$

and taking jumps of $\theta_t, \theta_x, q_t, q_x$, we use the Hadamard relation, (4.5), to find the speed, V, of a wave is given by $V^2 = \kappa(\theta^+)$. Taking the jump of equation (5.30) yields

$$[\theta_{tt}] = \kappa'[\theta_x]^2 + \kappa[\theta_{xx}]\,, \tag{5.32}$$

since $\theta_x^+ = 0$. Next, using the Hadamard relation, (4.5), we may show

$$[\theta_{tt}] = -2V\frac{\delta a}{\delta t} + V^2[\theta_{xx}]$$

and this in equation (5.32) together with the fact that $V^2 = \kappa$ allows us to deduce

$$-2V\frac{\delta a}{\delta t} = \kappa' a^2. \tag{5.33}$$

We consider the case where $\kappa = k_1 + k_2\theta$ and, in particular, take $k_1 = 1, k_2 = .7$, and $t_w = 1$. (These values are chosen so that amplitude blow - up which occurs at $t = t_\infty$, happens such that $t_\infty < t_w$.) Since $\theta^+ = 0$ the coefficients in equation (5.33) are constants and this equation may then be integrated to find

$$a(t) = \frac{1}{(a(0))^{-1} + (\kappa' t/2V)}\,. \tag{5.34}$$

In the present situation $a(0) = \theta_x^-(0) = -\pi/t_w$, $\kappa' = k_2$, and $V = 1$ at the wave. Thus, from (5.34) we find

$$a(t) = \theta_x^-(t) = \frac{1}{(-t_w/\pi) + (k_2 t/2)}\,.$$

This leads to amplitude blow-up at time $t_\infty = 1/.35\pi \approx 0.909456817$.

To investigate whether thermal shock formation actually occurs we solve the boundary initial value problem currently under study numerically. The acceleration wave analysis yields no information on the solution θ behind the wave, only information at the wavefront. We employ an explicit finite difference scheme, as suggested by (Jordan and Christov, 2005), (Jordan and Puri, 2005) and (Jordan, 2008b). Here we discretize equation (5.30) with the standard three point approximations for θ_{tt} and θ_{xx} and a centred difference for θ_x. Thus, if θ_m^k denotes the value of θ at the point (x_m, t_k),

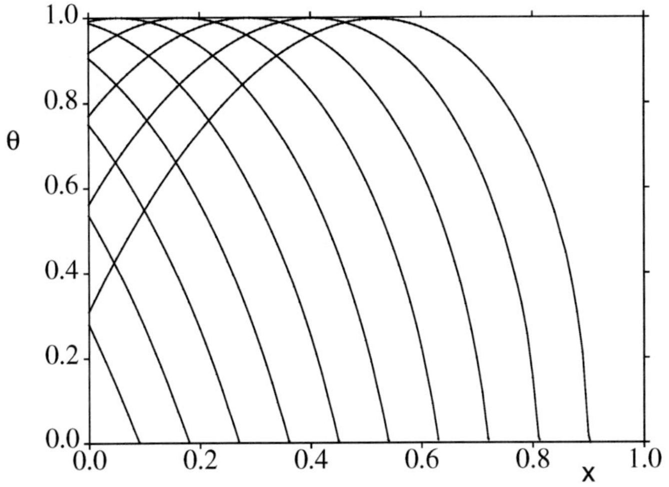

Figure 5.1. θ profile as t increases. The curves are for $t = .09, .18, \ldots, .9$, moving right. Note the steepening of θ_x^-.

then we use the finite difference scheme

$$\begin{aligned}\frac{\theta_m^{k+1} - 2\theta_m^k + \theta_m^{k-1}}{(\Delta t)^2} &= \kappa(\theta_m^k)\frac{\theta_{m+1}^k - 2\theta_m^k + \theta_{m-1}^k}{(\Delta x)^2} \\ &+ \kappa'(\theta_m^k)\frac{(\theta_{m+1}^k - \theta_{m-1}^k)^2}{4(\Delta x)^2}\,.\end{aligned} \tag{5.35}$$

This yields an explicit scheme for θ_m^{k+1} and θ_m^1, θ_m^0 are found from the initial data $\theta(x,0) = 0, \theta_t(x,0) = 0$, by standard means, cf. (Burden and Faires, 2001), pp. 719–721. Of course, the important point with this example is the boundary data equation (5.31). It is this which induces initial growth in θ_x. Even though (5.35) is an explicit scheme, the numerical results are surprisingly accurate if one chooses Δt and Δx carefully.

One clearly sees from figure 5.1 that the slope $|\theta_x^-|$ increases as $t(x)$ increases and this strongly indicates the formation of a thermal shock at $t = t_\infty$.

For a second example and also as a check on the numerical scheme we studied the system of (Christov and Jordan, 2010)

$$\begin{aligned}\frac{\partial \theta}{\partial t} &= -\frac{\partial q}{\partial x} \\ \frac{\partial q}{\partial t} + \frac{1}{\tau} q &= -(k_1 + k_2\theta)\theta_x\,.\end{aligned}$$

(Christov and Jordan, 2010) employ the relation $\tau^{-1} = k_1$ and then q may be eliminated and we find θ satisfies the equation

$$\frac{\partial^2 \theta}{\partial t^2} + k_1 \frac{\partial \theta}{\partial t} = \frac{\partial}{\partial x}\left(\kappa(\theta)\frac{\partial \theta}{\partial x}\right), \tag{5.36}$$

where $\kappa = k_1 + k_2\theta$. The boundary and initial conditions are as before, namely

$$\theta(0,t) = \big[H(t) - H(t-t_w)\big] \sin\Big(\frac{\pi t}{t_w}\Big), \qquad \theta(1,t) = 0,$$

$$\theta(x,0) = 0, \qquad \frac{\partial \theta}{\partial t}(x,0) = 0.$$

Defining the wave amplitude $a(t) = [\theta_x]$ one finds the wavespeed of an acceleration wave is given as $V^2 = \kappa(\theta^+)$. Then, taking the jump of equation (5.36) we find, in a manner similar to that leading to (5.33), that a satisfies the equation

$$\frac{\delta a}{\delta t} + \frac{k_1}{2}\, a + \frac{\kappa'}{2V}\, a^2 = 0.$$

This is integrated to yield the amplitude

$$a(t) = \frac{1}{e^{k_1 t/2}(a(0))^{-1} - (\kappa'/Vk_1)(1 - e^{k_1 t/2})}\,. \tag{5.37}$$

Blow-up of a now occurs if $a(0) < 0$ and $-a(0) > k_1^2/k_2$. The blow-up time is

$$t_\infty = \frac{2}{k_1} \log\Big(\frac{-\kappa' a(0)}{Vk_1 - a(0)\kappa'}\Big).$$

This is in agreement with (Christov and Jordan, 2010). For direct comparison with (Christov and Jordan, 2010), see their figure 6, we choose the same parameter values so that $k_1 = 1.13018161, k_2 = 0.95929815$ and $t_w = 0.940645282$.

We discretize (5.36) with a finite difference technique employing a centred difference for θ_t. Thus, we use

$$\begin{aligned}\frac{\theta_m^{k+1} - 2\theta_m^k + \theta_m^{k-1}}{(\Delta t)^2} &+ k_1\Big(\frac{\theta_m^{k+1} - \theta_m^{k-1}}{2\Delta t}\Big)\\ &= \kappa(\theta_m^k)\frac{\theta_{m+1}^k - 2\theta_m^k + \theta_{m-1}^k}{(\Delta x)^2} + \kappa'(\theta_m^k)\frac{(\theta_{m+1}^k - \theta_{m-1}^k)^2}{4(\Delta x)^2}\,.\end{aligned} \tag{5.38}$$

The solution is displayed in figure 5.2. (Christov and Jordan, 2010) plot θ for $t = .3, .5, .7$ and 0.9. The values chosen in figure 5.2 are different apart from $t = 0.9$, but show exactly the same trend of solution development. Again, moving right in figure 5.2 we see the steepening of $|\theta_x^-|$ and this strongly supports thermal shock formation at a value of $t_\infty \approx 0.9$, cf. (Christov and Jordan, 2010). Note that figure 5.2 shows the effect of dissipation present in equation (5.36) since θ_{max} is falling as the wave progresses. This does not happen in figure 5.1 since the dissipation term $k_1\theta_t$ is not present in Shablovskii's equation (5.30).

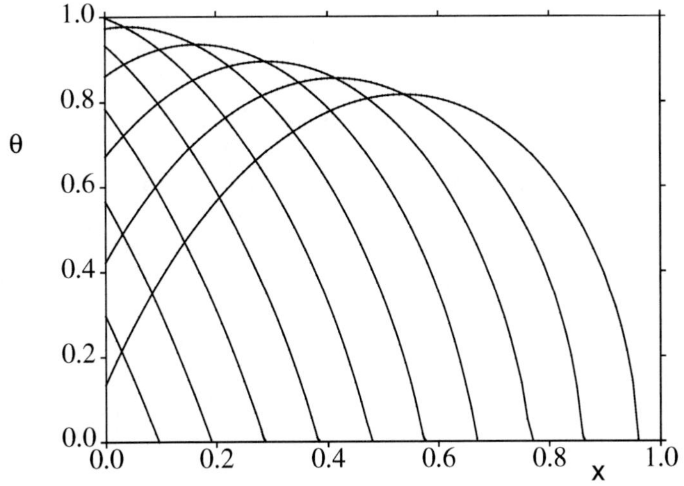

Figure 5.2. θ profile as t increases. The curves are for $t = .09, .18, \ldots, .9$, moving right. Note the steepening of θ_x^-.

5.2 Type II and type III thermoelasticity

(Bargmann and Steinmann, 2006; Bargmann and Steinmann, 2008), (Bargmann et al., 2008a), and (Bargmann et al., 2008b) are notable contributions dealing with aspects of type II and type III thermoelasticity and heat transfer.

(Bargmann and Steinmann, 2006) and (Bargmann and Steinmann, 2008) concentrate on developing finite element methods for solving the linearized equations of thermoelasticity according to Green-Naghdi theory of type II and type III, cf. sections 2.3 and 2.4, although they do also make comparison with what is found from classical thermelasticity employing Fourier's law. (Bargmann and Steinmann, 2006) and (Bargmann and Steinmann, 2008) essentially focus on developing accurate finite element methods for solving the equations, cf. equations (2.72), (2.73), (2.82), (2.83),

$$\begin{aligned} \rho\ddot{u}_i &= \big(E_{ijkh}\epsilon_{kh} - 3wK(T-T_0)\delta_{ij}\big)_{,j} + \rho b_i\,, \\ \rho c\ddot{\alpha} &= k_1\Delta\alpha - 3T_0wK\dot{\epsilon}_{ii} + \rho\theta s, \end{aligned} \tag{5.39}$$

for type II thermoelasticity, or

$$\begin{aligned} \rho\ddot{u}_i &= \big(E_{ijkh}\epsilon_{kh} - 3wK(T-T_0)\delta_{ij}\big)_{,j} + \rho b_i\,, \\ \rho c\ddot{\alpha} &= k_1\Delta\alpha + k_2\Delta\dot{\alpha} - 3T_0wK\dot{\epsilon}_{ii} + \rho\theta s, \end{aligned} \tag{5.40}$$

for type III thermoelasticity. In these equations u_i, α are elastic displacement and the temperature displacement variables, b_i and s are source terms. The tensor E_{ijkh} represents the elastic coefficients, ρ is the density, w, K are the thermal expansion coefficient and the bulk modulus, T_0 is a reference temperature, c is specific heat, ϵ_{ij} is the strain tensor, and θ and T

are equivalent functions for the temperature. The coefficients k_1 and k_2 are thermal coefficients, the difference between type III and type II being the presence of the k_2 term in (5.40).

(Bargmann and Steinmann, 2006) explain in detail how they approximate a solution to a boundary initial value problem for equations (5.39) or equations (5.40). Their approach is to use a Galerkin finite element method in both space and time using continuous elements in the discretization. Interestingly, they employ discontinuous Galerkin elements when discretizing the equations of classical thermoelasticity. They observe that the type II model requires the introduction of a stabilizing term in the temperature displacement equation $(5.39)_2$ to suppress numerical oscillations which otherwise arise.

(Bargmann and Steinmann, 2006) solve numerically the problem of a thermoelastic wave in one space dimension and particularly model the situation of a disturbance in NaF, as reported in the experiments of (Jackson et al., 1970). Their results are very revealing and certainly show they have successfully modelled numerically thermoelastic wave propagation in a realistic manner with type II and type III thermoelasticity. The numerical example of (Bargmann and Steinmann, 2008) considers thermoelastic wave propagation in a two-dimensional plate, in accordance with the experiments of (Narayanamurti and Dynes, 1972) on Bismuth. They show clearly how a heat pulse generated in the plate propagates outward.

The articles of (Bargmann and Steinmann, 2006) and (Bargmann and Steinmann, 2008) represent fundamental contributions to the numerical modelling of thermoelasticity using Green-Naghdi type II and III theories. The paper of (Bargmann et al., 2008a) is a very stimulating one addressing heat propagation on Saturn's moon Enceladus, and this is discussed further in section 9.2.1.

(Bargmann et al., 2008b) is another important paper dealing with thermal wave propagation according to Green-Naghdi type II and III theory. While it mainly concentrates on type III theory, type II and type I (classical) are discussed as limiting cases. (Bargmann et al., 2008b) models the siutation where a finite thermal pulse is input into a half space at some time, $t = 0$ say. Hence, they analyse the mathematical problem,

$$\begin{aligned} &\theta_{tt} - \nu^2\theta_{xx} = \chi\theta_{txx}\,, \quad 0 < x < \infty,\ 0 < t < \infty, \\ &\theta(0,t) = H(t), \qquad \theta(\infty,t) = 0, \\ &\theta(x,0) = 0, \qquad \theta_t(x,0) = 0, \end{aligned} \tag{5.41}$$

θ being temperature, H the Heaviside function, ν^2, χ positive constants.

Problem (5.41) is solved by a dual integral transform method and then (Bargmann et al., 2008b) examine closely the small and large time limits. They show clearly how the temperature pulse evolves. They also analyse a nonlinear version of type II theory, especially investigating the problem

$$\begin{aligned}
&(1+\lambda T)T_{tt} - (1+\lambda T)^2 T_{xx} = \lambda (T_t)^2 , \quad 0 < x < 1,\ t > 0, \\
&T(0,t) = H(t)\sin \pi t, \qquad T(1,t) = 0, \quad t < 1, \\
&T(x,0) = 0, \qquad T_t(x,0) = 0.
\end{aligned} \tag{5.42}$$

Here T is temperature (related to θ by $\theta = a + bT$), and λ is a positive constant. They use acceleration wave theory to show that a temperature rate wave (temperature acceleration wave for which T_x is discontinuous) may blow up in a finite time. They also solve (5.42) numerically using finite differences and graph the evolution of the temperature profile. This clearly shows temperature shock formation, and by equating this to their acceleration wave analysis they are able to deduce a value for λ.

(Bargmann et al., 2008b) also analyse travelling waves for equation (5.42) by setting $f(\xi) = 1 + \lambda T(x,t)$ with $\xi = x - vt$, v a constant speed. They are able to obtain an analytical solution in terms of the Lambert W-function. From this they deduce that the wave amplitude is directly related to propagation speed, and a thermal shock may form.

(Jordan, 2007) presents an interesting analysis of a hyperbolic equation which arises from the parabolic Fisher-Kolmogorov equation

$$\rho_t - \nu \rho_{xx} = \gamma \rho \Big(1 - \frac{\rho}{\rho_s} \Big) , \tag{5.43}$$

ν, γ, ρ_s constants. He notes that (5.43) is equivalent to

$$\rho_t + q_x = \gamma \rho \Big(1 - \frac{\rho}{\rho_s} \Big) , \tag{5.44}$$

and

$$q = -\nu \rho_x \tag{5.45}$$

for a flux q. A Cattaneo modification of (5.45) would write

$$\tau q_t + q = -\nu \rho_x \, . \tag{5.46}$$

(Jordan, 2007), however, argues that a type II version of (5.45) could be written

$$q_t = -c_\infty^2 \rho_x \tag{5.47}$$

where c_∞ is a positive constant. (Jordan, 2007) then works with the system (5.44) and (5.47). This is equivalent to the partial differential equation

$$\rho_{tt} - c_\infty^2 \rho_{xx} = \gamma \rho_t \Big(1 - \frac{2}{\rho_s} \rho \Big) . \tag{5.48}$$

The paper of (Jordan, 2007) produces some very interesting results. He derives a travelling wave solution and shows how it can produce a "Taylor shock". He then shows that a shock wave, where ρ and q have a discontinuity, may be analysed and the solution to the amplitude obtained exactly. It is also shown that an acceleration wave behaves as a weaker wave. (Jordan, 2007) also uses a finite difference method to investigate how a sinusoidal input into equation (5.48) via a boundary will evolve in time. The numerical simulations show shock formation and he compares this to his shock wave analysis. Various conclusions are drawn by (Jordan, 2007) regarding application of equation (5.48) as a model in applied mathematics.

Interesting results on travelling waves and on shock propagation for a hyperbolic Burgers equation and for a hyperbolic kinematic wave equation have been derived by (Jordan, 2010a).

5.3 Temperature dependent thermal conductivity

The Maxwell-Cattaneo equations with temperature dependent thermal conductivity have been the subject of much investigation. The relevant equations have been presented as equations (4.1) and a detailed analysis of acceleration waves is considered in section 4.1.

In this section we focus on work primarily related to thermal shock waves and numerical solutions to equations (4.1) when the thermal conductivity is a linear function of temperature. Thus, the relevant equations are

$$\begin{aligned} c\frac{\partial\theta}{\partial t} &= -\frac{\partial q_i}{\partial x_i}, \\ \tau\frac{\partial q_i}{\partial t} + q_i &= -\kappa(\theta)\theta_{,i}, \end{aligned} \tag{5.49}$$

where

$$\kappa(\theta) = k(1+\beta\theta), \tag{5.50}$$

θ, q_i being temperature and heat flux, with c, τ, k, β positive constants.

(Glass et al., 1986) added a source term $g(x, t)$ in one space dimension to equation $(5.49)_1$ and devised a finite difference MacCormack numerical method for the solution of system (5.49). They also solved numerically the analogous parabolic equation. They showed that the hyperbolic system predicted larger temperatures with a distinct temperature front. (Cramer et al., 2001) study the same system and analyse numerically the general evolution of the thermal shock. They allow a square wave heat source and a sinusoidal heat source input in time. This is a very interesting paper which investigates the interaction of shock waves with smoother parts of the temperature distribution.

(Reverberi et al., 2008) also study system (5.49) numerically employing relation (5.50) but also allowing κ to be an exponential function of temperature, i.e. $\kappa = ke^{\beta\theta}$. Their paper contains a brief but useful review of numerical methods used in this area. Their numerical scheme is a Hartree hybrid finite difference method. They analyse evolution profiles where a thermal shock forms and also follow shock evolution.

(Christov and Jordan, 2010) also analyse system (5.49) with the relation (5.50). After using Rankine-Hugoniot discontinuity relations for the thermal shocks they use a Godunov numerical scheme to accurately capture the shock wave. This paper also analyses temperature rate acceleration waves generated by a pulse at the boundary of form

$$\theta(0,t) = \big[H(t) - H(t-t_w)\big] \sin\left(\frac{\pi t}{t_w}\right)$$

for a constant $t_w > 0$. The blow-up time of the acceleration wave is compared to what is found by numerical simulation. (Christov and Jordan, 2010) also develop a detailed analysis of travelling waves for the equation

$$\tau\frac{\partial^2\theta}{\partial t^2} + \frac{\partial\theta}{\partial t} = k\frac{\partial}{\partial x}\left[(1+\beta\theta)\,\frac{\partial\theta}{\partial x}\right]$$

which arises from (5.49) in one space dimension. They develop an exact solution via a travelling wave and are able to relate this in a very interesting way to acceleration wave and shock wave behaviour.

5.4 Exercises

Exercise 5.4.1 *Consider the equation*

$$u_t + uu_x = 0, \qquad x \in \mathbb{R},\ t > 0, \tag{5.51}$$

with

$$u(x,0) = f(x).$$

By considering characteristics, show that

$$u = f(\xi) \qquad \text{on} \quad x = \xi + ut = \xi + f(\xi)t.$$

Deduce that

$$u_x = \frac{f'(\xi)}{1 + tf'(\xi)}$$

and so u_x blows up at $t = (-\inf_x \partial u_0/\partial x)^{-1}$ where $u_0(x) \equiv f(x)$, if $u_{0x} < 0$.

Develop an acceleration wave analysis for equation (5.51) with the amplitude $a(t) = [u_x]$. Show that the wavespeed $V = u$ and

$$\frac{\delta a}{\delta t} + 2u_x^+ a + a^2 = 0.$$

If $u_x^+ = 0$ show that

$$a(t) = \frac{a_0}{1 + ta_0}$$

$a_0 = a(0)$, and deduce $a(t) \to \infty$ as $t \to -1/a_0 = -1/u_x^-(0)$, provided $u_x^-(0) < 0$. (Here $u_x^-(0)$ refers to the value when $t = 0$.)

Exercise 5.4.2 *Consider the equation*

$$u_t + uu_x + \mu u = 0, \qquad x \in \mathbb{R},\ t > 0, \tag{5.52}$$

with

$$u(x, 0) = u_0(x).$$

By considering characteristics, show that

$$u = u_0(\xi)e^{-\mu t} \qquad \text{on} \quad x = \xi + \frac{u_0(\xi)}{\mu}\left(1 - e^{-\mu t}\right).$$

Show that

$$u_x = \frac{u_0(\xi)e^{-\mu t}}{\{1 + (1 - \exp -\mu t)\, u_0'(\xi)/\mu\}}$$

and deduce that u_x blows up, if $u_0'(\xi) < -\mu < 0$, at time

$$t = -\frac{1}{\mu}\log\left(1 + \frac{\mu}{u_0'(\xi)}\right).$$

Develop an acceleration wave analysis for equation (5.52) with the amplitude $a(t) = [u_x]$. Show that the wavespeed $V = u$ and if $u_x^+ = 0$ then

$$\frac{\delta a}{\delta t} + \mu a + a^2 = 0.$$

Solve this equation to show that

$$u_x^- = [u_x] = a(t) = \frac{1}{e^{\mu t}a(0)^{-1} + \{(e^{\mu t} - 1)/\mu\}}$$

and deduce that if $u_{0x} < -\mu$ then u_x^- blows up at

$$t = \frac{1}{\mu}\log\left(\frac{a(0)}{\mu + a(0)}\right).$$

How does this compare with the analysis of solution breakdown for equation (5.52) by characteristics?

Exercise 5.4.3 *Consider the initial-value problem*

$$\begin{aligned} u_t + c(u)u_x &= 0, \qquad & t > 0, x \in \mathbb{R}, \\ u &= f(x), & t = 0, x \in \mathbb{R}, \end{aligned} \tag{5.53}$$

where c, f are smooth functions. Write down the differential equations that determine the characteristics of (5.53) and hence show that

$$u = f(\xi) \qquad \text{on} \qquad x = \xi + F(\xi)t,$$

where $F(\xi) = c(f(\xi))$. Show also that if $c(u)$ is a decreasing function in x then the derivatives of u become discontinuous at a time

$$T = \left[\frac{-1}{F'(\xi)} \right]_m,$$

where the m indicates the characteristic on which the minimum is achieved.

When $c(u) = u^2$ and $f(x)$ is defined by

$$f(x) = \begin{cases} 1, & x < 0, \\ e \exp\left[-1/(1-x^2)\right], & 0 \le x \le 1, \\ 0, & x > 1, \end{cases}$$

deduce the solution u, determine whether u_x and u_t become discontinuous, and if they do find the first time of breakdown. Make a rough sketch of what is happening.

Exercise 5.4.4 *Repeat the analysis of exercise 5.4.3 when $c(u) = u^2$, but when $f(x)$ is defined by*

$$f(x) = \begin{cases} 1, & x < 0, \\ 1 - 2x^2 + x^4, & 0 \le x \le 1, \\ 0, & x > 1, \end{cases}$$

deduce the solution u, show that u_x and u_t become discontinuous, and show the first time of breakdown is $T = 7^{7/2}/1728$. Make a rough sketch of what is happening.

Exercise 5.4.5 *Find (implicitly) the solution to the partial differential equation*

$$\frac{\partial u}{\partial t} + u^2 \frac{\partial u}{\partial x} = 0, \qquad t > 0,\ x \in \mathbb{R},$$

with

$$u(x,0) = f(x) = \begin{cases} 1, & x < 0, \\ 1 - 3x^2 + 2x^3, & 0 \le x \le 1, \\ 0, & x > 1. \end{cases}$$

Show that u_x and u_t become discontinuous at time

$$T = \frac{100\sqrt{10}}{12(\sqrt{10}-1)(7\sqrt{10}+2)}.$$

Exercise 5.4.6 *Draw the characteristics in the (x,t) plane, and determine the characteristic on which "blow-up" occurs and the first time of blow-up of u_x, u_t where u solves:*

$$\frac{\partial u}{\partial t} + u \frac{\partial u}{\partial x} = 0, \qquad t > 0,\ x \in \mathbb{R},$$

and

$$u(x.0) = f(x) = \begin{cases} 1, & x < 0, \\ \frac{1}{2} - \tanh^{-1}(x - \frac{1}{2})/\log 3\,, & x \in [0,1], \\ 0, & x \in (1,2), \\ x - 2, & x \in [2,3), \\ 1, & x \in [3,4), \\ 5 - x, & x \in [4,5], \\ 0, & x > 5. \end{cases}$$

Exercise 5.4.7 *Consider the system of equations*

$$\frac{\partial u_i}{\partial t} + a_{ij}\frac{\partial u_j}{\partial x} + b_i = 0. \tag{5.54}$$

What is meant by saying (5.54) is a hyperbolic system?
Consider now the system of equations

$$\begin{aligned} u_t + u_x + 3v_x + v = 0, &\qquad x \in \mathbb{R}, \quad t > 0, \\ v_t + 3u_x + v_x + u = 0, &\qquad x \in \mathbb{R}, \quad t > 0, \end{aligned}$$

with initial data

$$u(x,0) = u_0(x), \qquad v(x,0) = v_0(x), \quad x \in \mathbb{R}.$$

Show that this may be reduced to the form

$$\begin{aligned} u(x,t) - v(x,t) &= e^t\big[u_0(\xi_1) - v_0(\xi_1)\big] \qquad \text{on} \quad x = -2t + \xi_1, \\ u(x,t) + v(x,t) &= e^{-t}\big[u_0(\xi_2) + v_0(\xi_2)\big] \qquad \text{on} \quad x = 4t + \xi_2. \end{aligned}$$

Find u, v at $x = 3, t = 1$, in terms of $u_0(-1), u_0(5), v_0(-1)$ and $v_0(5)$.

Exercise 5.4.8 *The transmission of electricity along a cable may be described by the equations*

$$\begin{aligned} i_t + v_x + i &= 0, \\ v_t + i_x + v &= 0, \end{aligned}$$

where v is voltage and i is current. Write these as a hyperbolic system and show that they may be reduced to the Riemann invariant form

$$\begin{aligned} e^t(i + v) &= \text{constant, along } x = t + k_1, \\ e^t(i - v) &= \text{constant, along } x + t = k_2, \end{aligned}$$

where k_1, k_2 are constants to be determined by the initial conditions.

Exercise 5.4.9 *Consider the system*

$$a_t + b_x = \alpha a, \qquad b_t + a_x = \alpha b, \tag{5.55}$$

with the initial conditions

$$a(x,0) = \sin x, \qquad b(x,0) = \cos x.$$

Write (5.55) as a hyperbolic system and hence solve to show that

$$a+b=e^{\alpha t}(\sin k_1+\cos k_1), \quad \text{on} \quad x=t+k_1,$$
$$a-b=e^{\alpha t}(\sin k_2-\cos k_2), \quad \text{on} \quad x=-t+k_2.$$

Deduce values for $a(2\pi,\pi)$, $b(2\pi,\pi)$, *when* $\alpha=2$. *Deduce also values for* $a(3\pi,2\pi)$, $b(3\pi,2\pi)$, *when* $\alpha=3$.

Exercise 5.4.10 *Consider the system of equations*

$$\frac{\partial u}{\partial t}+3\frac{\partial u}{\partial x}+4\frac{\partial v}{\partial x}+\alpha u=0,$$
$$\frac{\partial v}{\partial t}+\frac{\partial u}{\partial x}+\alpha v=0,$$

α *constant, with initial data*

$$u(x,0)=\exp(x^2), \quad v(x,0)=\sin \pi x.$$

Find expressions for $u-4v$ *and* $u+v$ *on suitable characteristics. Use these to determine* u,v *at* $(x,t)=(8,3)$.

Exercise 5.4.11 *Reduce the system of equations*

$$\frac{\partial u}{\partial t}+2\frac{\partial u}{\partial x}+\frac{\partial v}{\partial x}+b_1=0,$$
$$\frac{\partial v}{\partial t}+\frac{\partial u}{\partial x}+2\frac{\partial v}{\partial x}+b_2=0,$$

to two ordinary differential equations on characteristics. Suppose $b_1=u$, $b_2=v$, *and the initial data are*

$$u(x,0)=e^x, \qquad v(x,0)=1+x^2.$$

Calculate u,v *at* $(x,t)=(9/2,3/2)$.

Exercise 5.4.12 *Repeat exercise 5.4.11 with*

$$b_1=u^2+v^2, \qquad b_2=2uv,$$

and calculate a formula for $u-v$ *and* $u+v$ *on suitable characteristics. In terms of the initial data functions* $u_0(x)$, $v_0(x)$, *calculate* $u(0,1), v(0,1)$.

Exercise 5.4.13 *For the problem*

$$\frac{\partial u}{\partial t}+u\frac{\partial u}{\partial x}=0, \qquad x\in\mathbb{R},\ t>0,$$
$$u=0, \quad x<0, \qquad u=1, \quad x>0,$$

construct two weak solutions (one which keeps the shock wave for all t *and one which is continuous for* $t>0$*) and hence demonstrate non-uniqueness of a weak solution.*

Exercise 5.4.14 *Let u be a solution to the partial differential equation*

$$(e^u + e^{-u})^2 \frac{\partial u}{\partial t} + 4\frac{\partial u}{\partial x} = 0, \qquad x \in \mathbb{R},\ t > 0,$$

with

$$u(x,0) = \begin{cases} 1, & x < 0, \\ 0, & x > 0. \end{cases}$$

Find a shock wave solution valid for all $t > 0$*, such that* $u^- \equiv 1, u^+ \equiv 0$.

Exercise 5.4.15 *The equation*

$$\frac{\partial u}{\partial t} + u^2 \frac{\partial u}{\partial x} = 0,$$

is defined together with an initial condition which contains a discontinuity in u (shock wave). Suppose $0 < u^- < u^+$*, and define*

$$\lambda = \frac{-3u_x^+(u^- + u^+)}{(u^+ + 2u^-)}\ .$$

Show that

(i) *if* $[u_x] > \lambda$, $[u]$ *increases locally in* t,
(ii) *if* $[u_x] < \lambda$, $[u]$ *decreases locally in* t.

6
Qualitative estimates

6.1 Decay in time

The generalized Maxwell - Cattaneo equations, known as the GMC equations, are described in chapter 1, p. 15, and have the form

$$
\begin{aligned}
c\frac{\partial T}{\partial t} &= -\frac{\partial q_i}{\partial x_i}, \\
\tau\frac{\partial q_i}{\partial t} &= -q_i - \kappa\frac{\partial T}{\partial x_i} + \mu\Delta q_i + \nu\frac{\partial^2 q_j}{\partial x_j \partial x_i},
\end{aligned}
\tag{6.1}
$$

where T, q_i are the temperature and heat flux fields, and c, τ, κ, μ and ν are positive constants.

One of the early articles dealing with qualitative results for this system of equations was that of (Morro et al., 1990). These writers studied decay in time of the solution, established lower bounds, demonstrated uniqueness for the MC system (where μ, ν are zero), proved uniqueness and continuous dependence on the initial data for (6.1), and examined how solutions grow in the backward in time problem. (Morro et al., 1990) studied equations (6.1) on a bounded spatial domain, Ω. Further results, including structural stability results, for equations (6.1) are given by (Franchi and Straughan, 1994a).

A study of continuous dependence on the initial-time geometry, where perturbations of the initial data set are considered over a time zone, was given by (Payne and Song, 1997a) when Ω is a domain exterior to a bounded set Ω_0 in $\mathbb{R}^3$. A study of continuous dependence on the initial-time geometry for a solution to (6.1) when Ω is bounded is given by (Payne and Song, 1997b) who analysed both the forward in time and backward in time problems.

(Payne and Song, 1997a) investigated continuous dependence on changes in the spatial geometry itself, i.e. how the solution behaves when the boundary Γ of Ω is actually changed. The system of partial differential equations analysed by (Payne and Song, 1997a) is an extended version of (6.1), namely,

$$\begin{aligned} c\frac{\partial T}{\partial t} &= -\frac{\partial q_i}{\partial x_i} + \zeta \Delta T, \\ \tau\frac{\partial q_i}{\partial t} &= -q_i - \kappa\frac{\partial T}{\partial x_i} + \mu\Delta q_i + \nu\frac{\partial^2 q_j}{\partial x_j \partial x_i}, \end{aligned} \tag{6.2}$$

where ζ is a positive constant. The question of continuous dependence of the solution to (6.1) on changes in the spatial geometry was addressed by (Lin and Payne, 2004a).

(Payne and Song, 2001) analysed the behaviour of a solution to equations (6.1) as the relaxation time τ tends to zero. Interesting bounds for a solution to (6.1) when the initial data are replaced by non-standard data involving a combination of data at time $t = 0$ and data at a later time $t = T$ are provided by (Payne et al., 2005) and by (Payne et al., 2004).

In the remainder of this section we report some results of (Payne and Song, 2004b) on decay in time for T and q_i satisfying equations (6.1). Let equations (6.1) be defined on the domain $\Omega \times (0, \infty)$ where $\Omega \subset \mathbb{R}^3$ is a bounded domain with boundary Γ. The functions T and q_i are subject to initial data

$$T(\mathbf{x}, 0) = T_0(\mathbf{x}), \qquad q_i(\mathbf{x}, 0) = f_i(\mathbf{x}), \tag{6.3}$$

and satisfy boundary data of form

$$T(\mathbf{x}, t) = 0, \qquad \epsilon_{ijk} q_j n_k = 0, \qquad \text{on } \Gamma \times (0, \infty). \tag{6.4}$$

The second of (6.4) essentially states that the components of heat flux in the directions tangential to the unit normal to Γ are zero.

6.1.1 Decay of temperature

To derive a decay bound for the temperature (Payne and Song, 2004b) eliminate q_i from (6.1) and show that T satisfies the equation

$$\left(\frac{\tau}{\mu + \nu}\right) \frac{\partial^2 T}{\partial t^2} + \frac{1}{(\mu + \nu)} \frac{\partial T}{\partial t} - \frac{\kappa}{c(\mu + \nu)} \Delta T - \Delta \frac{\partial T}{\partial t} = 0, \tag{6.5}$$

on $\Omega \times (0, \infty)$. They introduce the variable θ by $\theta = e^{at}T$ for a positive number a to be chosen. Equation (6.5) may be rewritten in terms of θ as

$$\begin{aligned}\Big(\frac{\tau}{\mu+\nu}\Big)\frac{\partial^2\theta}{\partial t^2} + \Big(\frac{1-2a\tau}{\mu+\nu}\Big)\frac{\partial\theta}{\partial t} - \Big[\frac{\kappa}{c(\mu+\nu)} - a\Big]\Delta\theta \\ - \Delta\frac{\partial\theta}{\partial t} + \frac{a(a\tau-1)}{(\mu+\nu)}\theta = 0.\end{aligned} \tag{6.6}$$

Since $T = 0$ on Γ, we also know $\theta = 0$ on Γ. (Payne and Song, 2004b) define the function $F(t)$ by

$$F(t) = \Big(\frac{\tau}{\mu+\nu}\Big)\|\theta_t\|^2 + \Big[\frac{\kappa}{c(\mu+\nu)} - a\Big]\|\nabla\theta\|^2 + \frac{a(a\tau-1)}{(\mu+\nu)}\|\theta\|^2, \tag{6.7}$$

where $\|\cdot\|$ is the norm on $L^2(\Omega)$. By multiplying equation (6.6) by θ_t, integrating over Ω and using the boundary conditions, one shows that

$$\frac{1}{2}\frac{dF}{dt} + \|\nabla\theta_t\|^2 + \Big(\frac{1-2a\tau}{\mu+\nu}\Big)\|\theta_t\|^2 = 0. \tag{6.8}$$

If λ_1 denotes the first eigenvalue in Poincaré's inequality for Ω, then (Payne and Song, 2004b) require

$$\frac{1-2a\tau}{\mu+\nu} + \lambda_1 \geq 0. \tag{6.9}$$

Then, from (6.8) they deduce that

$$\frac{dF}{dt} \leq 0.$$

Upon integration in time they find

$$F(t) \leq F(0). \tag{6.10}$$

Now, recall $\theta = e^{at}T$ and rewrite inequality (6.10) in terms of $T(\mathbf{x}, t)$. The result is

$$E(t) \leq E(0)e^{-2at}, \tag{6.11}$$

where $E(t)$ is the function defined by

$$\begin{aligned}E(t) =& \Big(\frac{\tau}{\mu+\nu}\Big)\|T_t + aT\|^2 + \Big[\frac{\kappa}{c(\mu+\nu)} - a\Big]\|\nabla T\|^2 \\ &+ \frac{a(a\tau-1)}{(\mu+\nu)}\|T\|^2.\end{aligned} \tag{6.12}$$

In addition to inequality (6.9), the coefficient a is selected now so that

$$\kappa > ac(\mu+\nu). \tag{6.13}$$

Poincaré's inequality is again used and a is further restricted in order that the following inequalities hold for a constant $\gamma > 0$,

$$\begin{aligned} &\Big[\frac{\kappa}{c(\mu+\nu)} - a\Big] \|\nabla T\|^2 + \frac{a(a\tau - 1)}{(\nu+\mu)} \|T\|^2 \\ &\qquad \geq \Big\{\Big[\frac{\kappa}{c(\mu+\nu)} - a\Big] \lambda_1 + \frac{a(a\tau-1)}{(\nu+\mu)}\Big\} \|T\|^2 \\ &\qquad \geq \gamma \|T\|^2 . \end{aligned} \tag{6.14}$$

(Payne and Song, 2004b) actually select $\gamma = \kappa\lambda_1/2c(\mu+\nu)$ and pick a to satisfy inequalities (6.9), (6.13) and (6.14). In this way (Payne and Song, 2004b) show that one may deduce from (6.11),

$$\|T(t)\|^2 \leq E(0)\, \frac{e^{-2at}}{\gamma} . \tag{6.15}$$

They also observe that decay results for $\|T_t\|$ and $\|\nabla T\|$ may be derived from inequality (6.11).

6.1.2 Decay of heat flux

(Payne and Song, 2004b) also produce decay bounds for the heat flux q_i. To do this they write $(6.1)_2$ as

$$-\tau q_{i,t} - q_i - \kappa T_{,i} + \mu(q_{i,j} - q_{j,i})_{,j} + (\mu+\nu) q_{j,ji} = 0.$$

This equation is multiplied by q_i and the result integrated over Ω to see that with integration by parts

$$\begin{aligned} \frac{\tau}{2}\frac{d}{dt}\|\mathbf{q}\|^2 = &- \|\mathbf{q}\|^2 + \kappa(T, q_{i,i}) \\ &+ \mu \oint_\Gamma q_i(q_{i,j} - q_{j,i}) n_j dS - \mu \int_\Omega q_{i,j}(q_{i,j} - q_{j,i}) dx \\ &+ (\mu+\nu) \oint_\Gamma q_{j,j} q_i n_i dS - (\mu+\nu)\|q_{i,i}\|^2 , \end{aligned} \tag{6.16}$$

where $(\cdot,\cdot)$ is the inner product on $L^2(\Omega)$. By applying equation $(6.1)_1$ on the boundary Γ, the second last term is rewritten as

$$+(\mu+\nu) \oint_\Gamma q_{j,j} q_i n_i dS = c(\mu+\nu) \oint_\Gamma T_t q_i n_i dS = 0,$$

where the fact that the integral is zero follows since $T = 0$ on Γ. The other boundary term in (6.16) is seen to be zero by appealing to the boundary condition $\epsilon_{ijk} q_j n_k = 0$. Hence, (6.16) reduces to

$$\begin{aligned} \frac{\tau}{2}\frac{d}{dt}\|\mathbf{q}\|^2 = &- \|\mathbf{q}\|^2 + \kappa(T, q_{i,i}) \\ &- \mu \int_\Omega q_{i,j}(q_{i,j} - q_{j,i}) dx - (\mu+\nu)\|q_{i,i}\|^2 . \end{aligned} \tag{6.17}$$

The second last term in (6.17) is rewritten as

$$-\mu \int_\Omega q_{i,j}(q_{i,j} - q_{j,i})dx = -\frac{\mu}{2} \int_\Omega (q_{i,j} - q_{j,i})(q_{i,j} - q_{j,i})dx.$$

Thus, this term may be discarded in (6.17) to find

$$\frac{\tau}{2} \frac{d}{dt} \|\mathbf{q}\|^2 \leq -\|\mathbf{q}\|^2 + \kappa(T, q_{i,i}) - (\mu + \nu)\|q_{i,i}\|^2 . \tag{6.18}$$

The arithmetic-geometric mean inequality is used in the form

$$\kappa(T, q_{i,i}) \leq \frac{\kappa^2}{\sigma} \|T\|^2 + \frac{\sigma}{4} \|q_{i,i}\|^2$$

to obtain

$$\frac{\tau}{2} \frac{d}{dt} \|\mathbf{q}\|^2 \leq -\|\mathbf{q}\|^2 + \frac{\kappa^2}{\sigma} \|T\|^2 + \Big[\frac{\sigma}{4} - (\mu + \nu)\Big] \|q_{i,i}\|^2 .$$

Pick now $\sigma = 4(\mu + \nu)$ and use estimate (6.15), i.e. $\|T\|^2 \leq E(0)\gamma^{-1}e^{-2at}$, and one finds

$$\frac{\tau}{2} \frac{dZ}{dt} \leq -Z + \frac{\kappa^2}{4(\mu + \nu)} \frac{E(0)}{\gamma} e^{-2at} ,$$

where we momentarily define $Z(t) = \|\mathbf{q}(t)\|^2$. Rearranging and using an integrating factor one finds

$$\frac{d}{dt}\big(e^{2t/\tau} Z\big) \leq \frac{\kappa^2 E(0)}{2\tau(\mu + \nu)\gamma} \exp(-2at + 2t/\tau).$$

This inequality is integrated and we divide by $e^{2t/\tau}$ to derive the (Payne and Song, 2004b) inequality for q_i, namely

$$\|\mathbf{q}(t)\|^2 \leq \|\mathbf{q}(0)\|^2 e^{-2t/\tau} + \frac{\kappa^2 E(0)}{4\gamma(\mu + \nu)(a\tau - 1)} (e^{-2t/\tau} - e^{-2at}). \tag{6.19}$$

(Payne and Song, 2004b) derive further decay estimates for $\Omega \subset \mathbb{R}^2$ when the boundary condition $(6.4)_2$ is replaced by $q_i = 0$ on Γ. We refer the reader to (Payne and Song, 2004b) for further details.

6.1.3 Decay with other effects

We have only discussed time decay of the temperature field and heat flux in a rigid body. However, an increasingly important topic is the question of decay in thermoelastic systems with second sound effects, e.g. Lord-Shulman, Green-Lindsay, Green-Naghdi type III effects. Additionally, if there is decay, what sort of decay can be expected, e.g. exponential, polynomial in time, or what?

The question of decay in thermoelasticity was addressed for the classical theory (type I) in the fundamental articles of (Dafermos, 1968), (Slemrod, 1981), (Racke and Shibata, 1991), (Muñoz Rivera, 1992) and (Lebeau and

Zuazua, 1999). Recent work has addressed such issues in "hyperbolic" theories of thermoelasticity, i.e where heat may travel as a wave. Key to the type of decay found is whether the spatial domain Ω is one-dimensional, or two or three - dimensional. Often, exponential decay is found in the one dimensional case. For the two and three - dimensional situations the geometry is important and often only polynomial time decay is witnessed. However, in addition to thermoelastic effects there are many other physically important effects to consider. Among these we mention elastic materials containing voids (a class of porous materials) cf. section 2.6, mixtures of elastic materials, the effect of microtemperatures. A combination of these effects is leading to some surprising results. For example, (Casas and Quintanilla, 2005a) note that effects which separately lead to polynomial decay may lead to exponential decay when present simultaneously. We point out that such effects in conjunction with type II thermoelasticity might also lead to decay, the dissipation being provided by friction in the case of voids or mixtures, for example. It is important to derive the required theory correctly from full continuum thermodynamic principles, cf. (Green and Naghdi, 1992; Green and Naghdi, 1993), (Iesan, 2008), (Iesan and Quintanilla, 2009), (Straughan, 2008), pp. 329–332, and then derive the linearized equations either by considering a small deformation superimposed on a large one, or by providing a suitable free energy function.

The following articles address very interesting decay results in thermoelasticity, some being for theories appropriate to second sound, others involving effects like mixtures, voids, microtemperatures, (Alves et al., 2009), (Casas and Quintanilla, 2005a; Casas and Quintanilla, 2005b), (Dharmawadane et al., 2010), (Fabrizio et al., 2007), (Iesan and Quintanilla, 2009), (Irmscher and Racke, 2006), (Leseduarte et al., 2010), (Magaña and Quintanilla, 2006b; Magaña and Quintanilla, 2006a; Magaña and Quintanilla, 2007), (Messaoudi and Said-Houari, 2008; Messaoudi and Said-Houari, 2009), (Muñoz Rivera, 1997), (Muñoz Rivera and Qin, 2002), (Muñoz Rivera and Quintanilla, 2008), (Pamplona et al., 2009), (Passarella and Zampoli, 2007), (Qin and Muñoz Rivera, 2004), (Quintanilla, 2002a; Quintanilla, 2003; Quintanilla, 2004; Quintanilla, 2007b; Quintanilla, 2007a), (Quintanilla and Racke, 2003; Quintanilla and Racke, 2006a; Quintanilla and Racke, 2006b; Quintanilla and Racke, 2007; Quintanilla and Racke, 2008), (Racke, 2002; Racke, 2003), (Racke and Wang, 2005; Racke and Wang, 2008), (Reissig and Wang, 2005), (Sare et al., 2008), (Sun et al., 2006), (Vila Bravo and Muñoz Rivera, 2009), (Wang and Yang, 2006), (Weinmann, 2009), (Yang and Wang, 2006), (Zhang and Zuazua, 2003). We do not give a detailed description of these articles since not all are dealing with second sound theories. Nevertheless, the methods developed will undoubtedly prove useful in future analyses of second sound thermoelastic models and this is an area with potential.

(Sare and Racke, 2009) have recently proved an interesting result for a system of equations appropriate to a Timoshenko beam. They show that one may expect exponential decay of the solution when Fourier's law is adopted. However, incorporation of a Cattaneo heat flux equation does not lead to exponential decay.

Other very interesting decay results in thermoelasticity are those for composite bodies, where one material is touching another. Into this category comes the interesting work of (Bonfanti et al., 2008), see also (Quintanilla, 2008a). Likewise, the study of boundary effects giving rise to solution decay, cf. (Lazzari and Nibbi, 2007; Lazzari and Nibbi, 2008), is of much interest. These classes of problem will be useful in future, especially in connection with new classes of materials, such as auxetic materials, see e.g. (Lakes, 2008), and functionally graded materials, cf. (Mian and Spencer, 1998). Studies of decay and stability in auxetic materials are of particular interest, especially since results are beginning to emerge which do not require positive definiteness of the elastic coefficients, one merely needs strong ellipticity, see (Chirita and Ciarletta, 2006), (Chirita, 2007), (Xinchun and Lakes, 2007).

6.2 Uniqueness in type II thermoelasticity

A striking result of (Knops and Payne, 1970) demonstrated how to develop a logarithmic convexity technique to achieve uniqueness and continuous dependence on the initial data for the classical theory of linear thermoelasticity. Their paper involves some ingenious estimates and is based on showing convexity of the logarithm of the L^2 norm of the elastic displacement. Particularly striking is the fact that (Knops and Payne, 1970) do not require the elasticity tensor to be sign-definite. All they require is that the elastic coefficients, a_{ijkh}, be symmetric in the sense that

$$a_{ijkh} = a_{khij}. \tag{6.20}$$

The work of (Knops and Payne, 1970) was extended by (Levine, 1970) to derive uniqueness and continuous depedence for the solution to an abstract system of differential equations which includes the equations of classical thermoelasticity as a special case.

Further results using logarithmic convexity in classical thermoelasticity are due to (Ames and Payne, 1991; Ames and Payne, 1994; Ames and Payne, 1995). They establish a series of results on continuous dependence for the backward in time problem, for a unilateral problem, and for the initial-time geometry problem, respectively.

In this section we describe work of (Quintanilla and Straughan, 2000) who establish uniqueness for *anisotropic* linearised thermoelasticity of type II without requiring any definiteness whatsoever of the elasticity tensor. They use a novel logarithmic convexity technique.

The relevant equations of anisotropic inhomogeneous thermoelasticity of type II for a body with a centre of symmetry are, cf. (Quintanilla, 1999;

Quintanilla, 2002b), see also section 2.3, equations (2.73):

$$\rho\ddot{u}_i = (a_{ijkh}u_{k,h})_{,j} - (a_{ij}\theta)_{,j} + \rho f_i\,, \tag{6.21}$$

$$c\ddot{\theta} = -a_{ij}\ddot{u}_{i,j} + (k_{ik}\theta_{,k})_{,i} + \rho r\,, \tag{6.22}$$

where u_i, θ are the displacement vector and temperature field, and a superposed dot denotes $\partial/\partial t$. The quantities ρ, f_i and r are density, body force and heat supply, while $c > 0$ is a constant. The tensor $a_{ij}(\mathbf{x})$, is a coupling tensor and the elastic coefficients, $a_{ijkh}(\mathbf{x})$, satisfy the symmetry condition (6.20). No sign-definiteness is required of a_{ijkh}.

In this section Ω is a bounded domain in $\mathbb{R}^3$ with boundary Γ smooth enough to allow applications of the divergence theorem. Again, $(\,,\,)$ and $\|\cdot\|$ denote the inner product and norm on $L^2(\Omega)$. Equations (6.21) and (6.22) hold on $\Omega \times (0,T)$ where $T(\leq \infty)$ is some time. The boundary and initial conditions we consider are

$$u_i(\mathbf{x},t) = u_i^B(\mathbf{x},t), \quad \theta(\mathbf{x},t) = \theta^B(\mathbf{x},t), \qquad \mathbf{x} \in \Gamma, \tag{6.23}$$

and

$$\begin{aligned} u_i(\mathbf{x},0) &= g_i(\mathbf{x}), \quad \dot{u}_i(\mathbf{x},0) = h_i(\mathbf{x}), && \mathbf{x} \in \Omega, \\ \theta(\mathbf{x},0) &= \theta_0(\mathbf{x}), \quad \dot{\theta}(\mathbf{x},0) = \zeta(\mathbf{x}), && \mathbf{x} \in \Omega. \end{aligned} \tag{6.24}$$

The logarithmic convexity functional of (Quintanilla and Straughan, 2000) does not consist of simply L^2 norms of u_i and θ. They devise a "natural" functional which is a combination of the L^2 norm of u_i and a weighted H_0^1 norm of $\psi = \int_0^t \int_0^s \theta\, dq\, ds$.

We suppose the thermal tensor k_{ik} is symmetric and positive semi - definite, in the sense that

$$k_{ik} = k_{ki}, \qquad k_{ik}\xi_i\xi_k \geq 0, \quad \forall \xi_i. \tag{6.25}$$

To consider uniqueness, let (u_i^1, θ^1) and (u_i^2, θ^2) be two solutions to (6.21), (6.22), (6.23) and (6.24) for the same boundary and initial data, and for the same body force and heat supply. Then, the difference solution

$$u_i = u_i^1 - u_i^2, \quad \theta = \theta^1 - \theta^2, \tag{6.26}$$

may be shown to satisfy the equations

$$\rho\ddot{u}_i = (a_{ijkh}u_{k,h})_{,j} - (a_{ij}\theta)_{,j}\,, \tag{6.27}$$

$$c\ddot{\theta} = -a_{ij}\ddot{u}_{i,j} + (k_{ik}\theta_{,k})_{,i}\,. \tag{6.28}$$

In terms of $u_i.\theta$ the boundary and initial data are

$$u_i(\mathbf{x},t) = 0, \quad \theta(\mathbf{x},t) = 0, \qquad \mathbf{x} \in \Gamma, \tag{6.29}$$

and

$$u_i(\mathbf{x},0) = 0, \;\; \dot{u}_i(\mathbf{x},0) = 0, \quad \theta(\mathbf{x},0) = 0, \;\; \dot{\theta}_i(\mathbf{x},0) = 0, \qquad \mathbf{x} \in \Omega. \tag{6.30}$$

(Quintanilla and Straughan, 2000) begin by introducing the functions η and ψ by

$$\eta(\mathbf{x},t)=\int_0^t\theta(\mathbf{x},s)\,ds,\qquad \psi(\mathbf{x},t)=\int_0^t\int_0^s\theta(\mathbf{x},q)\,dq\,ds. \tag{6.31}$$

Equation (6.28) is rewritten in terms of either η or ψ as

$$c\ddot{\eta}=-a_{ij}\dot{u}_{i,j}+(k_{ik}\eta_{,k})_{,i}\qquad c\ddot{\psi}=-a_{ij}u_{i,j}+(k_{ik}\psi_{,k})_{,i}\,. \tag{6.32}$$

The basic functional, $F(t)$, of (Quintanilla and Straughan, 2000) is

$$F(t)=(\rho u_i,u_i)+(k_{ik}\psi_{,i},\psi_{,k})\,. \tag{6.33}$$

To use F, differentiate to see that

$$F'=2(\rho u_i,\dot{u}_i)+2(k_{ik}\psi_{,i},\dot{\psi}_{,k})\,, \tag{6.34}$$

and after a further differentiation,

$$F''=2(\rho\dot{u}_i,\dot{u}_i)+2(k_{ik}\dot{\psi}_{,i},\dot{\psi}_{,k})+2(\rho u_i,\ddot{u}_i)+2(k_{ik}\psi_{,i},\ddot{\psi}_{,k})\,. \tag{6.35}$$

Multiply $(6.32)_1$ by $\dot{\eta}$ and integrate over Ω. Add this to equation (6.27) multiplied by $\dot{u}_i$ after integration over Ω. By using the boundary conditions one may thus derive a conservation of energy law, of form

$$E(t)=E(0)=0, \tag{6.36}$$

where the total "energy", $E(t)$, is given by

$$E(t)\equiv\frac{1}{2}(\rho\dot{u}_i,\dot{u}_i)+\frac{1}{2}c\|\theta\|^2+\frac{1}{2}(a_{ijkh}u_{k,h},u_{i,j})+\frac{1}{2}(k_{ik}\eta_{,i},\eta_{,k}). \tag{6.37}$$

Multiplication of (6.27) by u_i and integration over Ω yields

$$(\rho u_i,\ddot{u}_i)+(a_{ijkh}u_{k,h},u_{i,j})=(a_{ij}\theta,u_{i,j}). \tag{6.38}$$

Also, multiplication of $(6.27)_2$ by $\ddot{\psi}$ and integration over Ω leads to

$$c\|\theta\|^2+(k_{ik}\psi_{,k},\ddot{\psi}_{,i})=-(a_{ij}u_{i,j},\theta). \tag{6.39}$$

The terms $(\rho u_i,\ddot{u}_i)$ and $(k_{ik}\psi_{,i},\ddot{\psi}_{,k})$ are substituted in equation (6.35), adding (6.38) and (6.39) to remove the right hand sides. We then find

$$F''=2(\rho\dot{u}_i,\dot{u}_i)+2(k_{ik}\dot{\psi}_{,i},\dot{\psi}_{,k})-2(a_{ijkh}u_{k,h},u_{i,j})-2c\|\theta\|^2\,. \tag{6.40}$$

This expression is conveniently rewritten using the energy equation (6.36) to see that

$$F''=4(\rho\dot{u}_i,\dot{u}_i)+4(k_{ik}\dot{\psi}_{,i},\dot{\psi}_{,k}). \tag{6.41}$$

This representation of F'' is a particularly convenient form to apply the logarithmic convexity method. Thus, we form $FF''-(F')^2$ to observe that

$$\begin{aligned} FF''-(F')^2 &= 4\Big[(\rho u_i,u_i)+(k_{ik}\psi_{,i},\psi_{,k})\Big]\Big[(\rho\dot u_i,\dot u_i)+(k_{ik}\dot\psi_{,i},\dot\psi_{,k})\Big] \\ &\quad -4\Big[(\rho u_i,\dot u_i)+(k_{ik}\psi_{,i},\dot\psi_{,k})\Big]^2 \\ &\geq 0, \end{aligned} \tag{6.42}$$

where the Cauchy-Schwarz inequality has been employed in the last line.

Inequality (6.42) is equivalent to $\log F(t)$ being a convex function of time. From this one then deduces, see e.g. (Payne, 1975), p. 12,

$$F(t)\leq\big[F(0)\big]^{(1-t/T)}\big[F(T)\big]^{t/T},\qquad t\in(0,T). \tag{6.43}$$

Hence, $F(t)\equiv 0$ on $[0,T]$, and from the expression for $F(t)$, equation (6.33), it follows that $u_i\equiv 0$ on $\Omega\times[0,T]$. Then, from equation (6.28) θ satisfies the equation

$$c\ddot\theta=(k_{ik}\theta_{,k})_{,i}\,.$$

Multiplication of this equation by $\dot\theta$ and integration over Ω together with an integration by parts leads to

$$c\|\dot\theta\|^2+(k_{ik}\theta_{,i},\theta_{,k})=0. \tag{6.44}$$

It follows from this that $\theta\equiv 0$ on $\Omega\times[0,T]$, and so uniqueness of a solution to the problem comprised of (6.21), (6.22), (6.23) and (6.24) follows.

(Quintanilla and Straughan, 2000) observe that equations (6.21), (6.22) are invariant under time reversal and so uniqueness holds also in the backward in time problem. In addition, they note that one may modify the function F in (6.33) to establish continuous dependence on the initial data, and results of structural stability such as continuous dependence on body force and heat source. (Quintanilla and Straughan, 2000) also note that boundary conditions (6.23) could be replaced by mixed boundary conditions involving a combination of (6.23) on part of the boundary together with prescribed traction and entropy flux on the remainder.

6.3 Growth in type II thermoelasticity

To produce a growth result for a solution to equations (6.21) and (6.22) (Quintanilla and Straughan, 2000) assume the thermal coefficients k_{ik} satisfy the conditions

$$k_{ik}=k_{ki},\qquad k_{ik}\xi_i\xi_k\geq k_0\xi_i\xi_i,\quad\forall\xi_i,\ \text{and some } k_0>0. \tag{6.45}$$

Let now (u_i,θ) be a solution to equations (6.21) and (6.22) with $f_i\equiv r\equiv 0$. Let (u_i,θ) satisfy zero boundary conditions as in (6.29). Hence, (u_i,θ) satisfy equations (6.27) and (6.28) with the initial data being given by (6.24). The key to a growth result is to find a suitable functional to which logarithmic convexity is applicable. (Quintanilla and Straughan, 2000) again define η and ψ as in (6.31). However, the initial data are now non-zero, and so we find the differential equations for η and ψ are

$$c\ddot{\eta} = -a_{ij}\dot{u}_{i,j} + (k_{ik}\eta_{,k})_{,i} + c\zeta(\mathbf{x}) + a_{ij}h_{i,j}(\mathbf{x})\,, \tag{6.46}$$

and

$$c\ddot{\psi} = -a_{ij}u_{i,j}+(k_{ik}\psi_{,k})_{,i}+\big[c\zeta(\mathbf{x})+a_{ij}h_{i,j}(\mathbf{x})\big]t+c\theta_0(\mathbf{x})+a_{ij}g_{i,j}(\mathbf{x})\,. \tag{6.47}$$

The data terms are incorporated into the entropy term, $(k_{ik}\eta_{,k})_{,i}$, by defining $Q_1(\mathbf{x})$ and $Q_2(\mathbf{x})$ to be solutions to the generalised Poisson equations

$$\begin{aligned} (k_{ik}Q_{1,k})_{,i} &= c\zeta(\mathbf{x}) + a_{ij}h_{i,j}(\mathbf{x}), \qquad && \mathbf{x}\in\Omega,\\ (k_{ik}Q_{2,k})_{,i} &= c\theta_0(\mathbf{x}) + a_{ij}g_{i,j}(\mathbf{x}), \qquad && \mathbf{x}\in\Omega, \end{aligned} \tag{6.48}$$

where Q_α satisfy the boundary conditions

$$Q_\alpha = 0, \quad \mathbf{x}\in\Gamma,\ \alpha=1,2. \tag{6.49}$$

Existence of the functions Q_α is guaranteed by theorems 4.3 and 6.8 of (Gilbarg and Trudinger, 1977) p. 55 and p. 95. In fact, the positive-definiteness condition on k_{ij}, (6.45), is only required to ensure the existence of Q_α for system (6.48), (6.49). (Quintanilla and Straughan, 2000) define α and β by

$$\alpha(\mathbf{x},t) = \psi(\mathbf{x},t) + Q_1 t + Q_2, \qquad \beta(\mathbf{x},t) = \eta(\mathbf{x},t) + Q_1, \tag{6.50}$$

and in terms of the functions α and β, equations (6.46) and (6.47) may be written as

$$\begin{aligned} c\ddot{\beta} &= -a_{ij}\dot{u}_{i,j} + (k_{ik}\beta_{,k})_{,i},\\ c\ddot{\alpha} &= -a_{ij}u_{i,j} + (k_{ik}\alpha_{,k})_{,i}\,. \end{aligned} \tag{6.51}$$

(Quintanilla and Straughan, 2000) define the functional $G(t)$ by

$$G(t) = (\rho u_i, u_i) + (k_{ik}\alpha_{,i}, \alpha_{,k}) + \alpha_1(t+t_0)^2\,, \tag{6.52}$$

where α_1 and t_0 are constants to be selected. The α_1 term follows (Knops and Payne, 1971a).

By differentiation and use of equations (6.27) and $(6.51)_1$ one may show the identities

$$(\rho u_i, \ddot{u}_i) = -(a_{ijkh}u_{k,h}, u_{i,j}) + (a_{ij}\theta, u_{i,j}), \tag{6.53}$$

and

$$c\|\theta\|^2 = -(k_{ik}\alpha_{,k}, \ddot{\alpha}_{,i}) - (a_{ij}u_{i,j}, \theta). \tag{6.54}$$

Equations (6.53) and (6.54) are added to obtain

$$(\rho u_i, \ddot{u}_i) + (k_{ik}\ddot{\alpha}_{,i}, \alpha_{,k}) = -c\|\theta\|^2 - (a_{ijkh}u_{k,h}, u_{i,j}). \tag{6.55}$$

Upon multiplication of (6.27) by $\dot{u}_i$ and integration over Ω, followed by multiplication of $(6.51)_1$ by $\dot{\beta}$ and integration over Ω, we derive the identity

$$E(t) = E(0), \tag{6.56}$$

where the energy $E(t)$ is defined by

$$E(t) \equiv \frac{1}{2}(\rho\dot{u}_i, \dot{u}_i) + \frac{1}{2}c\|\theta\|^2 + \frac{1}{2}(a_{ijkh}u_{k,h}, u_{i,j}) + \frac{1}{2}(k_{ik}\beta_{,i}, \beta_{,k}). \tag{6.57}$$

(Quintanilla and Straughan, 2000) show that one may differentiate G and use equations (6.55) and (6.56) to see that

$$\begin{aligned} GG'' - (G')^2 = \Big\{ & 4\Big[(\rho u_i, u_i) + (k_{ik}\alpha_{,i}, \alpha_{,k}) + \alpha_1(t+t_0)^2\Big] \\ & \times \Big[(\rho\dot{u}_i, \dot{u}_i) + (k_{ik}\dot{\alpha}_{,i}, \dot{\alpha}_{,k}) + \alpha_1\Big] \\ & - 4\Big[(\rho u_i, \dot{u}_i) + (k_{ik}\alpha_{,i}, \dot{\alpha}_{,k}) + \alpha_1(t+t_0)\Big]^2\Big\} \\ & - 2G\big[2E(0) + \alpha_1\big]. \end{aligned} \tag{6.58}$$

The term in braces on the right of (6.58) is non-negative by virtue of the Cauchy-Schwarz inequality, and so from (6.58) one derives

$$GG'' - (G')^2 \geq -2G\big[2E(0) + \alpha_1\big]. \tag{6.59}$$

From this follows

Theorem 6.3.1 *If either (a) $E(0) < 0$; or (b) $E(0) = 0$ and $G'(0) > 0$; or (c) $E(0) > 0$ and $G'(0) > 2\big[2G(0)E(0)\big]^{1/2}$, then $G(t)$ is bounded below by an increasing exponential function of t.*

The proof of this is given in (Quintanilla and Straughan, 2000) and follows a similar result in (Knops and Payne, 1971a), see also (Payne, 1975), p. 21.

6.4 Uniqueness in type III thermoelasticity

The appropriate linear equations for type III thermoelasticity theory may be derived from (Green and Naghdi, 1992) and for a body with a centre of symmetry they are, see e.g. section 2.4, equations (2.83):

$$\rho\ddot{u}_i = (a_{ijkh}u_{k,h})_{,j} - (a_{ij}\theta)_{,j} + \rho f_i\,, \tag{6.60}$$

$$c\ddot{\theta} = -a_{ij}\ddot{u}_{i,j} + (k_{ik}\theta_{,k})_{,i} + (b_{ik}\dot{\theta}_{,k})_{,i} + \rho r\,. \tag{6.61}$$

The tensors k_{ij} and b_{ij} are always symmetric.

In this section we record a uniqueness result of (Quintanilla and Straughan, 2000) for a solution to equations (6.60), (6.61). The boundary and initial data are

$$u_i(\mathbf{x},t) = u_i^B(\mathbf{x},t), \quad \theta(\mathbf{x},t) = \theta^B(\mathbf{x},t), \qquad \mathbf{x}\in\Gamma, \tag{6.62}$$

and

$$\begin{aligned} u_i(\mathbf{x},0) &= g_i(\mathbf{x}), & \dot{u}_i(\mathbf{x},0) &= h_i(\mathbf{x}), \\ \theta(\mathbf{x},0) &= \theta_0(\mathbf{x}), & \dot{\theta}(\mathbf{x},0) &= \zeta(\mathbf{x}), \qquad \mathbf{x}\in\Omega. \end{aligned} \tag{6.63}$$

The major symmetry condition (6.20) holds, i.e.

$$a_{ijkh} = a_{khij} \tag{6.64}$$

and k_{ik} and b_{ik} satisfy

$$b_{ik}\xi_i\xi_k \geq 0 \quad \text{and} \quad k_{ik}\xi_i\xi_k \geq 0, \quad \forall \xi_i. \tag{6.65}$$

To establish uniqueness let (u_i^1, θ^1) and (u_i^2, θ^2) be two solutions to (6.60) - (6.65) for the same boundary and initial data, and for the same body force and heat supply. Then, the difference solution (u_i, θ) given by

$$u_i = u_i^1 - u_i^2, \quad \theta = \theta^1 - \theta^2, \tag{6.66}$$

satisfies the equations

$$\rho \ddot{u}_i = (a_{ijkh}u_{k,h})_{,j} - (a_{ij}\theta)_{,j}\,, \tag{6.67}$$

$$c\ddot{\theta} = -a_{ij}\ddot{u}_{i,j} + (k_{ik}\theta_{,k})_{,i} + (b_{ik}\dot{\theta}_{,k})_{,i}\,. \tag{6.68}$$

The boundary and initial conditions satisfied by the difference solution (u_i, θ) are

$$u_i(\mathbf{x},t) = 0, \quad \theta(\mathbf{x},t) = 0, \qquad \mathbf{x}\in\Gamma, \tag{6.69}$$

and

$$u_i(\mathbf{x},0) = 0,\ \dot{u}_i(\mathbf{x},0) = 0, \quad \theta(\mathbf{x},0) = 0,\ \dot{\theta}_i(\mathbf{x},0) = 0, \qquad \mathbf{x}\in\Omega. \tag{6.70}$$

Define η by $\eta(\mathbf{x},t) = \int_0^t \theta(\mathbf{x},s)ds$ and observe equation (6.68) can be written as

$$c\ddot{\eta} = -a_{ij}\dot{u}_{i,j} + (k_{ik}\eta_{,k})_{,i} + (b_{ik}\dot{\eta}_{,k})_{,i}\,. \tag{6.71}$$

To establish uniqueness (Quintanilla and Straughan, 2000) use a Lagrange identity method. They consider $t\in(0,T)$ fixed, and form the identities

$$\begin{aligned} &\int_0^t \Big(\rho\ddot{u}_i(\tau), \dot{u}_i(2t-\tau)\Big)\,d\tau + \int_0^t \Big(a_{ijkh}u_{k,h}(\tau), \dot{u}_{i,j}(2t-\tau)\Big)\,d\tau \\ &\qquad = \int_0^t \Big(a_{ij}\theta(\tau), \dot{u}_{i,j}(2t-\tau)\Big)\,d\tau, \end{aligned} \tag{6.72}$$

and

$$\begin{aligned} &\int_0^t c\Big(\ddot{\eta}(\tau), \dot{\eta}(2t-\tau)\Big)\,d\tau + \int_0^t \Big(k_{ik}\eta_{,k}(\tau), \dot{\eta}_{,i}(2t-\tau)\Big)\,d\tau \\ &+ \int_0^t \Big(b_{ik}\dot{\eta}_{,k}(\tau), \dot{\eta}_{,i}(2t-\tau)\Big)\,d\tau = -\int_0^t \Big(a_{ij}\dot{u}_{i,j}(\tau), \theta(2t-\tau)\Big)\,d\tau, \end{aligned} \tag{6.73}$$

and

$$\int_0^t \Big(\rho\ddot{u}_i(2t-\tau), \dot{u}_i(\tau)\Big)\,d\tau + \int_0^t \Big(a_{ijkh}u_{k,h}(2t-\tau), \dot{u}_{i,j}(\tau)\Big)\,d\tau \\ = \int_0^t \Big(a_{ij}\theta(2t-\tau), \dot{u}_{i,j}(\tau)\Big)\,d\tau, \tag{6.74}$$

and

$$\int_0^t c\Big(\ddot{\eta}(2t-\tau), \dot{\eta}(\tau)\Big)\,d\tau + \int_0^t \Big(k_{ik}\eta_{,k}(2t-\tau), \dot{\eta}_{,i}(\tau)\Big)\,d\tau \\ + \int_0^t \Big(b_{ik}\dot{\eta}_{,k}(2t-\tau), \dot{\eta}_{,i}(\tau)\Big)\,d\tau = -\int_0^t \Big(a_{ij}\dot{u}_{i,j}(2t-\tau), \theta(\tau)\Big)\,d\tau. \tag{6.75}$$

Equations (6.72) - (6.75) are added and subtracted in the combination (6.72)-(6.73)+(6.74)-(6.75). (Quintanilla and Straughan, 2000) show that this yields the identity

$$\big(\rho\dot{u}_i(t), \dot{u}_i(t)\big) - \big(a_{ijkh}u_{k,h}(t), u_{i,j}(t)\big) \\ - c\|\theta(t)\|^2 + \big(k_{ik}\eta_{,k}(t), \eta_{,i}(t)\big) = 0. \tag{6.76}$$

An energy equation is derived by multiplying (6.67) by $\dot{u}_i$, and then by multiplying (6.71) by $\dot{\eta}$, and then integrating over Ω, also integrating by parts. The resulting energy equation is

$$E(t) = E(0) = 0, \tag{6.77}$$

where $E(t)$ is defined by

$$E(t) \equiv \frac{1}{2}\|\dot{\eta}(t)\|^2 + \frac{1}{2}\big(k_{ik}\eta_{,i}(t), \eta_{,k}(t)\big) + \frac{1}{2}\big(\rho\dot{u}_i(t), \dot{u}_i(t)\big) \\ + \frac{1}{2}\big(a_{ijkh}u_{k,h}(t), u_{i,j}(t)\big) + \int_0^t \Big(b_{ik}\dot{\eta}_{,i}, \dot{\eta}_{,k}\Big)\,ds. \tag{6.78}$$

Equations (6.76) and (6.77) are suitably added recalling $\dot{\eta} = \theta$ to obtain

$$\big(\rho\dot{u}_i(t), \dot{u}_i(t)\big) + \big(k_{ik}\eta_{,i}(t), \eta_{,k}(t)\big) + \int_0^t \Big(b_{ik}\theta_{,i}, \theta_{,k}\Big)\,ds = 0.$$

From this identity (Quintanilla and Straughan, 2000) deduce with the aid of the boundary conditions that $u_i \equiv 0$ on $\Omega \times [0,T]$. Then, from (6.68) θ satisfies the equation

$$c\ddot{\theta} = (k_{ik}\theta_{,k})_{,i} + (b_{ik}\dot{\theta}_{,k})_{,i}\,. \tag{6.79}$$

Upon multiplication by $\dot{\theta}$ and integration over Ω we find

$$\frac{1}{2}c\|\dot{\theta}\|^2 + \frac{1}{2}(k_{ik}\theta_{,i}, \theta_{,k}) + \int_0^t (b_{ik}\dot{\theta}_{,i}, \dot{\theta}_{,k})ds = 0. \tag{6.80}$$

It follows from (6.80) that $\theta \equiv 0$ on $\Omega \times [0,T]$. Hence, uniqueness follows.

(Quintanilla and Straughan, 2000) remark that one could extend the above analysis to unbounded spatial regions and not impose decay at infinity on u_i and θ by using a weighted Lagrange identity technique. The details involve a combination of their proof and that of (Rionero and Chirita, 1987) who used a weighted Lagrange identity in classical thermoelasticity. In this manner one can derive uniqueness and structural stability results for type III thermoelasticity on unbounded spatial domains without requiring the imposition of decay constraints at infinity, (Quintanilla, 2002a).

6.5 Uniqueness on an unbounded domain

To investigate uniqueness on an unbounded domain we employ the model of type III heat flow in a rigid heat conductor as described in section 1.11. We consider the linear theory as given by equation (1.127) and since we are considering the uniqueness question we may set $r = 0$ (as it disappears in the difference equation anyhow). Thus, we consider a solution u to the equation

$$\frac{\partial^2 u}{\partial t^2} + k_1 \frac{\partial u}{\partial t} = k_2 \Delta u + k_3 \Delta \frac{\partial u}{\partial t}, \qquad \mathbf{x} \in \Omega, \; t > 0. \tag{6.81}$$

The spatial domain will be specified below, and k_1, k_2, k_3 are positive constants. Let us observe that in type III heat flow theory u corresponds to the thermal displacement α. We also note, as in section 1.11, that equation (6.81) arises in Guyer-Krumhansl theory, equation (1.51), and in dual phase lag theory, equation (1.60). To illustrate the ideas we begin with $\Omega \subset \mathbb{R}^3$ being a bounded domain with boundary Γ. On the boundary Γ we prescribe u, i.e.

$$u(\mathbf{x}, t) = u_\Gamma(\mathbf{x}, t), \qquad \mathbf{x} \in \Gamma, \; t > 0, \tag{6.82}$$

while at time $t = 0$ we give u and u_t, so

$$u(\mathbf{x}, 0) = u_0(\mathbf{x}), \qquad \frac{\partial u}{\partial t}(\mathbf{x}, 0) = v_0(\mathbf{x}), \qquad \mathbf{x} \in \Omega, \tag{6.83}$$

u_Γ, u_0 and v_0 being given functions.

Let the boundary - initial value problem comprised of equation (6.81) together with the boundary and initial conditions (6.82) and (6.83) be denoted by $\mathcal{P}$. To demonstrate uniqueness we suppose there are two solutions $u_1(\mathbf{x}, t)$ and $u_2(\mathbf{x}, t)$ which both satisfy $\mathcal{P}$ for the same boundary and initial data u_Γ, u_0 and v_0. Then, define the difference solution u by

$$u = u_1 - u_2$$

and one finds u satisfies the boundary - initial value problem

$$\begin{aligned} &\frac{\partial^2 u}{\partial t^2} + k_1 \frac{\partial u}{\partial t} = k_2 \Delta u + k_3 \Delta \frac{\partial u}{\partial t}, \qquad \mathbf{x} \in \Omega,\ t > 0, \\ &u(\mathbf{x}, t) = 0, \qquad \mathbf{x} \in \Gamma,\ t > 0, \\ &u(\mathbf{x}, 0) = 0, \qquad \frac{\partial u}{\partial t}(\mathbf{x}, 0) = 0, \qquad \mathbf{x} \in \Omega. \end{aligned} \tag{6.84}$$

Next, multiply equation $(6.84)_1$ by $\partial u/\partial t$ and integrate over Ω and use the boundary conditions to see that

$$\frac{d}{dt}\Big[\frac{1}{2}\int_\Omega u_t^2\,dx + \frac{k_2}{2}\int_\Omega |\nabla u|^2 dx\Big] + k_1 \int_\Omega u_t^2\,dx + k_3 \int_\Omega |\nabla u_t|^2 dx = 0, \tag{6.85}$$

where we have used integration by parts and the boundary data. We discard the two positive terms in equation (6.85) and define

$$E(t) = \frac{1}{2}\int_\Omega u_t^2\,dx + \frac{k_2}{2}\int_\Omega |\nabla u|^2 dx \tag{6.86}$$

to deduce from equation (6.85) that

$$\frac{dE}{dt} \le 0.$$

Upon integration this inequality yields

$$E(t) \le E(0) = 0$$

where $E(0) = 0$ due to the initial data. Since E is given by (6.86) it follows that

$$u_t \equiv 0, \quad u_{,i} \equiv 0, \qquad \text{in } \Omega \times \{t > 0\}$$

and hence $u \equiv 0$ in $\Omega \times \{t > 0\}$, since $u \equiv 0$ at $t = 0$. Therefore, $u_1 \equiv u_2$, and the solution to $\mathcal{P}$ is unique.

Next, we consider the case where Ω is unbounded. Let $\Omega \subset \mathbb{R}^3$ be the domain exterior to a bounded domain $\Omega_0 \subset \mathbb{R}^3$. The inner boundary of Ω we denote by Γ. We now wish to consider the uniqueness question for $\mathcal{P}$ but when Ω is the exterior domain just identified. If we assume u and its derivatives decay sufficiently rapidly as $r \to \infty$, ($r = \sqrt{x_i x_i}$), then we may still deduce (6.85) and uniqueness follows as above. However, we are interested in the situation where u is allowed to grow as $r \to \infty$, and in particular, we allow exponential growth. In this case the energy method employed above fails and we now consider two alternative methods of establishing uniqueness.

6.5.1 *The Graffi method*

Suppose now the origin $0 \in \Omega_0$ and let R_0 be the first point such that $B(0, R_0) \supset \bar{\Omega}_0$ where $B(0, R_0)$ is the ball centred at 0 of radius R_0, and

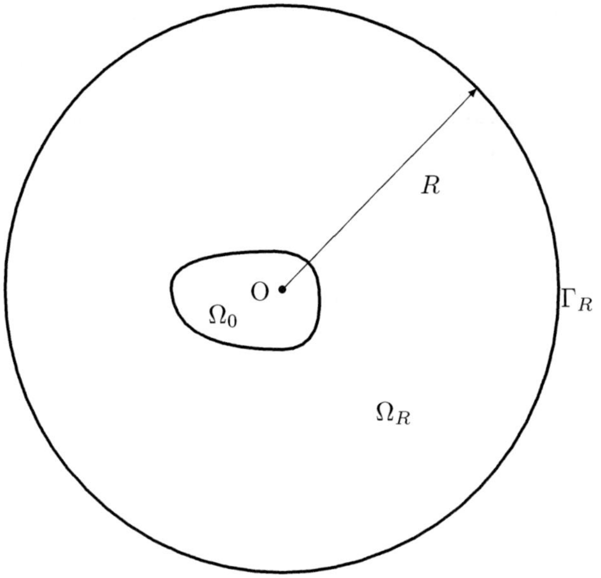

Figure 6.1. Geometry for uniqueness proof

$\bar{\Omega}_0$ is the closure of Ω. We begin by describing a technique due to (Graffi, 1960), which may also be conveniently found in (Graffi, 1999b).

To establish uniqueness of a solution to $\mathcal{P}$ we let u_1, u_2 be solutions as above and consider the boundary - initial value problem for the difference solution u, namely (6.84). To use Graffi's method we let Ω_R be the domain $\Omega_R = B(0,R) - \bar{\Omega}_0$, for $R \geq R_0$ and we let Γ_R be the outer boundary of Ω_R, as shown in figure 6.1.

We suppose the solution u satsifies the (growth) bounds

$$|u_t| \leq Ke^{\lambda r}, \quad |u_{,i}| \leq Ke^{\lambda r}, \quad |u_{,it}| \leq Ke^{\lambda r}, \qquad \text{as } r \to \infty, \tag{6.87}$$

$r = |\mathbf{x}| = \sqrt{x_i x_i}$, for some positive constants K, λ.

Multiply equation (6.84) by u_t and integrate over Ω_R for R fixed. Then, we obtain after integration by parts, since the terms on Γ vanish,

$$\begin{aligned} \frac{d}{dt}\Big[\frac{1}{2}\int_{\Omega_R} u_t^2\,dx + \frac{k_2}{2}\int_{\Omega_R} |\nabla u|^2 dx\Big] + k_1 \int_{\Omega_R} u_t^2\,dx \\ + k_3 \int_{\Omega_R} |\nabla u_t|^2 dx = k_2 \int_{\Gamma_R} u_t n_i u_{,i}\,dS + k_3 \int_{\Gamma_R} u_t n_i u_{,ti}\,dS, \end{aligned} \tag{6.88}$$

where n_i is the unit outward normal to Γ_R.

Next, employ the arithmetic - geometric mean inequality $ab \le a^2/2\alpha + \alpha b^2/2$ for $\alpha > 0$ on the right hand side of (6.88) and integrate twice in t over the fixed interval $[0, h]$. The term which arises on the right involving $|\nabla u|^2$ is reduced to an integral over $[0, h] \times \Gamma_R$ and we find for $\beta > 0$ another constant at our disposal,

$$\begin{aligned}
&\frac{1}{2}\int_0^h \int_{\Omega_R} u_t^2\, dx\, ds + \frac{k_2}{2}\int_0^h \int_{\Omega_R} |\nabla u|^2 dx\, ds \\
&\quad + k_1 \int_0^h \int_0^s \int_{\Omega_R} u_\eta^2\, dx\, d\eta\, ds + k_3 \int_0^h \int_0^s \int_{\Omega_R} |\nabla u_\eta|^2 dx\, d\eta\, ds \\
&\quad \le \frac{k_2 \alpha h}{2} \int_0^h \int_{\Gamma_R} |\nabla u|^2 dS\, ds \qquad (6.89)\\
&\quad + \Big(\frac{k_3}{2\beta} + \frac{k_2}{2\alpha}\Big) \int_0^h \int_0^s \int_{\Gamma_R} u_\eta^2\, dS\, d\eta\, ds \\
&\quad + \frac{k_3 \beta}{2} \int_0^h \int_0^s \int_{\Gamma_R} |\nabla u_\eta|^2\, dS\, d\eta\, ds.
\end{aligned}$$

We now drop the first term on the left of inequality (6.89) and define the function $F(R)$ by

$$\begin{aligned}
F(R) =& \frac{k_2}{2} \int_0^h \int_{\Omega_R} |\nabla u|^2 dx\, ds + k_1 \int_0^h \int_0^s \int_{\Omega_R} u_\eta^2\, dx\, d\eta\, ds \\
&+ k_3 \int_0^h \int_0^s \int_{\Omega_R} |\nabla u_\eta|^2 dx\, d\eta\, ds.
\end{aligned} \qquad (6.90)$$

Then, for a constant $A = A(h)$ we may obtain from (6.89)

$$F(R) \le AF'(R), \qquad (6.91)$$

where $F'(R) = dF/dR$. This inequality is integrated from R_0 to R to obtain

$$F(R) \ge \exp\{(R - R_0)/A\}\, F(R_0). \qquad (6.92)$$

Now $F(R_0)$ is a constant. We employ the bounds (6.87) on $F(R)$ in inequality (6.92) to obtain

$$\frac{4\pi}{3} R^3 k_4 e^{2\lambda R} \ge e^{R/A}\, F(R_0) \exp(-R_0/A), \qquad (6.93)$$

where k_4 is a constant depending on K, k_1, k_2 and k_3.

If $A^{-1} > 2\lambda$ then we let $R \to \infty$ in (6.93) and obtain a contradiction. Thus, F must be zero and uniqueness of a solution to $\mathcal{P}$ follows, on $\Omega \times [0, h]$. However, we may repeat this argument on $[h, 2h]$ etc., to obtain uniqueness for $t > 0$.

6.5.2 The weighted energy method

The next technique we describe to establish uniqueness allowing the solution to grow at infinity is an energy method, but one where a weight is added to control the growth at infinity. The idea is due to (Rionero and Galdi, 1976). This is a very versatile technique and expositions of this method in other contexts may be found in (Galdi and Rionero, 1985) and in (Flavin and Rionero, 1995).

Let Ω be an exterior domain as defined in section 6.5.1 and suppose u_1, u_2 are both solutions to $\mathcal{P}$ for the same boundary and initial data so that the difference solution u satisfies the boundary - initial value problem (6.84). We again suppose that the difference solution u is subject to the growth bounds (6.87). In this section we introduce the weight function g, where

$$g = e^{-\alpha r} \tag{6.94}$$

for $\alpha > 0$ to be selected. (More involved choices may be made depending on the outcome desired and many of these may be found in (Galdi and Rionero, 1985).)

The idea is now to multiply equation $(6.84)_1$ by gu_t and integrate over Ω. We select $\alpha > 2\lambda$ so that the integrals converge and one may then show after some integrations by parts

$$\begin{aligned}\frac{d}{dt}\Big[\frac{1}{2}\int_\Omega gu_t^2\,dx + \frac{k_2}{2}\int_\Omega gu_{,i}u_{,i}dx\Big]&\\ + k_1\int_\Omega gu_t^2\,dx + k_3\int_\Omega gu_{,it}u_{,it}\,dx&\\ = -k_2\int_\Omega g_{,i}u_tu_{,i}dx - k_3\int_\Omega g_{,i}u_tu_{,it}dx&\\ \le k_2\alpha\int_\Omega g|u_t|\,|u_{,i}|\,dx + k_3\alpha\int_\Omega g|u_t|\,|u_{,it}|\,dx,&\end{aligned} \tag{6.95}$$

where we have used the fact that $g_{,i} = -\alpha g x_i/r$. We now employ the arithmetic-geometric mean inequality with arbitrary positive constants ζ_1 and ζ_2 on the right hand side of (6.95) to find after dropping the k_1 term on the left,

$$\begin{aligned}\frac{d}{dt}\Big[\frac{1}{2}\int_\Omega gu_t^2\,dx + \frac{k_2}{2}\int_\Omega gu_{,i}u_{,i}dx\Big]&\\ + k_3\int_\Omega gu_{,it}u_{,it}\,dx&\\ \le \Big(\frac{k_2\alpha}{2\zeta_2} + \frac{k_3\alpha}{2\zeta_1}\Big)\int_\Omega g|u_t|^2\,dx&\\ + \frac{k_2\alpha\zeta_2}{2}\int_\Omega g|\nabla u|^2dx + \frac{k_3\alpha\zeta_1}{2}\int_\Omega g|\nabla u_t|^2dx.&\end{aligned} \tag{6.96}$$

Select $\zeta_1 = 2/\alpha$ and then we have freedom to pick ζ_2, for example, and here choose $\zeta_2 = 1/\alpha$. Then, from (6.96) we derive

$$\begin{aligned}\frac{d}{dt}\Big[\int_\Omega gu_t^2\,dx + k_2\int_\Omega gu_{,i}u_{,i}dx\Big]\\ \le \Big(k_2+\frac{k_3}{2}\Big)\alpha^2\int_\Omega g|u_t|^2\,dx + k_2\int_\Omega g|\nabla u|^2dx.\end{aligned} \tag{6.97}$$

Since we wish to allow λ large and we require $\alpha > 2\lambda$, inevitably the coefficient of gu_t^2 is the larger of the two coefficients on the right. Thus, if we define a weighted energy $E(t)$ by

$$E(t) = \int_\Omega gu_t^2dx + k_2\int_\Omega g|\nabla u|^2dx$$

then from inequality (6.97) we may derive

$$\frac{dE}{dt} \le mE,$$

where $m = (k_2 + k_3/2)\alpha^2$. Upon integration from 0 to t this inequality yields

$$E(t) \le e^{mt}E(0) = 0, \tag{6.98}$$

where $E(0) = 0$ follows from the initial data. Thus, from (6.98) we find $0 \le E(t) \le 0$ for all $t > 0$ and due to how $E(t)$ is defined and the fact that $u \equiv 0$ at $t = 0$ we see that $u \equiv 0\ \forall \mathbf{x} \in \Omega$, $\forall t > 0$. Thus, $u_1 \equiv u_2$ and uniqueness of a solution to $\mathcal{P}$ follows.

Note, this method allows the growth coefficient λ in (6.87) to be arbitrarily large.

6.6 Non-standard problems in thermoelasticity

Payne and his co-workers introduced a new class of non-standard problems which are relevant to many applied mathematical situations. Such non-standard problems are where the data are not given at time $t = 0$, but instead are given as a linear combination at times $t = 0$ and $t = T$, see (Payne and Schaefer, 2002), (Payne et al., 2004), (Ames et al., 2004a; Ames et al., 2004b). Such problems were originally introduced as a means to stabilize solutions to the improperly posed problem when the data is given at $t = T$ and one wishes to compute the solution backward in time, cf. (Ames et al., 1998), (Ames and Payne, 1999) and the references therein.

In this section we describe work of (Quintanilla and Straughan, 2005b) who obtain solution estimates in appropriate measures of the solution to thermoelasticity of type II or type III given data as a linear combination at $t = 0$ and $t = T$. They obtain solution estimates for the displacement, temperature, and strain, under a variety of conditions on the coupling constants. These estimates lead to continuous dependence on appropriate terms on the model and to uniqueness under the stated conditions. They also show that such problems do not always possess a unique solution, and thereby delimit the class of coupling constants for which the problems are physically useful. In addition to establishing *a priori* solution bounds for a solution to thermoelasticity of type II or III they show that the time region of (Payne et al., 2004) is sharp by demonstrating that there is non-uniqueness outside of this region. This result is extended to type III thermoelasticity. (Quintanilla and Straughan, 2005b) also consider non-homogeneous boundary conditions in thermoelasticity. (Quintanilla, 2010a) considers a non-standard problem in heat conduction where the heat flux constitutive equation actually has a delay.

(Quintanilla and Straughan, 2005b) commence with the linear equations of anisotropic inhomogeneous thermoelasticity of type II or type III . For a body with a centre of symmetry these are for a material of type II, cf. (Quintanilla, 2002b), or section 2.3,

$$\begin{aligned} \rho\ddot{u}_i =&(a_{ijkh}u_{k,h})_{,j} - (a_{ij}\theta)_{,j} + \rho f_i\,, \\ c\ddot{\theta} =& - a_{ij}\ddot{u}_{i,j} + (k_{ik}\theta_{,k})_{,i} + \rho r\,, \end{aligned} \tag{6.99}$$

where u_i, θ are the displacement vector and temperature field, and a superposed dot denotes $\partial/\partial t$. The quantities ρ, f_i and r are density, body force and heat supply, while $c > 0$ is a constant, and the tensor $a_{ij}(\mathbf{x})$, is a coupling tensor.

The appropriate equations for type III theory for a centrosymmetric body are,

$$\begin{aligned} \rho\ddot{u}_i =&(a_{ijkh}u_{k,h})_{,j} - (a_{ij}\theta)_{,j} + \rho f_i\,, \\ c\ddot{\theta} =& - a_{ij}\ddot{u}_{i,j} + (k_{ik}\theta_{,k})_{,i} + (b_{ik}\dot{\theta}_{,k})_{,i} + \rho r\,. \end{aligned} \tag{6.100}$$

The tensors k_{ij} and b_{ij} are symmetric.

Again, Ω is a bounded domain in $\mathbb{R}^3$ with boundary Γ smooth enough to allow applications of the divergence theorem, and (,) and $\|\cdot\|$ are the inner product and norm on $L^2(\Omega)$. Equations (6.99) or (6.100) hold on $\Omega \times (0,T)$ for some time $T(\leq \infty)$ and the boundary conditions have form

$$u_i(\mathbf{x},t) = 0, \quad \theta(\mathbf{x},t) = 0, \qquad \mathbf{x} \in \Gamma. \tag{6.101}$$

Non-homogeneous boundary conditions as also studied in (Quintanilla and Straughan, 2005b) are considered later.

Instead of initial conditions, the "non-standard" conditions imposed on the problem are

$$\begin{aligned} &\alpha u_i(\mathbf{x},0) + u_i(\mathbf{x},T) = g_i(\mathbf{x}), \qquad &&\beta\dot{u}_i(\mathbf{x},0) + \dot{u}_i(\mathbf{x},T) = h_i(\mathbf{x}), \\ &\alpha\theta(\mathbf{x},0) + \theta(\mathbf{x},T) = \alpha_0(\mathbf{x}), &&\beta\dot{\theta}(\mathbf{x},0) + \dot{\theta}(\mathbf{x},T) = \alpha_1(\mathbf{x}). \end{aligned} \tag{6.102}$$

Here, α, β are given real numbers with g_i, h_i, α_0 and α_1 being prescribed functions.

As in (Quintanilla and Straughan, 2005b) it is assumed that the elastic coefficients are symmetric and positive definite, i.e.

$$a_{ijkh} = a_{khij}, \qquad a_{ijkh}\xi_{ij}\xi_{kh} \geq a_0\xi_{ij}\xi_{ij}, \qquad (a_0 > 0) \tag{6.103}$$

for all second order tensors ξ_{ij}. The thermal conductivity k_{ij} is positive definite with b_{ij} non-negative, namely,

$$k_{ij}\xi_i\xi_j \geq k_0|\boldsymbol{\xi}|^2, \qquad b_{ij}\xi_i\xi_j \geq 0. \tag{6.104}$$

The interaction coefficients $a_{ij}(\mathbf{x})$ are assumed to remain bounded together with their derivatives, so

$$|a_{ij}| \leq a_1, \qquad |a_{ij,j}| \leq a_1 \tag{6.105}$$

where a_1 is a positive constant. It is further assumed that the functions ρ, c, f_i and r depend on $\mathbf{x}$ but not t, and $\rho, c > 0$.

(Quintanilla and Straughan, 2005b) denote by $A(u, v), K(\phi, \psi)$ the bilinear forms

$$A(u, v) = \int_\Omega a_{ijkh}u_{i,j}v_{k,h}dx, \qquad K(\phi, \psi) = \int_\Omega k_{ij}\phi_{,i}\psi_{,j}dx \tag{6.106}$$

and introduce the operator notation

$$A(u)_i = (a_{ijkh}u_{k,h})_{,j}\,. \tag{6.107}$$

To obtain estimates for an appropriate energy function (Quintanilla and Straughan, 2005b) work with a higher derivative "energy-like" function and so differentiate $(6.99)_1$ or $(6.100)_1$ to derive the partial differential equation

$$\rho u_{i,ttt} = (a_{ijkh}\dot{u}_{k,h})_{,j} - (a_{ij}\dot{\theta})_{,j}\,. \tag{6.108}$$

The introduction of an extra derivative requires a condition supplementary to the non-standard ones (6.102). This follows by evaluating $(6.99)_1$ or $(6.100)_1$ at $t = T$ and at $t = 0$ and employing (6.102) to see that

$$\ddot{u}_i(T) = -\alpha\ddot{u}_i(0) + F_i, \tag{6.109}$$

where the data function F_i is given by

$$F_i = (1+\alpha)f_i + \frac{1}{\rho}\big[(a_{ijkh}g_{k,h})_{,j} - (\alpha_0 a_{ij})_{,j}\big]. \tag{6.110}$$

(Quintanilla and Straughan, 2005b) also express $\ddot{u}_i(0)$ in terms of $\ddot{u}_i(T)$ as

$$\ddot{u}_i(0) = -\frac{1}{\alpha}\ddot{u}_i(T) + \frac{F_i}{\alpha}\,. \tag{6.111}$$

6.6.1 *Energy bounds*, $|\alpha|, |\beta| > 1$.

As (Quintanilla and Straughan, 2005b) observe, (Payne and Schaefer, 2002) split their analysis of the operator equation $u_{tt} + Au = F$ with conditions $\alpha u(0) + u(T) = g$, $\beta u_t(0) + u_t(T) = h$ into the cases $|\alpha|, |\beta| > 1$, and

$|\alpha|, |\beta| < 1$, and further show that $|\alpha| < 1, |\beta| > 1$ or $|\alpha| > 1, |\beta| < 1$ does not define a well posed problem. For type II thermoelasticity (Quintanilla and Straughan, 2005b) obtain results largely in agreement with this. For type III thermoelasticity, however, the situation for $|\alpha|, |\beta| < 1$ is very different.

In this section we assume $|\alpha|, |\beta| > 1$ and consider type III thermoelasticity. The analysis also holds for type II thermoelasticity. We follow (Quintanilla and Straughan, 2005b) and consider for simplicity the non-standard conditions (6.102) for the same constants α and β. One could repeat the analysis here, mutatis mutandis, with the general non-standard boundary conditions of $(6.102)_1$ and $(6.102)_2$ having α replaced by γ, and having β replaced by δ, for $\alpha, \beta, \gamma, \delta$ all different.

(Quintanilla and Straughan, 2005b) begin by multiplying (6.108) by $\ddot{u}_i$ and integrating over Ω, using the boundary conditions to derive

$$\frac{d}{dt}\Big[\frac{1}{2}\int_\Omega \rho\ddot{u}_i\ddot{u}_i dx + \frac{1}{2}A(\dot{u}, \dot{u})\Big] - \int_\Omega a_{ij}\dot{\theta}\ddot{u}_{i,j}dx = 0. \tag{6.112}$$

They also multiply $(6.100)_2$ by $\dot{\theta}$ and integrate to derive

$$\begin{aligned}\frac{d}{dt}\Big[\frac{1}{2}\int_\Omega c\dot{\theta}^2 dx + \frac{1}{2}K(\theta, \theta)\Big] &+ \int_\Omega b_{ij}\dot{\theta}_{,i}\dot{\theta}_{,j}dx\\ &+ \int_\Omega a_{ij}\dot{\theta}\ddot{u}_{i,j}dx = \int_\Omega \rho r\dot{\theta}dx.\end{aligned} \tag{6.113}$$

Define $E_1(t)$ by

$$E_1(t) = \frac{1}{2}\int_\Omega \rho\ddot{u}_i\ddot{u}_i dx + \frac{1}{2}A(\dot{u}, \dot{u}) + \frac{1}{2}\int_\Omega c\dot{\theta}^2 dx + \frac{1}{2}K(\theta, \theta), \tag{6.114}$$

and note that upon addition of (6.112) and (6.113) one obtains

$$\frac{dE_1}{dt} + \int_\Omega b_{ij}\dot{\theta}_{,i}\dot{\theta}_{,j}dx = \int_\Omega \rho r\dot{\theta}dx. \tag{6.115}$$

This equation is integrated from 0 to t to find for any $0 \le t \le T$,

$$\begin{aligned}E_1(t) + \int_0^t \int_\Omega b_{ij}\dot{\theta}_{,i}\dot{\theta}_{,j}dx\, ds =&E_1(0) + \int_\Omega \rho r\theta(\mathbf{x}, t)dx\\ &- \int_\Omega \rho r\theta(\mathbf{x}, 0)dx.\end{aligned} \tag{6.116}$$

Choosing $t = T$, from (6.116) one may drop a positive term to see that

$$E_1(T) \le E_1(0) + \int_\Omega \rho r\theta(\mathbf{x}, T)dx - \int_\Omega \rho r\theta(\mathbf{x}, 0)dx. \tag{6.117}$$

The idea is now to use (6.102) together with (6.109) to remove terms evaluated at T and replace them with terms evaluated at $t = 0$. This procedure yields, after use of the arithmetic-geometric mean inequality on the $\int_\Omega \rho r\theta(\mathbf{x},0)dx$ term, the Poincaré inequality, and inequalities (6.104),

$$\begin{aligned}
&\frac{1}{2}\int_\Omega \rho F_iF_i dx + \frac{1}{2}\alpha^2\int_\Omega \rho\ddot{u}_i(0)\ddot{u}_i(0)dx - \alpha\int_\Omega \rho F_i\ddot{u}_i(0)dx \\
&\quad + \frac{1}{2}A(h,h) + \frac{1}{2}\beta^2 A(\dot{u}(0),\dot{u}(0)) - \beta A(h,\dot{u}(0)) \\
&\quad + \frac{1}{2}\int_\Omega c\alpha_1^2 dx + \frac{1}{2}\beta^2\int_\Omega c\big[\dot{\theta}(0)\big]^2 dx - \beta\int_\Omega c\alpha_1\dot{\theta}(0)dx \\
&\quad + \frac{1}{2}K(\alpha_0,\alpha_0) + \frac{1}{2}\alpha^2 K(\theta(0),\theta(0)) - \alpha K(\alpha_0,\theta(0)) \qquad (6.118)\\
&\quad \le \frac{1}{2}\int_\Omega \rho\ddot{u}_i(0)\ddot{u}_i(0)dx + \frac{1}{2}A(\dot{u}(0),\dot{u}(0)) + \frac{1}{2}\int_\Omega c\big[\dot{\theta}(0)\big]^2 dx \\
&\quad + \frac{1}{2}K(\theta(0),\theta(0)) + \int_\Omega \rho r\alpha_0 dx + \frac{(1+\alpha)^2}{2\gamma_1}\int_\Omega \rho^2 r^2 dx \\
&\quad + \frac{\gamma_1}{2k_0\lambda_1}K(\theta(0),\theta(0)),
\end{aligned}$$

where the explicit dependence on $\mathbf{x}$ in terms evaluated at $t=0$ has been suppressed.

Next we use the arithmetic-geometric mean inequality, for arbitrary constants $\gamma_2,\gamma_3,\gamma_4,\gamma_5$ at our disposal, on the third, sixth, ninth and twelth terms. As an example, we write

$$\alpha\int_\Omega \rho F_i\ddot{u}_i(0)dx \le \frac{\alpha^2}{2\gamma_2}\int_\Omega \rho F_iF_i dx + \frac{\gamma_2}{2}\int_\Omega \rho\ddot{u}_i(0)\rho\ddot{u}_i(0)dx.$$

Then one shows from (6.118)

$$\begin{aligned}
&\frac{1}{2}(\alpha^2-\gamma_2-1)\int_\Omega \rho\ddot{u}_i(0)\ddot{u}_i(0)dx + \frac{1}{2}(\beta^2-\gamma_3-1)A(\dot{u}(0),\dot{u}(0)) \\
&+ \frac{1}{2}(\beta^2-\gamma_4-1)\int_\Omega c\big[\dot{\theta}(0)\big]^2 dx + \frac{1}{2}(\alpha^2-\gamma_5-1-\gamma_1/2\lambda_1k_0)\,K(\theta(0),\theta(0)) \\
&\le \Big(\frac{\alpha^2}{2\gamma_2}-\frac{1}{2}\Big)\int_\Omega \rho F_iF_i dx + \Big(\frac{\beta^2}{2\gamma_3}-\frac{1}{2}\Big)A(h,h) \\
&+ \Big(\frac{\beta^2}{2\gamma_4}-\frac{1}{2}\Big)\int_\Omega c\alpha_1^2 dx + \Big(\frac{\alpha^2}{2\gamma_5}-\frac{1}{2}\Big)K(\alpha_0,\alpha_0) \\
&+ \int_\Omega \rho r\alpha_0 dx + \frac{(1+\alpha)^2}{2\gamma_1}\,\|\rho r\|^2. \qquad (6.119)
\end{aligned}$$

Since $|\alpha|>1$, $|\beta|>1$, the numbers $\gamma_1,\gamma_2,\gamma_3,\gamma_4,\gamma_5$ are selected small enough that

$$\begin{aligned}
&\alpha^2-\gamma_2-1>0, \quad \beta^2-\gamma_3-1>0, \\
&\beta^2-\gamma_4-1>0, \quad \alpha^2-\gamma_5-1-\frac{\gamma_1}{2\lambda_1k_0}>0.
\end{aligned}$$

Thus, the coefficients of the terms on the left of (6.119) are positive. In this way we determine computable constants $k_1, \ldots, k_6$ such that

$$\begin{aligned} E_1(0) \leq & k_1 \int_\Omega \rho F_i F_i dx + k_2 A(h,h) + k_3 \int_\Omega c\alpha_1^2 dx \\ & + k_4 K(\alpha_0, \alpha_0) + k_5 \int_\Omega \rho r \alpha_0 dx + k_6 \|\rho r\|^2 \equiv A_0, \end{aligned} \tag{6.120}$$

where A_0 is defined as indicated and is a data term.

At this point we return to (6.116) and drop the b_{ij} term to derive

$$E_1(t) \leq E_1(0) + \int_\Omega \rho r \theta(t) dx - \int_\Omega \rho r \theta(0) dx. \tag{6.121}$$

The second term on the right is estimated with the arithmetic-geometric mean inequality for arbitrary $\gamma_6, \gamma_7 > 0$ to derive

$$\begin{aligned} \int_\Omega \rho r \theta(t) dx & \leq \frac{1}{2\gamma_6} \|\rho r\|^2 + \frac{\gamma_6}{2} \|\theta\|^2 \\ & \leq \frac{1}{2\gamma_6} \|\rho r\|^2 + \frac{\gamma_6}{2\lambda_1 k_0} K(\theta, \theta) \end{aligned}$$

with a similar calculation for $-\int_\Omega \rho r \theta(0) dx$. Pick $\gamma_6 < \lambda_1 k_0$ and then from (6.121) one may determine constants k_7 and k_8 such that

$$E_1(t) \leq k_7 E_1(0) + k_8 \|\rho r\|^2. \tag{6.122}$$

Estimate (6.120) is employed in inequality (6.122) to obtain

$$\begin{aligned} E_1(t) \leq & c_1 \int_\Omega \rho F_i F_i dx + c_2 A(h,h) + c_3 \int_\Omega c\alpha_1^2 dx \\ & + c_4 K(\alpha_0, \alpha_0) + c_5 \int_\Omega \rho r \alpha_0 dx + c_6 \|\rho r\|^2 = B_0, \end{aligned} \tag{6.123}$$

for computable constants $c_1, \ldots, c_6$. The term B_0 is data and so inequality (6.123) is an *a-priori* bound for the function $E_1(t)$ for $0 \leq t \leq T$.

(Quintanilla and Straughan, 2005b) note that (6.123) also yields an estimate for $\|\theta(t)\|^2$ and $\|\nabla\theta(t)\|^2$ by using (6.104) together with Poincaré's inequality. To estimate $\|\mathbf{u}(t)\|$ and $\|\nabla \mathbf{u}(t)\|$ multiply $(6.99)_1$ by $\dot{u}_i$ and integrate over Ω to see that

$$\frac{dU}{dt} = \int_\Omega \rho f_i \dot{u}_i dx + \int_\Omega a_{ij} \dot{u}_{i,j} \theta \, dx, \tag{6.124}$$

where the function $U(t)$ is given by

$$U(t) = \frac{1}{2} \int_\Omega \rho \dot{u}_i \dot{u}_i dx + \frac{1}{2} A(u,u). \tag{6.125}$$

Integrating (6.124) we see that, for any $0 \leq t \leq T$,

$$U(t) = U(0) + \int_\Omega \rho f_i u_i(t) dx - \int_\Omega \rho f_i u_i(0) dx + \int_0^t \int_\Omega a_{ij} \dot{u}_{i,j} \theta \, dx \, ds. \tag{6.126}$$

Evaluate this equation at $t = T$ and use the non-standard conditions (6.102), the bounds (6.105) and the arithmetic-geometric mean inequality to obtain

$$\begin{aligned}
&\frac{1}{2}\int_\Omega \rho(h_i-\beta\dot{u}_i(0))(h_i-\beta\dot{u}_i(0))dx+\frac{1}{2}A(g-u(0),g-u(0))\\
&\qquad\le\frac{1}{2}\int_\Omega \rho\dot{u}_i(0)\dot{u}_i(0)dx+\frac{1}{2}A(u(0),u(0))+\int_\Omega \rho f_i g_i dx \qquad (6.127)\\
&\qquad -\int_\Omega \rho f_i(1+\alpha)u_i(0)dx+\frac{a_1}{2}\int_0^T(\|\nabla\dot{\mathbf{u}}\|^2+\|\theta\|^2)ds.
\end{aligned}$$

As in (Quintanilla and Straughan, 2005b) the terms on the left are expanded and the arithmetic-geometric mean inequality is employed on the $\dot{u}_i(0)h_i$ and $u(0)g$ terms, for arbitrary positive constants δ_1 and δ_2. Then use the arithmetic-geometric mean inequality in the following manner, for $\delta_3 > 0$ at our disposal,

$$\begin{aligned}
-\int_\Omega \rho(1+\alpha)f_i u_i(0)dx &\le \frac{(1+\alpha)^2}{2\delta_3}\|\rho\mathbf{f}\|^2+\frac{\delta_3}{2}\|\mathbf{u}(0)\|^2\\
&\le \frac{(1+\alpha)^2}{2\delta_3}\|\rho\mathbf{f}\|^2+\frac{\delta_3}{2a_0\lambda_1}A(u(0),u(0)),
\end{aligned}$$

where restrictions (6.103) and Poincaré's inequality have been employed. With the aid of this inequality one thus shows from inequality (6.121)

$$\begin{aligned}
&\frac{1}{2}(\beta^2-\delta_1-1)\int_\Omega \rho\dot{u}_i(0)\dot{u}_i(0)dx+\frac{1}{2}\Big(\alpha^2-\delta_2-1-\frac{\delta_3}{2a_0\lambda_1}\Big)A(u(0),u(0))\\
&\qquad\le\frac{1}{2}(1+\beta^2/\delta_1)\|\rho^{1/2}\mathbf{h}\|^2+\frac{1}{2}(1+\alpha^2/\delta_2)A(g,g)+\int_\Omega \rho f_i g_i dx\\
&\qquad +\frac{(1+\alpha)^2}{2\delta_3}\|\rho\mathbf{f}\|^2+\frac{a_1}{2a_0}\int_0^T A(\dot{u},\dot{u})ds\\
&\qquad +\frac{a_1}{2\lambda_1 k_0}\int_0^T K(\theta,\theta)ds. \qquad (6.128)
\end{aligned}$$

The last two terms are bounded by a piece of $E_1(T)$. Then, selecting $\beta^2 > 1+\delta_1$, $\alpha^2 > 1+\delta_2+\delta_3/2a_0\lambda_1$ one calculates computable constants $d_1, \ldots, d_4$ such that

$$\begin{aligned}
U(0) &\le d_1\|\rho^{1/2}\mathbf{h}\|^2+d_2A(g,g)+d_3\int_\Omega \rho f_i g_i dx+d_4\|\rho\mathbf{f}\|^2+E_1(T)\\
&\le d_1\|\rho^{1/2}\mathbf{h}\|^2+d_2A(g,g)\\
&\quad +d_3\int_\Omega \rho f_i g_i dx+d_4\|\rho\mathbf{f}\|^2+B_0=C_0, \qquad (6.129)
\end{aligned}$$

where C_0 is defined as indicated and is a data term.

From equation (6.126) we estimate

$$U(t) \leq U(0) + \frac{1}{2\zeta_1}\|\rho\mathbf{f}\|^2 + \frac{\zeta_1}{2\lambda_1 a_0}A(u,u) + \frac{1}{2\zeta_2}\|\rho\mathbf{f}\|^2 + \frac{\zeta_2}{2\lambda_1 a_0}A(u(0),u(0)) + \frac{a_1}{2a_0}\int_0^t A(\dot{u},\dot{u})ds + \frac{a_1}{2\lambda_1 k_0}\int_0^t K(\theta,\theta)ds,$$

for constants $\zeta_1, \zeta_2 > 0$ at our disposal. The number ζ_1 is chosen so small that $\zeta_1 < \lambda_1 a_0$ and we bound the $A(u(0),u(0))$ term by $U(0)$, then bound the last two terms by $E_1(t)$, to find

$$U(t) \leq \zeta_3 U(0) + \zeta_3\|\rho\mathbf{f}\|^2 + \zeta_4 E_1(t). \tag{6.130}$$

$E_1(t)$ is estimated using (6.123), $U(0)$ is bounded by means of (6.129) and then from (6.130) we find for computable constants $\ell_1, \ldots, \ell_{10}$,

$$\begin{aligned} U(t) \leq & \ell_1 \int_\Omega \rho F_i F_i dx + \ell_2 A(h,h) + \ell_3 \int_\Omega c\alpha_1^2 dx + \ell_4 K(\alpha_0,\alpha_0) \\ & + \ell_5 \int_\Omega \rho r \alpha_0 dx + \ell_6\|\rho r\|^2 + \ell_7\|\rho^{1/2}\mathbf{h}\|^2 + \ell_8 A(g,g) \\ & + \ell_9 \int_\Omega \rho f_i g_i dx + \ell_{10}\|\rho\mathbf{f}\|^2 \equiv D_0. \end{aligned} \tag{6.131}$$

The term D_0 is defined as shown. It is important to note that D_0 involves only data and thus inequality (6.131) is an *a priori* bound for $U(t)$. From this inequality we obtain an *a priori* bound for either $\|\mathbf{u}(t)\|^2$ or $\|\nabla\mathbf{u}(t)\|^2$.

6.6.2 Energy bounds, $|\alpha|, |\beta| < 1$.

When $|\alpha|, |\beta| < 1$ (Quintanilla and Straughan, 2005b) are able to progress only with type II thermoelasticity.

They observe that equations (6.99) are invariant under time reversal, $t \to -t$. They rewrite the non-standard conditions (6.102), to place the coefficients α, β etc. in front of the terms involving T, e.g. $u_i(0)+(1/\alpha)u_i(T) = g_i/\alpha$. In this manner, this is essentially the problem of the previous subsection but with 0 and T reversed. The roles of 0 and T are interchanged and $E_1(0)$ is bounded by $E_1(T)$. Then use the fact that $1/|\alpha| > 1$, $1/|\beta| > 1$ to bound $E_1(T)$ by data. In this way one finds a bound for $E_1(t)$ in terms of data for all $0 \leq t \leq T$.

To handle U (Quintanilla and Straughan, 2005b) write

$$U(0) = U(T) - \int_\Omega \rho f_i u_i(T)dx + \int_\Omega \rho f_i u_i(0)dx - \int_0^T\int_\Omega a_{ij}\dot{u}_{i,j}\theta\, dx\, ds.$$

They then eliminate the quantities at $t = 0$ in favour of those at $t = T$. In this way they derive a bound of form $U(T) \leq$ data. This in turn leads to a bound $U(0) \leq$ data. Then from (6.126) they derive an estimate of form $U(t) \leq$ data, $0 \leq t \leq T$.

(Quintanilla and Straughan, 2005b) observe that when $|\alpha| < 1$, $|\beta| < 1$ in type III theory they were unable to find a suitable bound. Since (Payne and Schaefer, 2002), (Payne et al., 2005) did find bounds for a wave equation with dissipation, they investigated why things break down. (Payne and Schaefer, 2002) establish an *a priori* bound for a solution to the equation

$$u_{tt} + au_t + Au = 0$$

for A a densely defined symmetric linear operator and $a > 0$ constant. They require the restriction $|\alpha|, |\beta| < e^{-aT}$. If one inspects their (clever) proof carefully, it hinges on being able to bound the dissipation function $\int_t^T \|u_s(s)\|^2 ds$ by $\int_t^T E(s)ds$ where in their case $E(s) = \|u_s\|^2/2 + (Au(s), u(s))$, the norm and inner product being on an appropriate function space. For type III thermoelasticity if one works with $E_1(t)$ then one encounters a dissipation of form

$$\int_t^T \int_\Omega b_{ij}\dot{\theta}_{,i}\dot{\theta}_{,j}\, dx\, ds \tag{6.132}$$

whereas the θ part in E_1 essentially involves $\|\dot{\theta}\|^2$ and $\|\nabla\theta\|^2$. This prevents (Quintanilla and Straughan, 2005b) from bounding the dissipation term in (6.132) by $\int_t^T E_1 ds$ since the function in (6.132) is effectively $\int_t^T \|\nabla\dot{\theta}\|^2 ds$. Thus, the analogue of the (Payne and Schaefer, 2002) proof, in going from equation (6.4) to inequality (6.5) of their paper, breaks down.

(Quintanilla and Straughan, 2005b) further observe that one cannot expect to find a bound for $E_1(t)$ for $|\alpha|, |\beta| < e^{-aT}$ in type III thermoelasticity, (where a is a constant related to the first eigenfunction in the membrane problem). They show that there is non-uniqueness in this range and hence one cannot expect such a bound to hold.

6.6.3 Non-homogeneous boundary conditions

(Quintanilla and Straughan, 2005b) show that the analysis of the non-standard problems required $u_i = 0$ and $\theta = 0$ on the spatial boundary, Γ, but this may be overcome. If one requires inhomogeneous boundary conditions of the form

$$u_i(\mathbf{x}, t) = u_i^B(\mathbf{x}, t), \qquad \theta(\mathbf{x}, t) = \theta^B(\mathbf{x}, t), \qquad \mathbf{x} \in \Gamma, \tag{6.133}$$

the procedure is as follows.

Let u_i, θ be a solution to (6.99) or (6.100) with conditions (6.102), but with the zero boundary conditions replaced by (6.133). To derive estimates for suitable norms of u_i and θ in the inhomogeneous problem one introduces functions v_i, ψ which solve the system

$$\begin{aligned} \rho\ddot{v}_i &= A(v)_i - (a_{ij}\psi)_{,j} + \rho f_i \\ c\ddot{\psi} &= -a_{ij}\ddot{v}_{i,j} + (k_{ij}\psi_{,j})_{,i} + (b_{ij}\dot{\psi}_{,j})_{,i} + \rho r \end{aligned} \tag{6.134}$$

in $\Omega \times (0, T]$, with boundary conditions

$$v_i(\mathbf{x}, t) = u_i^B(\mathbf{x}, t), \quad \psi(\mathbf{x}, t) = \theta^B(\mathbf{x}, t), \qquad \mathbf{x} \in \Gamma. \tag{6.135}$$

The functions v_i, ψ satisfy *standard* initial conditions

$$\begin{aligned} v_i(\mathbf{x}, 0) &= k_i(\mathbf{x}), \quad \dot{v}_i(\mathbf{x}, 0) = \tilde{k}_i(\mathbf{x}), \\ \psi(\mathbf{x}, 0) &= \ell(\mathbf{x}), \quad \dot{\psi}(\mathbf{x}, 0) = \tilde{\ell}(\mathbf{x}). \end{aligned} \tag{6.136}$$

Next, introduce the difference functions $w_i = u_i - v_i$, $\phi = \theta - \psi$. By calculation w_i, ϕ solves the problem

$$\begin{aligned} \rho \ddot{w}_i &= A(w)_i - (a_{ij}\phi)_{,j} \\ c\ddot{\phi} &= -a_{ij}\ddot{w}_{i,j} + (k_{ij}\phi_{,j})_{,i} + (b_{ij}\dot{\phi}_{,j})_{,i} \end{aligned}$$

in $\Omega \times (0, T)$ with $w_i = \phi = 0$ on Γ, and the non-standard conditions

$$\begin{aligned} &\alpha w_i(\mathbf{x}, 0) + w_i(\mathbf{x}, T) = g_i(\mathbf{x}) - \alpha k_i(\mathbf{x}) - v_i(\mathbf{x}, T) \\ &\beta \dot{w}_i(\mathbf{x}, 0) + \dot{w}_i(\mathbf{x}, T) = h_i(\mathbf{x}) - \beta \tilde{k}_i(\mathbf{x}) - \dot{v}_i(\mathbf{x}, T) \\ &\alpha \phi(\mathbf{x}, 0) + \phi(\mathbf{x}, T) = \alpha_0(\mathbf{x}) - \alpha \ell(\mathbf{x}) - \psi(\mathbf{x}, T) \\ &\beta \dot{\phi}(\mathbf{x}, 0) + \dot{\phi}(\mathbf{x}, T) = \alpha_1(\mathbf{x}) - \beta \tilde{\ell}(\mathbf{x}) - \dot{\psi}(\mathbf{x}, T). \end{aligned}$$

To obtain an estimate for u_i, (Quintanilla and Straughan, 2005b) let $\|u\|$ be a suitable norm for u_i, e.g. in $L^2(\Omega)$ or $\sqrt{A(u, u)}$, then from the triangle inequality

$$\|\mathbf{u}(t)\| \leq \|\mathbf{w}(t)\| + \|\mathbf{v}(t)\|.$$

The quantity $\|\mathbf{v}(t)\|$ is known and $\|\mathbf{w}(t)\|$ may be found in terms of the data functions $g_i, h_i, \alpha_0, \alpha_1, k_i, \tilde{k}_i, \ell, \tilde{\ell}$ and the functions $v_i(T)$ and $\psi(T)$. Since these solve a standard boundary-initial value problem they are known and so we can find bounds for $\|\mathbf{u}(t)\|$. Likewise $\|\dot{\mathbf{u}}(t)\|$, $\|\theta(t)\|$, $\|\dot{\theta}(t)\|$ may be estimated in terms of data. The boundary data u_i^B and θ^B are involved in the bounds through the functions $\mathbf{v}(t)$, $\mathbf{v}(T)$, etc.

6.7 Explosive instabilities in heat transfer

6.7.1 *Third order theory*

In this section we describe a result of (Quintanilla and Straughan, 2002) which focusses on a theory of (Ghaleb and El-Deen Mohamedein, 1989) who derive a third order in time theory for heat propagation.

Exponential growth and connected results for the linearised (Ghaleb and El-Deen Mohamedein, 1989) theory are given by (Franchi and Straughan, 1994a) and by (Quintanilla, 1997). (Payne and Song, 2006) study an interesting class of problems for the (Ghaleb and El-Deen Mohamedein, 1989) theory in in that they consider the temperature T and its time derivative

T_t prescribed at time $t = 0$ but also prescribe the value of T_t at a later time $t = t_1$. They derive interesting bounds for the solution and also investigate decay in the spatial variable.

(Quintanilla and Straughan, 2002) concentrate on a nonlinear equation arising from the (Ghaleb and El-Deen Mohamedein, 1989) theory. They argue that the thermal conductivity is a function of temperature and they include this effect. This then leads to a nonlinear theory of heat conduction. They observe that the inclusion of temperature - dependent thermal conductivity has disastrous consequences in that it predicts blow-up of the temperature field in finite time for certain parameters.

The theory of (Ghaleb and El-Deen Mohamedein, 1989) is based on three equations which govern the behaviour of the temperature field, T, the heat flux, q_i, and the entropy, which they denote by s. They have an entropy production equation

$$\rho\frac{\partial s}{\partial t} = -\frac{1}{T_0}\frac{\partial q_i}{\partial x_i}, \tag{6.137}$$

in which ρ, T_0 are positive constants. For the heat flux they propose the following law,

$$h\frac{\partial q_i}{\partial t} + \frac{1}{T_0}q_i = -K\frac{\partial T}{\partial x_i} - j\frac{\partial^2 T}{\partial x_i \partial t}, \tag{6.138}$$

in which h and j are positive constants. (Quintanilla and Straughan, 2002) allow the thermal conductivity K be a function of temperature, $K = K(T) > 0$. In fact, in reality K does depend on T. In addition to equations (6.137), (6.138) (Ghaleb and El-Deen Mohamedein, 1989) have the following constitutive equation for the entropy,

$$\rho s = \hat{a}T - \ell\frac{\partial T}{\partial t}, \tag{6.139}$$

in which $\hat{a}$ and ℓ are positive constants.

(Quintanilla and Straughan, 2002) eliminate the variables s and q_i to obtain a single equation for T which has form

$$\ell T_0\frac{\partial^3 T}{\partial t^3} + \Big(\frac{\ell}{h} - \hat{a}T_0\Big)\frac{\partial^2 T}{\partial t^2} - \frac{\hat{a}}{h}\frac{\partial T}{\partial t} = -\frac{j}{h}\Delta\frac{\partial T}{\partial t} - \frac{1}{h}\big(K(T)T_{,i}\big)_{,i}. \tag{6.140}$$

Let us set $K(T) = K\,f(T)$ where K is a positive constant and

$$f(T) = 1 + \gamma' T^{\epsilon} \tag{6.141}$$

for positive constants γ' and ϵ. (Quintanilla and Straughan, 2002) non-dimensionalise equation (6.140) with the variables

$$\begin{aligned} &t' = t/T_0h, \qquad x_i' = x_i/(KT_0h/\hat{a})^{1/2}, \qquad uT_0 = T,\\ &\alpha = \ell/\hat{a}T_0h, \qquad \beta = j/KT_0h, \qquad a = \gamma' T_0\epsilon, \end{aligned}$$

where u is the non-dimensional temperature. Then, equation (6.140) is transformed to

$$\alpha \frac{\partial^3 u}{\partial t^3} - (1-\alpha)\frac{\partial^2 u}{\partial t^2} - \frac{\partial u}{\partial t} = -\beta\Delta\frac{\partial u}{\partial t} - \big[f(u)u_{,i}\big]_{,i}, \tag{6.142}$$

where $f(u)$ is given by

$$f(u) = 1 + au^{\epsilon}. \tag{6.143}$$

(Quintanilla and Straughan, 2002) consider the situation where $u \geq 0$, and also

$$0 < \alpha < 1. \tag{6.144}$$

As usual Ω is a bounded domain in three-space, with boundary Γ. Equation (6.142) is defined on $\Omega \times (0, \mathcal{T})$ for some time $\mathcal{T}$. The function u is assumed zero on the boundary Γ and initial values are prescribed for $u(\mathbf{x},0)$, $\dot{u}(\mathbf{x},0)$, and $\ddot{u}(\mathbf{x},0)$. In other words, u satisfies the boundary-initial value problem $\mathcal{P}$ defined by equation (6.142) with the boundary and initial conditions,

$$u(\mathbf{x},t) = 0, \qquad \mathbf{x} \in \Gamma, \tag{6.145}$$

$$u(\mathbf{x},0) = u_0(\mathbf{x}), \quad \frac{\partial u}{\partial t}(\mathbf{x},0) = v_0(\mathbf{x}), \quad \frac{\partial^2 u}{\partial t^2}(\mathbf{x},0) = a_0(\mathbf{x}), \tag{6.146}$$

for prescribed functions u_0, v_0 and a_0.

Throughout, as usual, $\|\cdot\|$ and $(\,,\,)$ are the norm and inner product on $L^2(\Omega)$.

6.7.2 Nonexistence of a solution

We report on work of (Quintanilla and Straughan, 2002) who establish an upper bound for the interval of existence of a solution to $\mathcal{P}$. (Quintanilla and Straughan, 2002) remark that there are many blow-up results for first and second order in time partial differential equations occurring in the physical literature but they have not seen such an example for a third order derivative in time equation. In connection with this (Goldstein, 1985) shows that the abstract equation

$$\frac{d^n u}{dt^n} = Au, \qquad n \geq 3, \tag{6.147}$$

is well posed if and only if A is a bounded linear operator. We note that equation (6.142) possesses the Δu_t term and so the linearised version does not come into the category of the theory of (Goldstein, 1985). In an interesting article, (Dreher et al., 2009) consider the abstract operator equation

$$b_0 u_t + b_1 u_{tt} + \cdots + b_j \frac{\partial^{j+1} u}{\partial t^{j+1}} = c_0 Au + c_1 Au_t + \cdots + c_m A\frac{\partial^m u}{\partial t^m}, \tag{6.148}$$

with A being an appropriate operator in a suitable Banach space. (A has to possess a sequence of real eigenvalues λ_k such that $0 > \lambda_k \to -\infty$ as $k \to \infty$.) They show that the initial value problem for (6.148) is not well - posed if $j + 1 - m \geq 3$.

The proof of global nonexistence given in (Quintanilla and Straughan, 2002) begins by multiplying equation (6.142) by u and integrating over Ω. They define

$$G(t) = \|u(t)\|^2 \qquad \text{and} \qquad K(t) = \|u_t(t)\|^2. \tag{6.149}$$

Then one shows

$$G''' - \mu G'' = 3K' - 2\mu K + \frac{1}{\alpha}\frac{d}{dt}(G + \beta\|\nabla u\|^2) + \frac{2}{\alpha}(f\nabla u, \nabla u), \tag{6.150}$$

where $\mu = (1-\alpha)/\alpha > 0$ and superscript prime denotes differentiation with respect to t. Upon use of an integrating factor and an integration one may find

$$\begin{aligned} G'' =&3K + \frac{1}{\alpha}(G + \beta\|\nabla u\|^2) + \int_0^t e^{\mu(t-s)}\Big[\mu K + \frac{\mu}{\alpha}(G + \beta\|\nabla u\|^2)\Big]\, ds \\ &+ \frac{2}{\alpha}\int_0^t e^{\mu(t-s)}(fu_{,i}, u_{,i})ds \\ &+ \Big(G''(0) - 3K(0) - \frac{1}{\alpha}G(0) - \frac{\beta}{\alpha}\|\nabla u_0\|^2\Big)e^{\mu t}. \end{aligned} \tag{6.151}$$

The function $F(t)$ is defined by

$$F(t) = \int_0^t \|u(s)\|^2 ds. \tag{6.152}$$

It is supposed the initial data are such that

$$G''(0) - 3K(0) - \frac{1}{\alpha}G(0) - \frac{\beta}{\alpha}\|\nabla u_0\|^2 \geq 0\,. \tag{6.153}$$

The Poincaré inequality $\|\nabla u\|^2 \geq \lambda_1\|u\|^2$, is next used on the nonlinear term involving f, and using the definition of f, one may use Hölder's inequality and then discard some non-negative terms on the right of inequality (6.151) to obtain

$$F''' \geq \tilde{k}\int_0^t e^{\mu(t-s)}\|u\|^{2+\epsilon} ds\,. \tag{6.154}$$

Here the constant $\tilde{k}$ is given by $\tilde{k} = 8a\lambda_1/\alpha m^{\epsilon/2}(2+\epsilon)^2$, where m is the volume of Ω.

By using Hölder's inequality one may show

$$\begin{aligned}\int_0^t \|u\|^2 ds &\leq \Big(\frac{\epsilon}{2\mu}\Big)^{\epsilon/(2+\epsilon)} [1-e^{-2\mu t/\epsilon}]\Big(\int_0^t e^{\mu(t-s)}\|u\|^{2+\epsilon}ds\Big)^{2/(2+\epsilon)}\\ &\leq \Big(\frac{\epsilon}{2\mu}\Big)^{\epsilon/(2+\epsilon)} \Big(\int_0^t e^{\mu(t-s)}\|u\|^{2+\epsilon}ds\Big)^{2/(2+\epsilon)}. \qquad (6.155)\end{aligned}$$

Upon using (6.155) in inequality (6.154) one may show

$$F''' \geq kF^{1+\epsilon/2}, \qquad (6.156)$$

where now k is the constant $k = (2\mu/\epsilon)^{\epsilon/2}\,\tilde{k}$.

It is assumed that $\|u_0\|^2 > 0$ and thus $F'(0) > 0$. Inequality (6.156) is multiplied by F' and one finds

$$(F'F'')' \geq (F'')^2 + \Big(\frac{2k}{4+\epsilon}\Big)\frac{d}{dt}F^{2+\epsilon/2}. \qquad (6.157)$$

After dropping the $(F'')^2$ term and integrating in time we find

$$F'F'' \geq F'(0)F''(0) + \Big(\frac{2k}{4+\epsilon}\Big)F^{2+\epsilon/2}.$$

Now multiply by F' and integrate from 0 to t to find

$$(F')^3 \geq \big[F'(0)\big]^3 + 3F'(0)F''(0)F + \zeta F^{3+\epsilon/2}, \qquad (6.158)$$

where the constant ζ is given by

$$\zeta = \frac{12k}{[(4+\epsilon)(6+\epsilon)]}.$$

One now starts with inequality (6.158) and argues that u exists for all time. Separate variables and integrate to find

$$\begin{aligned} t &\leq \int_0^{F(t)} \frac{dF}{(\alpha_1+\beta_1 F+\zeta F^{3+\epsilon/2})^{1/3}}\\ &\leq \int_0^{\infty} \frac{dF}{(\alpha_1+\beta_1 F+\zeta F^{3+\epsilon/2})^{1/3}} < \infty, \end{aligned} \qquad (6.159)$$

where we have put $\alpha_1 = \big[F'(0)\big]^3$ and $\beta_1 = 3F'(0)F''(0)$. Inequality (6.159) leads to a contradiction and so the solution cannot exist in a classical sense for all time. One sees that an upper bound for the interval of existence is given by

$$T_u = \int_0^{\infty} \frac{dF}{(\alpha_1+\beta_1 F+\zeta F^{3+\epsilon/2})^{1/3}}.$$

Thus, nonexistence of a solution to $\mathcal{P}$ has been established provided $\|u_0\|^2 > 0$ and

$$2(u_0, a_0) \geq \|v_0\|^2 + \frac{1}{\alpha}\|u_0\|^2 + \frac{\beta}{\alpha}\|\nabla u_0\|^2.$$

As noted in (Quintanilla and Straughan, 2002), the expected behaviour is blow-up in a finite time T with $T \leq T_u$.

6.8 Qualitative results for fluids

As we observed in chapter 3, section 3.1 a viscous fluid model incorporating the Maxwell-Cattaneo law employing Fox's derivative for the heat flux was presented by (Straughan and Franchi, 1984). Their equations may be written

$$
\begin{aligned}
&\frac{\partial v_i}{\partial t} + v_j \frac{\partial v_i}{\partial x_j} = -\frac{\partial p}{\partial x_i} + b_i T + \nu \Delta v_i,\\
&\frac{\partial v_i}{\partial x_i} = 0,\\
&c\Big(\frac{\partial T}{\partial t} + v_i \frac{\partial T}{\partial x_i}\Big) = -\frac{\partial q_i}{\partial x_i},\\
&\tau\Big(\frac{\partial q_i}{\partial t} + v_j \frac{\partial q_i}{\partial x_j} - \epsilon_{ijk}\omega_j q_k\Big) = -q_i - \kappa T_{,i}\,,
\end{aligned}
\tag{6.160}
$$

where without loss of generality for the class of problems of concern here we have taken the density to be 1, and b_i is a body force vector (in (Straughan and Franchi, 1984) $b_i = \alpha g k_i$, α, g being thermal expansion coefficient and gravity, and $\mathbf{k} = (0, 0, 1)$). Here v_i, T and p are the velocity, temperature and pressure, c, ν, τ are positive constants, and $\boldsymbol{\omega} = \mathrm{curl}\, \mathbf{v}/2$.

If one works with the generalized Maxwell-Cattaneo equations, or equivalently the Guyer-Krumhansl equations, again employing Fox's derivative, then the corresponding system of equations is, cf. (Franchi and Straughan, 1994b) and section 3.1, specifically section 3.1.3,

$$
\begin{aligned}
&\frac{\partial v_i}{\partial t} + v_j \frac{\partial v_i}{\partial x_j} = -\frac{\partial p}{\partial x_i} + b_i T + \nu \Delta v_i,\\
&\frac{\partial v_i}{\partial x_i} = 0,\\
&c\Big(\frac{\partial T}{\partial t} + v_i \frac{\partial T}{\partial x_i}\Big) = -\frac{\partial q_i}{\partial x_i},\\
&\tau\Big(\frac{\partial q_i}{\partial t} + v_j \frac{\partial q_i}{\partial x_j} - \epsilon_{ijk}\omega_j q_k\Big) = -q_i - \kappa T_{,i} + \tilde{\tau}(\Delta q_i + 2q_{k,ki})\,,
\end{aligned}
\tag{6.161}
$$

where c and $\tilde{\tau}$ are further positive constants.

6.8.1 Decay for a solution to (6.160)*?*

Suppose we consider the system of equations (6.160) defined on a bounded domain $\Omega \subset \mathbb{R}^3$ with on the boundary Γ,

$$v_i = 0, \quad T = T_B, \quad \epsilon_{ijk} n_j q_k = 0 \qquad \text{on} \qquad \Gamma, \tag{6.162}$$

for $T_B > 0$ a constant. We may replace T in $(6.160)_3$ and $(6.160)_4$ by $\theta = T - T_B$ and then absorb T_B into the pressure in equation $(6.160)_1$. This allows us to replace equations (6.160) by the system

$$\begin{aligned}
&\frac{\partial v_i}{\partial t} + v_j \frac{\partial v_i}{\partial x_j} = -\frac{\partial p}{\partial x_i} + b_i\theta + \nu \Delta v_i, \\
&\frac{\partial v_i}{\partial x_i} = 0, \\
&c\Big(\frac{\partial \theta}{\partial t} + v_i \frac{\partial \theta}{\partial x_i}\Big) = -\frac{\partial q_i}{\partial x_i}, \\
&\tau\Big(\frac{\partial q_i}{\partial t} + v_j \frac{\partial q_i}{\partial x_j} - \epsilon_{ijk}\omega_j q_k\Big) = -q_i - \kappa\theta_{,i}\, .
\end{aligned} \tag{6.163}$$

These equations hold in $\Omega \times \{t > 0\}$ and on the boundary Γ we now have

$$v_i = 0, \quad \theta = 0, \quad \epsilon_{ijk} n_j q_k = 0 \qquad \text{on} \qquad \Gamma. \tag{6.164}$$

Then, multiplying each of $(6.163)_1$, $(6.163)_3$, $(6.163)_4$ in turn by v_i, θ and q_i and integrating over Ω we find

$$\frac{d}{dt}\frac{1}{2}\|\mathbf{v}\|^2 = (\theta, v_i b_i) - \nu\|\nabla \mathbf{v}\|^2, \tag{6.165}$$

$$\frac{d}{dt}\frac{c}{2}\|\theta\|^2 = -(q_{i,i}, \theta), \tag{6.166}$$

$$\frac{d}{dt}\frac{\tau}{2}\|\mathbf{q}\|^2 = -\|\mathbf{q}\|^2 - \kappa(\theta_{,i}, q_i). \tag{6.167}$$

Thus, from (6.166) and (6.167), upon use of the boundary conditions and an integration by parts we see that

$$\frac{d}{dt}\frac{c\kappa}{2}\|\theta\|^2 + \frac{\tau}{2}\|\mathbf{q}\|^2 = -\|\mathbf{q}\|^2. \tag{6.168}$$

Hence,

$$\frac{c\kappa}{2}\|\theta(t)\|^2 + \frac{\tau}{2}\|\mathbf{q(t)}\|^2 + \int_0^t \|\mathbf{q(s)}\|^2 ds = \frac{c\kappa}{2}\|\theta(0)\|^2 + \frac{\tau}{2}\|\mathbf{q(0)}\|^2. \tag{6.169}$$

From this we may deduce $\|\mathbf{q(t)}\|^2 \in L^1(0,\infty)$ which suggests possible decay of q_i although it does not prove it. We also deduce $\|\theta(t)\|^2$ is bounded for all t. From equation (6.165) we may use Poincaré's inequality and the arithmetic-geometric mean inequality to find

$$\frac{d}{dt}\|\mathbf{v}\|^2 + \lambda_1\nu\|\mathbf{v}\|^2 \le \frac{\|\theta\|^2}{\mu},$$

where $\mu = \lambda_1\nu/|\mathbf{b}|^2$. Given $\|\theta(t)\|^2$ is bounded, from this inequality we may easily deduce $\|\mathbf{v}(t)\|$ is bounded. However, I have not seen how to show $\|\mathbf{v}(t)\|, \|\mathbf{q}(t)\|$ and $\|\theta(t)\|$ decay, if they do. The nonlinear system (6.160) appears difficult to treat.

If instead we work with the equivalent of (6.161) then instead of system (6.163) we have

$$\begin{aligned}
&\frac{\partial v_i}{\partial t} + v_j\frac{\partial v_i}{\partial x_j} = -\frac{\partial p}{\partial x_i} + b_i\theta + \nu\Delta v_i,\\
&\frac{\partial v_i}{\partial x_i} = 0,\\
&c\Big(\frac{\partial \theta}{\partial t} + v_i\frac{\partial \theta}{\partial x_i}\Big) = -\frac{\partial q_i}{\partial x_i},\\
&\tau\Big(\frac{\partial q_i}{\partial t} + v_j\frac{\partial q_i}{\partial x_j} - \epsilon_{ijk}\omega_j q_k\Big) = -q_i - \kappa\theta_{,i} + \tilde{\tau}\Delta q_i + 2\tilde{\tau}q_{k,ki}\,.
\end{aligned} \tag{6.170}$$

The boundary equations are again (6.164).

To derive the equivalent of equation (6.167) we write the last two terms in $(6.170)_4$ as

$$3\tilde{\tau}q_{k,ik} + \tilde{\tau}(q_{i,j} - q_{j,i})_{,j}\,.$$

Then, upon multiplication of $(6.170)_4$ by q_i and integration over Ω we find

$$\begin{aligned}
\frac{\tau}{2}\frac{d}{dt}\|\mathbf{q}\|^2 = &-\|\mathbf{q}\|^2 + \kappa(\theta, q_{i,i}) + \tilde{\tau}\oint_\Gamma n_j q_i(q_{i,j} - q_{j,i})dS\\
&+ 3\tilde{\tau}\oint_\Gamma n_i q_i q_{k,k}dS - \tilde{\tau}\int_\Omega q_{i,j}(q_{i,j} - q_{j,i})dx\\
&- 3\tilde{\tau}\int_\Omega (q_{i,i})^2 dx.
\end{aligned} \tag{6.171}$$

Since $\epsilon_{ijk}n_j q_k = 0$ on Γ we have $\oint_\Gamma n_j q_i(q_{i,j} - q_{j,i})dS = 0$. Also, using $(6.170)_3$,

$$\oint_\Gamma n_i q_i q_{k,k}dS = -c\oint_\Gamma n_i q_i \theta_{,t}dS - c\oint_\Gamma n_i q_i v_r\theta_{,r}dS = 0$$

since $\theta = 0$ and $v_i = 0$ on Γ. Whence, (6.171) leads to

$$\begin{aligned}
\frac{\tau}{2}\frac{d}{dt}\|\mathbf{q}\|^2 = &-\|\mathbf{q}\|^2 + \kappa(\theta, q_{i,i})\\
&- \tilde{\tau}\int_\Omega q_{i,j}(q_{i,j} - q_{j,i})dx - 3\tilde{\tau}\|q_{i,i}\|^2.
\end{aligned} \tag{6.172}$$

Equation (6.166) still holds here and so we may derive the following equation, using (6.172)

$$\frac{d}{dt}\Big(\frac{c\kappa}{2}\|\theta\|^2+\frac{\tau}{2}\|\mathbf{q}\|^2\Big)=-\|\mathbf{q}\|^2-\tilde{\tau}\int_\Omega q_{i,j}(q_{i,j}-q_{j,i})dx \\ -3\tilde{\tau}\|q_{i,i}\|^2. \tag{6.173}$$

From this equation we deduce that $\|\mathbf{q}(t)\|^2$, $\|q_{[i,j]}\|^2$ and $\|q_{i,i}\|^2$ are $L^1(0,\infty)$ where $q_{[i,j]}=(q_{i,j}-q_{j,i})/2$. Furthermore,

$$\frac{c\kappa}{2}\|\theta(t)\|^2+\frac{\tau}{2}\|\mathbf{q}(t)\|^2\le\frac{c\kappa}{2}\|\theta(0)\|^2+\frac{\tau}{2}\|\mathbf{q}(0)\|^2. \tag{6.174}$$

Use the arithmetic-geometric mean inequality on the $(\theta, q_{i,i})$ term in (6.171) and then for a constant $\beta>0$ we find

$$\frac{\tau}{2}\frac{d}{dt}\|\mathbf{q}\|^2\le-\|\mathbf{q}\|^2+\frac{\kappa}{2\beta}\|\theta\|^2+\Big(\frac{\kappa\beta}{2}-3\tilde{\tau}\Big)\|q_{i,i}\|^2 \\ -\tilde{\tau}\int_\Omega q_{i,j}(q_{i,j}-q_{j,i})dx. \tag{6.175}$$

Pick $\beta\ge 6\tilde{\tau}/\kappa$ and from (6.175) we see that

$$\frac{d}{dt}\|\mathbf{q}\|^2\le\frac{\kappa}{\tau\beta}\|\theta(t)\|^2\le\frac{\kappa}{\tau\beta}\|\theta(0)\|^2+\frac{1}{c\beta}\|\mathbf{q}(0)\|^2<\infty. \tag{6.176}$$

Thus we have shown $\|\mathbf{q}(t)\|^2\in L^1(0,\infty)$ and $d\|\mathbf{q}\|^2/dt\le K<\infty$ and so $\|\mathbf{q}(t)\|\to 0$ as $t\to\infty$.

For the GMC fluid system of equations with Fox's derivative, we have decay of q_i but we only have bounds for v_i and θ. It is easy to derive decay, continuous dependence and uniqueness results for the linearized systems (3.11), (3.12), (3.15), but such results for the fully nonlinear systems (6.160) or (6.161) would appear substantially more difficult.

For the Cattaneo - Christov equations given in section 3.1.2, or the Guyer - Krumhansl extension given in section 3.1.4, we are unaware of general qualitative results on the complete nonlinear systems.

6.9 Exercises

Exercise 6.9.1 *(See (Quintanilla, 2002d), and also (Ames and Straughan, 1997), p. 26, and (Russo, 1987).) Consider the following boundary - initial value problem for equation* (1.127) *for type III heat flow in a rigid solid,*

$$\begin{aligned}&\frac{\partial^2 u}{\partial t^2}+k_1\frac{\partial u}{\partial t}=k_2\Delta u+k_3\Delta\frac{\partial u}{\partial t}, && t>0,\mathbf{x}\in\Omega\\ &u(\mathbf{x},t)=g(\mathbf{x},t), \qquad t>0,\mathbf{x}\in\Gamma\\ &u(\mathbf{x},0)=h(\mathbf{x}),\quad \frac{\partial u}{\partial t}(\mathbf{x},0)=j(\mathbf{x}), && \mathbf{x}\in\Omega,\end{aligned} \tag{6.177}$$

where Ω *is a domain exterior to a bounded domain* $\Omega_0 \subset \mathbb{R}^3$, *and* Γ *is the boundary of* Ω_0.

By using a weighted energy method with a suitable weight show the solution u *is unique when* u *satisfies the following growth conditions*

$$|u_t|,\ |u_{,i}|,\ |u_{,it}| \le K\exp(\lambda r^2),$$

as $r \to \infty$. *(Note, this result is optimal in the sense that the growth condition cannot be increased to* $\exp(\lambda r^{2+\epsilon})$.*)*

Exercise 6.9.2 *Let* Ω *be a bounded domain with boundary* Γ *smooth enough to apply the divergence theorem. Consider the Green - Lindsay linearized equations of thermoelasticity, equations* (2.57), *with* $b_i = 0$ *and redefine the thermal conductivity term* k_{ij} *to be* k_{ij}/ρ. *(If* $\rho = \rho(\mathbf{x})$ *then the method outlined below needs to be trivially modified.) Then the displacement* u_i *and temperature* θ *satisfy the equations*

$$\begin{aligned} \rho\ddot{u}_i &= \rho F_i + (c_{ijkh}u_{k,h})_{,j} + \big[a_{ij}(\theta + \alpha\dot{\theta})\big]_{,j}, \\ h\ddot{\theta} + d\dot{\theta} - a_{ij}\dot{u}_{i,j} &= R + (k_{ij}\theta_{,j})_{,i} \end{aligned} \tag{6.178}$$

where $R = r/\theta_0$. *Consider equations* (6.178) *defined on* $\Omega \times \{t > 0\}$ *together with the boundary conditions*

$$u_i(\mathbf{x},t) = g_i(\mathbf{x},t), \quad \theta(\mathbf{x},t) = \theta_\Gamma(\mathbf{x},t), \qquad \mathbf{x} \in \Gamma,\ t > 0, \tag{6.179}$$

and the initial data

$$\begin{aligned} &u_i(\mathbf{x},0) = h_i(\mathbf{x}), \quad \dot{u}_i(\mathbf{x},0) = j_i(\mathbf{x}), \qquad \mathbf{x} \in \Omega, \\ &\theta(\mathbf{x},0) = \theta_0(\mathbf{x}), \quad \dot{\theta}(\mathbf{x},0) = \theta_1(\mathbf{x}), \qquad \mathbf{x} \in \Omega. \end{aligned} \tag{6.180}$$

Let the boundary - initial value problem comprised of equations (6.178) *-* (6.180) *be denoted by* $\mathcal{P}$. *The symmetries on the coefficients are as in* (2.58).

Prove uniqueness of a solution to $\mathcal{P}$ *with the assumptions*

$$\alpha d - h \ge 0,\ \alpha > 0,\ h > 0,\ k_{ij}\xi_i\xi_j \ge 0,\ c_{ijkh}\xi_{ij}\xi_{kh} \ge 0,$$

for all ξ_i, ξ_{ij}, *where the coefficients may depend on* $\mathbf{x}$.
Hint. Let (u_i^1, θ^1) *and* (u_i^2, θ^2) *be solutions to* $\mathcal{P}$ *for the same functions* F_i *and* R, *and for the same boundary data,* g_i, θ_Γ, *and initial data,* h_i, j_i, θ_0 *and* θ_1. *Show the difference solution* $u_i = u_i^1 - u_i^2, \theta = \theta^1 - \theta^2$, *satisfies the boundary - initial value problem,*

$$\begin{aligned} &\rho\ddot{u}_i = (c_{ijkh}u_{k,h})_{,j} + \big[a_{ij}(\theta + \alpha\dot{\theta})\big]_{,j}, \\ &h\ddot{\theta} + d\dot{\theta} - a_{ij}\dot{u}_{i,j} = (k_{ij}\theta_{,j})_{,i}, \qquad \text{in } \Omega \times \{t > 0\}, \\ &u_i(\mathbf{x},t) = 0, \quad \theta(\mathbf{x},t) = 0, \qquad \mathbf{x} \in \Gamma,\ t > 0, \\ &u_i(\mathbf{x},0) = 0, \quad \dot{u}_i(\mathbf{x},0) = 0, \qquad \mathbf{x} \in \Omega, \\ &\theta(\mathbf{x},0) = 0, \quad \dot{\theta}(\mathbf{x},0) = 0, \qquad \mathbf{x} \in \Omega. \end{aligned} \tag{6.181}$$

Multiply equation $(6.181)_1$ *by* u_i *and integrate over* Ω, *and multiply equation* $(6.181)_2$ *by* $\theta + \alpha\dot{\theta}$ *and integrate over* Ω. *Show that the energy* $E(t)$ *is conserved, i.e.*

$$E(t) = E(0) \qquad \forall t > 0,$$

where

$$\begin{aligned} E(t) =& \frac{1}{2}\int_\Omega \alpha h\Big(\dot{\theta} + \frac{1}{\alpha}\theta\Big)^2 dx + \frac{1}{2}\int_\Omega \Big(d - \frac{h}{\alpha}\Big)\theta^2 dx \\ &+ \frac{1}{2}\int_\Omega \alpha k_{ij}\theta_{,i}\theta_{,j} dx + \frac{1}{2}\int_\Omega \rho\dot{u}_i\dot{u}_i dx \\ &+ \frac{1}{2}\int_\Omega c_{ijkh}u_{i,j}u_{k,h} dx + \int_0^t\int_\Omega k_{ij}\theta_{,i}\theta_{,j}\, dx\, ds \\ &+ \int_0^t\int_\Omega (d\alpha - h)\dot{\theta}^2\, dx\, ds. \end{aligned}$$

Hence, deduce that a solution to $\mathcal{P}$ *is unique.*

Exercise 6.9.3 *(See (Straughan, 1974)). Consider the uniqueness question for the Green - Lindsay thermoelasticity problem as posed in exercise 6.9.2, but when the definiteness condition on the* c_{ijkh} *is removed. Instead assume*

$$\alpha d - h > 0, \;\; \alpha > 0, \;\; h > 0, \;\; k_{ij}\xi_i\xi_j \geq k_0\xi_i\xi_i\,,$$

for all ξ_i, *for a constant* $k_0 > 0$ *and* $c_{ijkh} = c_{khij}$ *(no definiteness).*

Show that the function

$$\begin{aligned} F(t) = \int_\Omega \rho u_i u_i\, dx + \alpha\int_\Omega k_{ij}\eta_{,i}\eta_{,j}\, dx \\ + \int_0^t\int_\Omega (\alpha m\theta^2 + k_{ij}\eta_{,i}\eta_{,j})\, dx\, ds \end{aligned} \tag{6.182}$$

is a logarithmically convex function of t, *where*

$$\eta(\mathbf{x}, t) = \int_0^t \theta(\mathbf{x}, s) ds, \qquad m = d - \frac{h}{\alpha}.$$

Hence, deduce the solution to $\mathcal{P}$ *is unique, where* $\mathcal{P}$ *is the boundary - initial value problem* (6.178) - (6.180) *but now with the restrictions of this question as stated above*

Hint. Let $\phi = \theta + \alpha\dot{\theta}$ *and show that equation* $(6.181)_2$ *is equivalent to*

$$\frac{h}{\alpha}\dot{\phi} + m\dot{\theta} - a_{ij}\dot{u}_{i,j} = (k_{ij}\theta_{,j})_{,i}\,. \tag{6.183}$$

Use the initial data and integrate equation (6.183) *in time to show*

$$\frac{h}{\alpha}\phi + m\theta - a_{ij}u_{i,j} = (k_{ij}\eta_{,j})_{,i}\,. \tag{6.184}$$

Multiply $(6.181)_1$ *by* u_i *and integrate over* Ω. *Multiply* (6.184) *by* ϕ *and integrate over* Ω. *Add the resulting equations and expand to see that*

$$\begin{aligned}\int_\Omega \rho u_i \ddot{u}_i \, dx &+ \int_\Omega k_{ij}\eta_{,j}\theta_{,i}\, dx \\ &+ \alpha \int_\Omega k_{ij}\eta_{,i}\ddot{\eta}_{,j}\, dx + \int_\Omega m\alpha\theta\dot{\theta}\, dx \\ &= -\int_\Omega m\theta^2 dx - \int_\Omega \frac{h}{\alpha}\phi^2 dx - \int_\Omega c_{ijkh}u_{i,j}u_{k,h}dx.\end{aligned} \tag{6.185}$$

Calculate $F''(t)$ *and use* (6.185) *to see that*

$$\begin{aligned}F'' =& 2\int_\Omega \rho\dot{u}_i\dot{u}_i\, dx + 2\alpha\int_\Omega k_{ij}\theta_{,j}\theta_{,i}\, dx \\ &- 2\int_\Omega m\theta^2 dx - 2\int_\Omega c_{ijkh}u_{i,j}u_{k,h}dx - 2\int_\Omega \frac{h}{\alpha}\phi^2\, dx.\end{aligned} \tag{6.186}$$

Next, multiply equation $(6.181)_1$ *by* $\dot{u}_i$ *and integrate over* Ω. *Add this to equation* (6.183) *multiplied by* ϕ *and integrated over* Ω. *Integrate the result from* 0 *to* t *in time and show that*

$$4E(t) + 4\int_0^t\int_\Omega k_{ij}\theta_{,i}\theta_{,j}\, dx\, ds + 4\int_0^t\int_\Omega m\alpha\dot{\theta}^2\, dx\, ds = 4E(0), \tag{6.187}$$

where $E(t)$ *is the energy function,*

$$\begin{aligned}E(t) =& \frac{1}{2}\bigg[\int_\Omega \frac{h}{\alpha}\phi^2 dx + \int_\Omega \alpha k_{ij}\theta_{,i}\theta_{,j}dx + \int_\Omega m\theta^2 dx \\ &+ \int_\Omega c_{ijkh}u_{i,j}u_{k,h}dx + \int_\Omega \rho\dot{u}_i\dot{u}_i dx\bigg].\end{aligned}$$

Next, use the initial data to see $E(0) = 0$ *and then substitute in* (6.186) *using* (6.187) *to derive*

$$\begin{aligned}F'' =& 4\int_\Omega \rho\dot{u}_i\dot{u}_i dx + 4\alpha\int_\Omega k_{ij}\theta_{,i}\theta_{,j}dx \\ &+ 4\int_0^t\int_\Omega k_{ij}\theta_{,i}\theta_{,j}\, dx\, ds + 4\int_0^t\int_\Omega m\alpha\dot{\theta}^2\, dx\, ds.\end{aligned} \tag{6.188}$$

Show by differentiating (6.182) *that*

$$\begin{aligned}F' =& 2\int_\Omega \rho u_i\dot{u}_i dx + 2\alpha\int_\Omega k_{ij}\eta_{,i}\theta_{,j}dx \\ &+ \int_\Omega m\alpha\theta^2 dx + \int_\Omega k_{ij}\eta_{,i}\eta_{,j}dx \\ =& 2\int_\Omega \rho u_i\dot{u}_i dx + 2\alpha\int_\Omega k_{ij}\eta_{,i}\theta_{,j}dx \\ &+ 2\int_0^t\int_\Omega m\alpha\theta\dot{\theta}\, dx\, ds + 2\int_0^t\int_\Omega k_{ij}\theta_{,i}\eta_{,j}\, dx\, ds.\end{aligned} \tag{6.189}$$

Now, form the expression $FF''-(F')^2$ *using* (6.182), (6.188) *and* (6.189) *and show*

$$FF'' - (F')^2 \geq 0 \tag{6.190}$$

by means of the Cauchy-Schwarz inequality.

Deduce that the solution to $\mathcal{P}$ *is unique from the inequality* (6.190).

Exercise 6.9.4 *Let* Ω *be a domain exterior to a bounded domain* $\Omega_0 \subset \mathbb{R}^3$ *with inner boundary* Γ. *Suppose* (u_i, θ) *satisfy the Green-Lindsay equations of thermoelasticity as in* (6.178) - (6.180) *but where* Ω *is now the exterior region.*

Suppose

$$\alpha d - h > 0,\ \alpha > 0,\ h > 0,\ \rho > 0,\ k_{ij}\xi_i\xi_j \geq k_0\xi_i\xi_i\,,$$
$$c_{ijkh}\xi_{ij}\xi_{kh} \geq a_0\xi_{ij}\xi_{kh}\,,$$

for positive constants k_0, a_0. *Let the corresponding boundary - initial value problem be denoted by* $\mathcal{P}$. *Show that a solution to* $\mathcal{P}$ *is unique employing the Graffi method of section 6.5.1 provided*

$$|u_{k,h}|,\ |\dot{u}_i|,\ |\theta|,\ |\dot{\theta}|,\ |\theta_{,i}|\ \leq\ Ke^{\lambda r}$$

for some $K, \lambda > 0$.
Hint. Multiply equation (6.181) *by* $\dot{u}_i$ *and integrate over* Ω_R *and multiply equation* (6.181) *by* $\phi = \theta + \alpha\dot{\theta}$ *and integrate over* Ω_R. *Show that*

$$\begin{aligned}
&\frac{d}{dt}\Big[\frac{1}{2}\int_{\Omega_R}\rho\dot{u}_i\dot{u}_i\,dx + \frac{1}{2}\int_{\Omega_R}c_{ijkh}u_{i,j}u_{k,h}\,dx + \frac{1}{2}\int_{\Omega_R}\alpha k_{ij}\theta_{,i}\theta_{,j}\,dx\\
&\quad + \frac{1}{2}\int_{\Omega_R}\frac{h}{\alpha}\phi^2\,dx + \frac{1}{2}\int_{\Omega_R}\Big(d - \frac{h}{\alpha}\Big)\theta^2\,dx\Big]\\
&\quad + \int_{\Omega_R}k_{ij}\theta_{,i}\theta_{,j}\,dx + \int_{\Omega_R}(d\alpha - h)\dot{\theta}^2\,dx\\
&= \int_{\Gamma_R}c_{ijkh}u_{k,h}\dot{u}_i n_j\,dS + \int_{\Gamma_R}a_{ij}\phi\dot{u}_i n_j\,dS + \int_{\Gamma_R}k_{ij}\theta_{,j}\phi n_i\,dS\\
&\leq a_1\int_{\Gamma_R}c_{ijkh}u_{i,j}u_{k,h}\,dS + a_2\int_{\Gamma_R}\rho\dot{u}_i\dot{u}_i\,dS + a_3\int_{\Gamma_R}\theta^2\,dS\\
&\quad + a_4\int_{\Gamma_R}\dot{\theta}^2\,dS + a_5\int_{\Gamma_R}k_{ij}\theta_{,i}\theta_{,j}\,dS
\end{aligned}$$

for suitable constants $a_1, \ldots, a_5$. *Integrate this inequality twice over a fixed time interval* $[0, \ell]$ *and select a suitable function* $F(R)$ *to use the Graffi method with.*

Exercise 6.9.5 *Let* Ω *be a domain exterior to a bounded domain* $\Omega_0 \subset \mathbb{R}^3$ *with inner boundary* Γ. *Suppose* (u_i, θ) *satisfy the Green-Lindsay equations of thermoelasticity as in* (6.178) - (6.180) *but where* Ω *is now the exterior region.*

Suppose

$$\alpha d - h > 0,\ \alpha > 0,\ h > 0,\ \rho > 0,\ k_{ij}\xi_i\xi_j \geq k_0\xi_i\xi_i\,,$$
$$c_{ijkh}\xi_{ij}\xi_{kh} \geq a_0\xi_{ij}\xi_{kh}\,.$$

Let the corresponding boundary - initial value problem be denoted by $\mathcal{P}$. Show that a solution to $\mathcal{P}$ is unique employing the weighted energy method of section 6.5.2 provided

$$|u_{k,h}|,\ |\dot{u}_i|,\ |\theta|,\ |\dot{\theta}|,\ |\theta_{,i}| \ \leq\ Ke^{\lambda r}$$

for some $K, \lambda > 0$.

Exercise 6.9.6 *Let Ω be the domain exterior to a bounded domain $\Omega_0 \subset \mathbb{R}^3$. Let Γ be the inner boundary of Ω. Use the Graffi method of section 6.5.1 to show that a solution to the linear equations of type III thermoelasticity, equations (2.83), is unique provided*

$$\rho > 0,\ c > 0,\ k_{ij}\xi_i\xi_j \geq k_0\xi_i\xi_i\,,$$
$$c_{ijkh}\xi_{ij}\xi_{kh} \geq c_0\xi_{ij}\xi_{kh}\,,\ b_{ij}\xi_i\xi_j \geq b_0\xi_i\xi_i$$

and

$$|u_{k,h}|,\ |\dot{u}_i|,\ |\theta|,\ |\eta_{,i}|,\ |\theta_{,i}| \ \leq\ ke^{\lambda r},$$

as $r \to \infty$, for $k, \lambda > 0$, where $\eta = \int_0^t \theta\, ds$.
Hint. Let the boundary - initial value problem for (2.83) on the exterior domain Ω be denoted by $\mathcal{P}$. Suppose $(u_i^1, \theta^1), (u_i^2, \theta^2)$ are two solutions to $\mathcal{P}$ for the same boundary and initial data. Define the difference solution $u_i = u_i^1 - u_i^2, \theta - \theta^1 - \theta^2$ and find the boundary - initial value problem for this solution. Define $\eta(\mathbf{x}, t) = \int_0^t \theta(\mathbf{x}, s) ds$ for the difference solution and show from (2.83) that η satisfies the equation

$$c\ddot{\eta} = -a_{ij}\dot{u}_{i,j} + (k_{ij}\eta_{,j})_{,i} + (b_{ij}\theta_{,j})_{,i}\,.$$

Show that

$$\frac{d}{dt}\left[\frac{1}{2}\int_{\Omega_R} \rho\dot{u}_i\dot{u}_i\, dx + \frac{1}{2}\int_{\Omega_R} c_{ijkh}u_{i,j}u_{k,h}\, dx\right]$$
$$= \int_{\Omega_R} a_{ij}\theta\dot{u}_{i,j}\, dx - \int_{\Gamma_R} a_{ij}\theta\dot{u}_i n_j\, dS + \int_{\Gamma_R} c_{ijkh}u_{k,h}\dot{u}_i n_j\, dS$$

and

$$\frac{d}{dt}\left[\frac{1}{2}\int_{\Omega_R} c\theta^2\, dx + \frac{1}{2}\int_{\Omega_R} k_{ij}\eta_{,i}\eta_{,j}\, dx\right] + \int_{\Omega_R} b_{ij}\theta_{,i}\theta_{,j}\, dx$$
$$= -\int_{\Omega_R} a_{ij}\theta\dot{u}_{i,j}\, dx + \int_{\Gamma_R} k_{ij}\eta_{,j}\theta n_i\, dS + \int_{\Gamma_R} b_{ij}\theta_{,j}\theta n_i\, dS.$$

Then proceed from these two equations.

Exercise 6.9.7 *Let Ω be the domain exterior to a bounded domain $\Omega_0 \subset \mathbb{R}^3$. Let Γ be the inner boundary of Ω. Use the weighted energy method of section 6.5.2 to show that a solution to the linear equations of type III thermoelasticity, equations* (2.83), *is unique provided*

$$\rho > 0, \ c > 0, \ k_{ij}\xi_i\xi_j \geq k_0\xi_i\xi_i \,,$$
$$c_{ijkh}\xi_{ij}\xi_{kh} \geq c_0\xi_{ij}\xi_{kh} \,, \ b_{ij}\xi_i\xi_j \geq b_0\xi_i\xi_i$$

and

$$|u_{k,h}|, \ |\dot{u}_i|, \ |\theta|, \ |\eta_{,i}|, \ |\theta_{,i}| \ \leq \ ke^{\lambda r},$$

as $r \longrightarrow \infty$, for $k, \lambda > 0$, where $\eta = \int_0^t \theta\, ds$.

Exercise 6.9.8 *(See (Ciarletta and Straughan, 2010)). For a compressible fluid the Cattaneo - Christov equations, see section 3.1.2, are*

$$\begin{aligned} &\rho c_p\Big(\frac{\partial T}{\partial t} + v_i\frac{\partial T}{\partial x_i}\Big) = -\frac{\partial Q_i}{\partial x_i}\,,\\ &\tau\Big(\frac{\partial Q_i}{\partial t} + v_j\frac{\partial Q_i}{\partial x_j} - Q_j\frac{\partial v_i}{\partial x_j} + \frac{\partial v_m}{\partial x_m}Q_i\Big) + Q_i = -\kappa\frac{\partial T}{\partial x_i}\,. \end{aligned} \tag{6.191}$$

Suppose equations (6.191) *are defined on a bounded domain Ω for $t > 0$, and the velocity* $\mathbf{v}$ *is given. Suppose the temperature T is prescribed on the boundary Γ of Ω and T and Q_i are prescribed at $t = 0$. Suppose further $v_in_i = 0$ on $\Gamma \times \{t > 0\}$. Consider the difference solution $\theta = T^1 - T^2, q_i = Q_i^1 - Q_i^2$, where (T^1, Q_i^1) and (T^2, Q_i^2) are solutions to* (6.191) *for the same boundary and initial data. By deriving a differential inequality for the function*

$$\frac{1}{2}\|\theta\|^2 + \frac{\tau}{2}\|\mathbf{q}\|^2 \,,$$

$\|\cdot\|$ being the norm on $L^2(\Omega)$, show that the solution is unique if $|v_{i,i}|$ and $|v_{i,j}|$ are bounded in $\bar{\Omega} \times [0, \mathcal{T}]$, for some number $\mathcal{T} < \infty$.

7
Spatial decay

The topic of how a solution to a problem in continuum mechanics decays in space, including those equations which involve second sound, has been one of immense interest over the last few years. The first articles to deal with spatial decay in thermoelasticity would appear to be those of (Chirita, 1995a; Chirita, 1995b), (Chirita, 1997) and (Horgan and Payne, 1997). It would appear that the first articles dealing with spatial decay in second sound theories were those of (Quintanilla, 1996), who derived estimates for a solution to a damped wave equation, (Payne and Song, 1996), who established spatial decay bounds for a generalized Maxwell-Cattaneo theory (Guyer-Krumhansl model), and (Chirita and Quintanilla, 1996), who obtained spatial decay for a suitable functional measure of a solution to the Green-Lindsay equations of thermoelasticity.

The area of spatial decay estimates is still at the forefront of research in elasticity, thermoelasticity, and thermoelastic theories admitting heat waves. In particular, with the advent of auxetic materials where Poisson's ratio may be negative there has been a surge of interest in spatial decay in elasticity requiring only strong ellipticity of the elastic coefficients as opposed to requiring positive definiteness. This aspect is discussed in section 7.5.

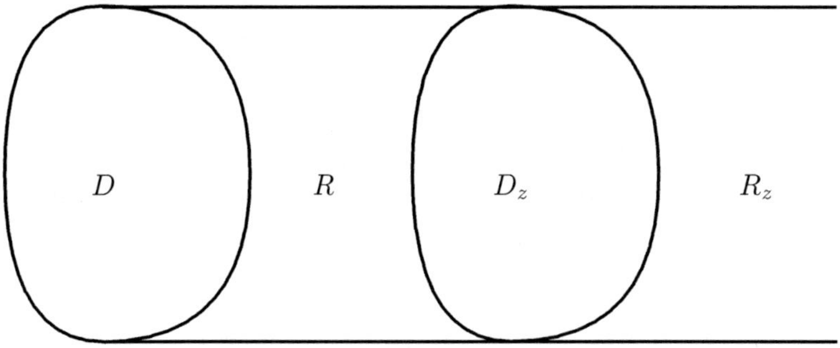

Figure 7.1. Spatial cylinder domain

7.1 Generalized Maxwell - Cattaneo theory

Spatial decay estimates for the generalized Maxwell - Cattaneo equations, see section 1.3, have been provided by (Payne and Song, 1996; Payne and Song, 2004a; Payne and Song, 2005) and by (Lin and Payne, 2004a). Here we review the results of (Payne and Song, 2005).

The basic equations are the GMC equations (or Guyer-Krumhansl equations) given in section 1.3, namely

$$\begin{aligned} c\frac{\partial T}{\partial t} &= -\frac{\partial q_i}{\partial x_i}, \\ \tau\frac{\partial q_i}{\partial t} &= -q_i - \kappa\frac{\partial T}{\partial x_i} + \mu\Delta q_i + \nu\frac{\partial^2 q_j}{\partial x_i \partial x_j}, \end{aligned} \tag{7.1}$$

where T, q_i are temperature and heat flux, c, τ, κ, μ and ν are positive constants.

Let D be a domain in $\mathbb{R}^2$, with boundary ∂D, and then we consider the semi-infinite cylinder $R \subset \mathbb{R}^3$ which is formed by the domain D running from $z = 0$ to $z = \infty$. The domain $D \times \{z\}$ we denote by D_z and R_z is the domain $D \times (z, \infty)$, as shown in figure 7.1

We denote the boundary of R by ∂R. Observe that ∂R is composed of $D(z = 0)$ together with the curved boundary of the cylinder which we denote by ∂R_c, and the limit boundary of D as $z \to \infty$.

The initial conditions of (Payne and Song, 2005) are

$$T(\mathbf{x}, 0) = 0, \quad q_i(\mathbf{x}, 0) = 0, \qquad \text{in } R \times \{t = 0\}, \tag{7.2}$$

whereas the boundary conditions may be of two types. Either

$$T(\mathbf{x}, t) = 0, \quad \epsilon_{ijk} q_j n_k = 0, \qquad \text{on } \partial R_c \times \{t > 0\}, \tag{7.3}$$

together with

$$\begin{aligned} T(x_1, x_2, 0, t) &= g(x_1, x_2, t), \qquad \text{on } D \times \{t > 0\}, \\ q_\alpha(x_1, x_2, 0, t) &= f_\alpha(x_1, x_2, t), \qquad \text{on } D \times \{t > 0\}, \end{aligned} \tag{7.4}$$

where n_i is the unit outward normal to ∂R_c, $\alpha = 1, 2$, and g and f_α are prescribed functions. Alternatively,

$$q_i(\mathbf{x}, t) = 0, \qquad \text{on } \partial R_c \times \{t > 0\}, \tag{7.5}$$

together with

$$q_i(x_1, x_2, 0, t) = f_i(x_1, x_2, t), \qquad \text{on } D \times \{t > 0\}, \tag{7.6}$$

where f_i, $i = 1, 2, 3$, are given functions.

7.1.1 Temperature spatial decay

To establish a spatial decay estimate for the temperature under the boundary conditions (7.3) and (7.4) (Payne and Song, 2005) remove q_i from (7.1) to find that T satisfies the equation

$$\Big(\frac{\tau}{\mu+\nu}\Big)\frac{\partial^2 T}{\partial t^2} + \frac{1}{(\mu+\nu)}\frac{\partial T}{\partial t} = r\Delta T + \Delta\frac{\partial T}{\partial t}, \quad \text{in } R \times \{t > 0\},$$

where $r = \kappa/[c(\mu+\nu)]$. They make use of a solution measure over the cross section D_z, namely,

$$E(z, t) = \int_0^t \int_{D_z} (T_s + rT)^2 dA\, ds, \tag{7.7}$$

where dA denotes the integration element on D_z. After some calculation they show

$$\frac{\partial^2 E}{\partial z^2} - \Big(\frac{\tau}{\mu+\nu}\Big)\frac{\partial E}{\partial t} \geq 2\lambda_1 E + 2a\int_0^t \int_{D_z} T_s(T_s + rT) dA\, ds, \tag{7.8}$$

where λ_1 is the first eigenvalue in the problem

$$\frac{\partial^2 w}{\partial x_\alpha \partial x_\alpha} + \lambda w = 0, \quad x_\alpha \in D, \qquad w = 0 \text{ on } \partial D,$$

and $a = (1 - r\tau)/(\mu+\nu)$. Regardless of the sign of a, they then show that one may deduce from (7.8),

$$\frac{\partial^2 E}{\partial z^2} - \Big(\frac{\tau}{\mu+\nu}\Big)\frac{\partial E}{\partial t} \geq 2\tilde{\lambda} E, \tag{7.9}$$

where $\tilde{\lambda} = \lambda_1$ for $a \geq 0$ and $\tilde{\lambda} = \lambda_1 - a$ when $a < 0$.

(Payne and Song, 2005) put $\beta = 2(\mu+\nu)\tilde{\lambda}/\tau$ and show that the function $P = Ee^{\beta t}$ satisfies from (7.9),

$$\frac{\partial^2 P}{\partial z^2} - k\frac{\partial P}{\partial t} \geq 0, \tag{7.10}$$

for $k = \tau/(\mu+\nu)$. They then compare P to the following solution $\chi(z,t)$ which solves the one-dimensional heat equation,

$$\chi(z,t) = \left(\frac{k}{t+t_0}\right) \exp\left(\frac{-kz^2}{4(t+t_0)}\right). \tag{7.11}$$

Provided M and t_0 are constants selected so that $P(0,t) \leq M\chi(0,t)$ (Payne and Song, 2005) show that

$$P(z,t) \leq M\chi(z,t), \quad z \geq 0, t > 0.$$

From this estimate they are able to deduce that

$$E(z,t) \leq E(0,t) \exp\left[\frac{-kz^2}{4(t+t_0)}\right], \tag{7.12}$$

where $E(0,t)$ is the data term

$$E(0,t) = \int_0^t ds \int_D (g_s + rg)^2 dA.$$

Inequality (7.12) is the spatial decay estimate for the temperature T.

7.1.2 Spatial decay of heat flux

To derive an estimate for the spatial decay of the heat flux under the boundary conditions (7.3) and (7.4), (Payne and Song, 2005) employ the weighted volume measure

$$F(z,t) = \int_0^t ds \int_{R_z} (\xi - z)^2 q_i q_i dA\, d\xi\,. \tag{7.13}$$

After much calculation (Payne and Song, 2005) show that for the constant $\Gamma = 4c^2\mu(2+\nu/\mu)^2$, F satisfies the inequality

$$\frac{\partial^2 F}{\partial z^2} - k\frac{\partial F}{\partial t} \geq -\Gamma E(0,t) \int_z^\infty \exp\left[\frac{-k\xi^2}{4(t+t_0)}\right] d\xi\,. \tag{7.14}$$

From this inequality (Payne and Song, 2005) use a comparison argument to show that

$$F(z,t) \leq \left[F(0,t) + 2\Gamma E(0,t) t \frac{(t+t_0)}{k^2 z}\right] \exp\left(\frac{-kz^2}{4(t+t_0)}\right). \tag{7.15}$$

Inequality (7.15) is not an *a priori* spatial decay estimate becuase $F(0,t)$ is not in terms of boundary data on the cylinder end D. In fact, bounding $F(0,t)$ in terms of known boundary data is not easy. Nevertheless (Payne and Song, 2005) show that

$$F(0,t) \leq K\sqrt{\pi}\sqrt{\frac{t+t_0}{k}}\, E(0,t) + 4\sqrt{2\mu}\, G(0,t)\,, \tag{7.16}$$

where $K = 8c^2[(\mu + \nu)^2 + (2\mu + \nu)^2]$ and $G(0,t)$ is an explicit function involving the data terms $f_\alpha, g, \partial g/\partial x_\alpha$ and $\partial g/\partial t$.

When estimate (7.16) is employed in (7.15) this yields an *a priori* spatial decay estimate for a functional of the heat flux q_i.

7.1.3 Spatial decay with heat flux prescribed

(Payne and Song, 2005) also derive spatial decay estimates for both the temperature field T and the heat flux q_i when the heat flux boundary conditions (7.5) and (7.6) are considered. These estimates are somewhat complicated and we refer to the original article for complete details. Nevertheless, we point out that the solution measures involved are

$$\phi(z,t) = \int_0^t (t-s)ds \int_{R_z} (\xi - z)^4 \Big(\frac{\partial T}{\partial s}\Big)^2 dA\, d\xi$$

and

$$\psi(z,t) = \int_0^t (t-s)ds \int_{R_z} (\xi - z)^2 q_i q_i\, dA\, d\xi\,.$$

7.2 MC theory backward in time

The Maxwell - Cattaneo equations involve temperature T and heat flux q_i, see section 1.2. If one eliminates the heat flux then the temperature satisfies a damped wave equation of form

$$\tau \frac{\partial^2 T}{\partial t^2} + \frac{\partial T}{\partial t} = \Delta T. \tag{7.17}$$

Spatial decay results for T were derived by (Quintanilla, 1996).

We here recollect some interesting results of (Ames and Payne, 1998) who dealt with the equation (7.17), but *backward in time*. The backward in time problem may be regarded as a forward in time problem by reversing time and then we study the equation

$$\tau \frac{\partial^2 T}{\partial t^2} - \frac{\partial T}{\partial t} = \Delta T. \tag{7.18}$$

This equation is here studied on the cylinder $R \times \{t > 0\}$, the region defined in section 7.1. For τ very small, equation (7.18) may be regarded as a regularization to the backward heat equation $T_t + \Delta T = 0$ which is well known to yield an improperly posed problem.

(Ames and Payne, 1998) actually deal with (7.18) in a cylinder R in $\mathbb{R}^N$. Here we restrict attention to $\mathbb{R}^3$. Thus, they consider equation (7.18) defined on the domain $R \times \{t > 0\}$ with the boundary and initial data

$$
\begin{aligned}
&T(\mathbf{x},t) = g_1(x_1,x_2,x_3,t) \qquad \text{on } \partial R_c \times \{t>0\},\\
&T(x_1,x_2,0,t) = g_0(x_1,x_2,t) \qquad \text{on } D \times \{t>0\},\\
&T(\mathbf{x},0) = f_0(\mathbf{x}), \quad T_t(\mathbf{x},0) = h_0(\mathbf{x}),
\end{aligned}
\tag{7.19}
$$

where $\mathbf{x} = (x_1,x_2,x_3) = (x_1,x_2,z)$.

In fact, (Ames and Payne, 1998) consider two solutions T_1 and T_2 which satisfy (7.18) and (7.19) for the same functions g_1, f_0 and h_0, but allow different g_0. Defining the function

$$w = T_1 - T_2,$$

this leads to w satisfying the equation and conditions

$$
\begin{aligned}
&\Delta w + \frac{\partial w}{\partial t} - \tau \frac{\partial^2 w}{\partial t^2} = 0, \qquad \text{in } R \times \{t>0\},\\
&w = 0, \qquad \text{on } \partial R_c \times \{t>0\},\\
&w(x_1,x_2,0,t) = g(x_1,x_2,t), \qquad \text{on } D \times \{t>0\},\\
&w(\mathbf{x},0) = 0, \qquad w_t(\mathbf{x},0) = 0.
\end{aligned}
\tag{7.20}
$$

The analysis of (Ames and Payne, 1998) is interesting and begins by showing the function $\Phi(z,t)$ given by

$$\Phi = \int_0^t e^{-\gamma\eta/\tau}\, d\eta \int_{D_z} w_{,z} w_{,\eta} dA$$

is non-positive for $\gamma > 2$. To do this they show Φ satisfies the following differential inequalities,

$$\Phi_{,z} \mp \tau^{1/2}\Phi_{,t} \geq \pm\sqrt{\frac{\gamma(\gamma-2)}{\tau}}\,\Phi, \tag{7.21}$$

where we refer to the upper (lower) signs as $(7.21)_1$ and $(7.21)_2$. They first integrate $(7.21)_1$ along characteristics of form

$$t + \sqrt{\tau}\, z = \text{constant},$$

to conclude $\Phi \leq 0$. They then integrate $(7.21)_2$ along characteristics of form

$$t - \sqrt{\tau}\, z = \text{constant},$$

to find that

$$-\Phi(z,t) \leq -\Phi(0,t_0)\exp\Big\{-\sqrt{\frac{\gamma(\gamma-2)}{\tau}}\, z\Big\}, \tag{7.22}$$

for a suitable number t_0.

Inequality (7.22) is a spatial decay estimate. However, it remains to be shown that $\Phi(0,t_0)$ can be bounded in terms of the data function g. In fact, this is tricky but (Ames and Payne, 1998) show that one may derive an inequality of form

$$\begin{aligned}
-\Phi(0,t) \le & \frac{2}{\tau}\int_R e^{-\gamma t/\tau} v_{,s}^2\,dx + \frac{2\tau}{\gamma}\int_0^t e^{-\gamma s/\tau}\,ds\int_R v_{,is}v_{,is}dx \\
& + \frac{2}{(\gamma-2)}\int_0^t e^{-\gamma s/\tau}\,ds\int_R \big[(\gamma-1)v_{,\eta} - \tau v_{,\eta\eta}\big]^2 dx.
\end{aligned} \tag{7.23}$$

The function v may be selected, one such choice being

$$v(x_1,x_2,x_3,t) = g(x_1,x_2,t)e^{-\delta z}$$

for $\delta > 0$ a constant to be chosen optimally.

In addition to the spatial decay bound (7.22) combined with (7.23), (Ames and Payne, 1998) also derive an either/or Saint-Venant result for a solution w and also relax the assumption of $w(\mathbf{x},0) = 0$, $w_t(\mathbf{x},0) = 0$, to allow non-zero initial data values.

7.3 Green-Lindsay thermoelasticity

The linear equations according to the theory of thermoelasticity derived by (Green and Lindsay, 1972), see section 2.2, may be written

$$\begin{aligned}
\rho\ddot{u}_i &= (c_{ijkh}u_{k,h})_{,j} + \big[a_{ij}(\theta + \alpha\dot{\theta})\big]_{,j}\,, \\
h\ddot{\theta} + d\dot{\theta} - a_{ij}\dot{u}_{i,j} - b_i\dot{\theta}_{,i} - (b_i\dot{\theta})_{,i} &= (k_{ij}\theta_{,j})_{,i}\,,
\end{aligned} \tag{7.24}$$

where u_i and θ are the displacement and temperature, respectively. We follow the analysis of (Payne and Song, 2002) who adopt the conditions given below on the coefficients. The symmetry conditions,

$$c_{ijkh} = c_{khij}\,, \quad a_{ij} = a_{ji}\,, \quad k_{ij} = k_{ji}\,,$$

and for all arbitrary tensors ξ_{ij} and vectors ξ_i, the bounds

$$0 < c_0\xi_{ij}\xi_{ij} \le c_{ijkh}\xi_{ij}\xi_{kh} \le c_1\xi_{ij}\xi_{ij},$$

$$0 < k_0\xi_i\xi_i \le k_{ij}\xi_i\xi_j,$$

and

$$\begin{aligned}
&0 < k_{33}(\mathbf{x}) \le k_1, \quad 0 < \rho_0 \le \rho(\mathbf{x}), \quad 0 \le h_0 \le h(\mathbf{x}), \\
&0 < \alpha_0 \le \alpha(\mathbf{x}) \le \alpha_1, \quad -d_0 \le d(\mathbf{x}),
\end{aligned}$$

for all $\mathbf{x} \in R$, R being a domain as defined in section 7.1, with $|a_{ij}|$, $|a_{ij,k}|$, $|\nabla\alpha|$ and $|\mathbf{b}|$ being bounded.

The domain R is as in section 7.1 with the notation there in use here. In fact, a Saint-Venant type of result for (7.24) was given by (Chirita and Quintanilla, 1996). These writers chose the body to have a centre of symmetry and so $b_i = 0$, and their domain R was bounded, but in some ways more general. We here report on the results derived by (Payne and Song, 2002) on solution bounds. In fact, the results of (Payne and Song, 2002) depend strongly on the boundary conditions imposed on the lateral boundary ∂R_c.

7.3.1 Dirichlet boundary conditions

The first conditions of (Payne and Song, 2002) involve initial conditions,

$$u_i(\mathbf{x},0)=0,\ \dot{u}_i(\mathbf{x},0)=0,\ \theta(\mathbf{x},0)=0,\ \dot{\theta}(\mathbf{x},0)=0,\quad \mathbf{x}\in R, \tag{7.25}$$

together with the boundary conditions on the lateral wall of the cylinder ∂R_c,

$$u_i=0,\quad \theta=0,\qquad \text{on } \partial R_c\times\{t>0\}, \tag{7.26}$$

and the conditions on the end of the cylinder, D,

$$u_i(x_1,x_2,0,t)=f_i(x_1,x_2,t),\qquad \theta(x_1,x_2,0,t)=g(x_1,x_2,t), \tag{7.27}$$

with f_i and g being prescribed functions. For each t, the functions u_i,θ and their first derivatives are assumed to decay uniformly in x_1 and x_2 as $x_3=z\to\infty$.

The solution measure of (Payne and Song, 2002) under the above initial and boundary conditions is of form

$$E(z,t)=\int_z^\infty\int_{D_\xi}\big[\rho\dot{u}_i\dot{u}_i+c_{ijkh}u_{i,j}u_{k,h}+\alpha(h\dot{\theta}^2+k_{ij}\theta_{,i}\theta_{,j})\big]dA\,d\xi.$$

After extensive computation and clever use of inequalities (Payne and Song, 2002) show that E satisfies the partial differential inequality, for constants K_1 and K_2,

$$\frac{\partial E}{\partial t}\le -K_1\frac{\partial E}{\partial z}+K_2E. \tag{7.28}$$

Inequality (7.28) is then rearranged after multiplication by e^{-K_2t} as

$$\frac{\partial}{\partial t}(e^{-K_2t}E)+K_1\frac{\partial}{\partial z}(e^{-K_2t}E)\le 0. \tag{7.29}$$

Inequality (7.29) is next integrated along the characteristic $z-z_0=K_1(t-t_0)$, so that for $z>z_0\ge 0$, $t>t_0\ge 0$,

$$E(z,t)\le e^{K_2(t-t_0)}E(z_0,t_0). \tag{7.30}$$

(Payne and Song, 2002) then show that if (z,t) lie on a line $z=K_1t+z_0$, for $z_0\ge 0$, $E(z,t)\equiv 0$, whereas for $z<K_1t$ inequality (7.30) is a bound for $E(z,t)$ provided $E(z_0,t_0)$ can be estimated. The estimation of $E(z_0,t_0)$ is non-trivial. Nevertheless, (Payne and Song, 2002) are able to derive an estimate for $E(0,t)$ in terms of data. This bound is very technical but has

the form

$$E(0,t) \leq \frac{Q}{1-\epsilon}\, e^{\delta t/(1-\epsilon)}$$

for Q, δ, ϵ completely computable in terms of data. Combined with inequality (7.30) this yields an *a priori* bound for the function $E(z,t)$ of form

$$E(z,t) \leq \frac{Q}{1-\epsilon}\, e^{\delta t/(1-\epsilon)}\, e^{K_2(t-t_0)}\,. \tag{7.31}$$

7.3.2 Neumann boundary conditions

(Payne and Song, 2002) also investigate the situation where the initial conditions are those of (7.25) but the lateral boundary conditions (7.26) and end conditions (7.27) are replaced by conditions on the stress tensor and heat flux vector. Then, define the stress tensor σ_{ij} and heat flux vector q_i by

$$\begin{aligned} \sigma_{ij} &= c_{ijkh}e_{kh} + a_{ij}(\theta + \alpha\dot{\theta}),\\ q_i &= -b_i\dot{\theta} - k_{ij}\theta_{,j}\,,\\ e_{kh} &= \frac{1}{2}(u_{k,h} + u_{h,k})\,, \end{aligned}$$

and (Payne and Song, 2002) replace the boundary conditions (7.26) and (7.27) by

$$\sigma_{ij}n_j = 0, \quad q_in_i = 0 \qquad \text{on } \partial R_c \times \{t > 0\}, \tag{7.32}$$

and

$$\begin{aligned} \sigma_{i3}(x_1,x_2,0,t) &= \tilde{f}_i(x_1,x_2,t),\\ q_3(x_1,x_2,0,t) &= \tilde{g}(x_1,x_2,t), \end{aligned} \tag{7.33}$$

with $\tilde{f}_i$ and $\tilde{g}$ being prescribed functions. The functions u_i, θ and their first derivatives are again required to decay uniformly in x_1 and x_2 as $z \to \infty$.

The solution measure adopted by (Payne and Song, 2002) for Neumann boundary conditions has form

$$\Phi(z,t) = \int_0^t \int_{R_z} \{\rho\dot{u}_i\dot{u}_i + c_{ijkh}u_{k,h}u_{i,j} + \alpha(h\dot{\theta}^2 + k_{ij}\theta_{,i}\theta_{,j})\}\, dA\, dz\, ds.$$

After much involved computation and bounding of terms (Payne and Song, 2002) show that for constants M_1, M_2 and M_3, the function Φ satisfies the partial differential inequality

$$\frac{\partial \Phi}{\partial t} \leq -M_1\frac{\partial \Phi}{\partial z} + (M_2 + M_3t)\Phi. \tag{7.34}$$

After multiplication by a suitable integrating factor this inequality is rewritten as

$$\frac{\partial}{\partial t}\Big[\Phi e^{-(M_2t+M_3t^2/2)}\Big] + M_1\frac{\partial}{\partial z}\Big[\Phi e^{-(M_2t+M_3t^2/2)}\Big] \le 0. \tag{7.35}$$

(Payne and Song, 2002) then integrate inequality (7.35) along the characteristic $z - z_0 = M_1(t - t_0)$ with $z > z_0 \ge 0$. They deduce that

$$\Phi(z,t) \le \Phi(z_0,t_0)\,\exp\Big[M_2(t-t_0) + M_3\Big(\frac{t^2-t_0^2}{2}\Big)\Big]$$

and conclude that if $z \ge M_1 t$ then $\Phi(z,t) \equiv 0$ whereas when $z < M_1 t$

$$\Phi(z,t) \le \Phi(0,t_0)\,\exp\Big[M_2(t-t_0) + M_3\Big(\frac{t^2-t_0^2}{2}\Big)\Big]. \tag{7.36}$$

Since $\Phi(0,t_0)$ is not directly in terms of data it is necessary to bound this term. This (Payne and Song, 2002) do by showing

$$\frac{\partial\Phi(0,t)}{\partial t} \le 2\tilde{Q}(t) + 2(R_5 + \kappa t)\Phi(0,t)$$

for data terms $\tilde{Q}, R_5$ and κ. Upon integration this yields the bound

$$\Phi(0,t) \le 2\int_0^t \tilde{Q}(s)\,\exp\big[2R_5(t-s) + \kappa(t^2-s^2)\big]\,ds. \tag{7.37}$$

Inequality (7.36) employed in conjunction with the bound (7.37) yields the desired explicit bound for the function $\Phi(z,t)$.

7.4 Type III thermoelasticity

We turn now to estimates of spatial decay type for the Green-Naghdi theories of thermoelasticity. For type II theory without dissipation, see section 2.3, interesting spatial decay bounds have been developed by (Nappa, 1998).

In this section we report a spatial decay result for thermoelasticity of type III, see section 2.4, produced by (Quintanilla, 2001a). The linear equations of type III thermoelasticity for an anisotropic body with a centre of symmetry are, cf. section 2.4,

$$\begin{aligned} \rho\ddot{u}_i &= (c_{ijkh}u_{k,h})_{,j} - (a_{ij}\theta)_{,j}\,, \\ c\ddot{\theta} &= -a_{ij}\ddot{u}_{i,j} + (b_{ij}\dot{\theta}_{,i})_{,j} + (k_{ij}\theta_{,i})_{,j} \end{aligned} \tag{7.38}$$

where u_i, θ are displacement and temperature, respectively. The terms c_{ijkh}, b_{ij} and k_{ij} are symmetric, c_{ijkh} being symmetric in the sense of the major symmetry in that $c_{ijkh} = c_{khij}$.

(Quintanilla, 2001a) considers equations (7.38) on the domain $R \times \{t > 0\}$ together with the initial conditions

$$u_i(\mathbf{x},0) = 0, \; u_{i,t}(\mathbf{x},0) = 0, \quad \theta(\mathbf{x},0) = 0, \; \theta_{,t}(\mathbf{x},0) = 0, \tag{7.39}$$

and the boundary conditions

$$u_i = 0, \; \theta = 0, \quad \text{on } \partial R_c \times \{t > 0\}. \tag{7.40}$$

The notation for R, D is as in section 7.1. (Quintanilla, 2001a) introduces two solution measures, namely $E_0(z,t)$ and $E_1(z,t)$ which are defined by

$$E_0(z,t) = -\int_0^t \int_{D_z} [\tau_{i1} u_{i,ss} + (k_{1j}\theta_{,j} + b_{1j}\dot{\theta}_{,j})\theta_s] dA\, ds \tag{7.41}$$

and

$$E_1(z,t) = \int_z^\infty E_0(\xi,t)\, d\xi. \tag{7.42}$$

(Quintanilla, 2001a) also requires the restriction

$$\lim_{z\to\infty} E_0(z,t) = 0.$$

In (7.41), τ_{ij} is the function given by $\tau_{ij} = c_{ijkh}\dot{u}_{k,h} - a_{ij}\dot{\theta}$.

The decay estimate of (Quintanilla, 2001a) proceeds by showing that the function E_1 satisfies the differential inequality

$$\frac{\partial E_1}{\partial t} \le -\beta_1 \frac{\partial E_1}{\partial z} + \beta_2 \frac{\partial^2 E_1}{\partial z^2}, \tag{7.43}$$

for constants β_1, β_2. Then for the constant $a = \beta_1/2\beta_2$ he shows the function $F_1(z,t) = e^{-az}E_1(z,t)$ satisfies the inequality

$$\frac{\partial F_1}{\partial t} + a^2\beta_2 F_1 \le \beta_2 \frac{\partial^2 F_1}{\partial z^2}. \tag{7.44}$$

The transformation $v(z,t) = \exp(a^2\beta_2 t)\, F_1(z,t)$ leads to v satisfying the inequality

$$\frac{\partial^2 v}{\partial z^2} - \frac{1}{\beta_2}\frac{\partial v}{\partial t} \ge 0, \tag{7.45}$$

for $z \ge 0, t \ge 0$, where $v(z,0) = 0$ and with $v(z,t) \to 0$ as $z \to \infty$. To bound v (Quintanilla, 2001a) appeals to a comparison result comparing v to the solution w to the differential equation

$$\frac{\partial^2 w}{\partial z^2} - \frac{1}{\beta_2}\frac{\partial w}{\partial t} = 0$$

with initial value $w(z,0) = 0$ and boundary values

$$w(0,t) = \exp(a^2\beta_2 t)\, E_1(0,t) = g(t) \qquad \text{and} \qquad w(z,t) \to 0 \quad \text{as } z \to \infty.$$

He gives this solution as

$$w(z,t) = \frac{z}{\sqrt{4\pi\beta_2}} \int_0^t \frac{g(s)}{(t-s)^{3/2}} \exp\left[\frac{-z^2}{4\beta_2(t-s)}\right] ds\,.$$

In this way (Quintanilla, 2001a) is able to bound v directly in terms of w and he deduces that

$$\begin{aligned} E_1(z,t) \leq & \frac{\exp(az - z^2/4t\beta_2)}{z} \sqrt{4t\beta_2}\, \exp(-a^2\beta_2 t) \\ & \times \sup_{0\leq s\leq t} \left[\exp(a^2\beta_2 s)\, E_1(0,s)\right]. \end{aligned} \tag{7.46}$$

Inequality (7.46) is the decay estimate for $E_1(z,t)$ for type III thermoelasticity. An explicit bound for $E_1(0,s)$ is not given in (Quintanilla, 2001a), although he gives references as to how such a bound may be derived.

Further spatial decay results in type III thermoelasticity may be found in (Quintanilla, 2010b).

7.5 Strong ellipticity in thermoelasticity

In recent years the use of auxetic materials, see (Lakes, 2008), has become increasingly important. These are typically foam like structures and for linear elasticity, the elasticity tensor does not define a positive-definite quadratic form. Very interesting stability results for an isotropic body with only strong ellipticity are given by (Xinchun and Lakes, 2007), see also (Lakes and Wojciechowski, 2008).

For the spatial decay problem (Chirita and Ciarletta, 2003a) have relaxed the condition of positive-definiteness of the elasticity tensor in isotropic thermoelasticity. (Chirita, 2007) has also derived spatial decay results in linearized anisotropic thermoelastodynamics requiring only strong ellipticity. Both of these articles were for the equations of classical thermoelasticity, i.e. without any second sound effects. Further very interesting articles dealing with strong ellipticity are those of (Chirita, 2006; Chirita, 2009), (Chirita et al., 2007), (Chirita and Danescu, 2008), (Chirita and Ciarletta, 2008; Chirita and Ciarletta, 2010b; Chirita and Ciarletta, 2010c), (Passarella and Zampoli, 2010), (Tibullo and Vaccaro, 2008), and (Chirita and Ghiba, 2010b; Chirita and Ghiba, 2010a).

A very interesting article of (Chirita and Ciarletta, 2006) derives beautiful spatial decay bounds for the equations of static linear elasticity requiring only strong ellipticity of the elastic coefficients. We here generalize their results to include temperature, albeit in a thermostatic configuration. Since we are in the realms of stationary thermoelasticity, our results apply, for example, to the stationary equations of Green and Naghdi thermoelasticity of type II or of type III.

The notation of R, D, D_z etc., is as in section 7.1. The basic equations of stationary thermoelasticity we consider are

$$\begin{aligned} &c_{ijkh}u_{k,hj} + a_{ij}\theta_{,j} = 0, \\ &k_{ij}\theta_{,ji} = 0. \end{aligned} \tag{7.47}$$

In these equations u_i, θ are displacement and temperature and the coefficients c_{ijkh}, a_{ij} and k_{ij} are assumed constant. (It is not difficult to allow them to depend on x_1 and x_2, but we follow (Chirita and Ciarletta, 2006) and maintain them constant.) These coefficients satisfy the symmetries

$$c_{ijkh} = c_{khij} = c_{jikh} \quad \text{and} \quad k_{ij} = k_{ji}\,. \tag{7.48}$$

We suppose c_{ijkh} are strongly elliptic, i.e.

$$c_{ijkh}\xi_i\xi_k\eta_j\eta_h > 0, \;\forall \xi_i, \eta_i \neq 0.$$

Also, a_{ij} are bounded and k_{ij} is positive-definite, so $k_{33} > 0$, and

$$k_{ij}\xi_i\xi_j \geq k_0\xi_i\xi_i\,, \quad k_0 > 0\,.$$

As noted by (Chirita and Ciarletta, 2006), if we consider an isotropic body then

$$c_{ijkh} = \lambda\delta_{ij}\delta_{kh} + \mu(\delta_{ik}\delta_{jh} + \delta_{ih}\delta_{jk}).$$

Positive-definiteness of the tensor c_{ijkh} requires

$$\mu > 0, \; 3\lambda + 2\mu > 0,$$

whereas strong ellipticity only needs

$$\mu > 0, \; \lambda + 2\mu > 0.$$

The boundary conditions considered are

$$u_i = 0, \; \theta = 0, \quad \text{on } \partial R_c \times \{t > 0\}, \tag{7.49}$$

$$u_i(x_1, x_2, 0) = g_i(x_1, x_2), \quad \theta(x_1, x_2, 0) = \theta_0(x_1, x_2), \qquad \text{on } D. \tag{7.50}$$

(Chirita and Ciarletta, 2006) note that strong ellipticity shows

$$c_{i3k3}\xi_i\xi_k > 0, \;\forall \xi_i \neq 0,$$

and

$$c_{i\alpha k\beta}\xi_i\xi_k\eta_\alpha\eta_\beta > 0, \;\forall \xi_i, \eta_\alpha \neq 0.$$

Throughout this section a repeated Roman index sums from 1 to 3 whereas a repeated Greek index sums from 1 to 2. Hence ξ_i stands for a vector in 3 - dimensions whereas η_α denotes a vector in 2 - dimensions (e.g. in a crosss section D_z). (Chirita and Ciarletta, 2006) also denote by k_m, k_M and $\hat{k}_m, \hat{k}_M$ the minimum and maximum eigenvalues of the tensors c_{i3k3}

and $c_{\alpha 3\beta 3}$, so that

$$k_m \xi_i \xi_i \le c_{i3k3} \xi_i \xi_k \le k_M \xi_i \xi_i \,, \tag{7.51}$$

$$\hat{k}_m \eta_\alpha \eta_\alpha \le c_{\alpha 3\beta 3} \eta_\alpha \eta_\beta \le \hat{k}_M \eta_\alpha \eta_\alpha \,. \tag{7.52}$$

To commence with a decay estimate we define the function F by

$$F(z) = \int_{D_z} k_{33} \theta^2 dA. \tag{7.53}$$

By differentiation

$$F''(z) = 2 \int_{D_z} k_{33} \theta \theta_{zz} dA + 2 \int_{D_z} k_{33} \theta_z^2 dA. \tag{7.54}$$

Employing equation $(7.47)_2$,

$$k_{33} \theta_{zz} = -k_{3\alpha} \theta_{,3\alpha} - k_{\alpha 3} \theta_{,3\alpha} - k_{\alpha\beta} \theta_{,\alpha\beta}$$

and so using this in (7.54) we find

$$\begin{aligned} F''(z) =& 2 \int_{D_z} k_{33} \theta_z^2 dA - 2 \int_{D_z} k_{3\alpha} \theta \theta_{,3\alpha} dA \\ & - 2 \int_{D_z} k_{\alpha 3} \theta \theta_{,3\alpha} dA - 2 \int_{D_z} k_{\alpha\beta} \theta \theta_{,\alpha\beta} dA \\ =& 2 \int_{D_z} k_{33} \theta_z^2 dA + 2 \int_{D_z} k_{3\alpha} \theta_{,\alpha} \theta_{,3} dA \\ & + 2 \int_{D_z} k_{\alpha 3} \theta_{,\alpha} \theta_{,3} dA + 2 \int_{D_z} k_{\alpha\beta} \theta_{,\alpha} \theta_{,\beta} dA, \end{aligned}$$

where in deriving the last line we have integrated by parts in x_α and used the boundary conditions on ∂D (on ∂R_c). Then,

$$F''(z) = 2 \int_{D_z} k_{ij} \theta_{,i} \theta_{,j} dA \ge 2k_0 \int_{D_z} \theta_{,i} \theta_{,i} dA \tag{7.55}$$

$$\ge 2k_0 \lambda_1 \int_{D_z} \theta^2 dA, \tag{7.56}$$

where the last line follows on using Poincaré's inequality on D_z, i.e. for functions $\phi = 0$ on ∂D_z,

$$\int_{D_z} \phi_{,\alpha} \phi_{,\alpha} dA \ge \lambda_1 \int_{D_z} \phi^2 dA.$$

To proceed we now follow (Chirita and Ciarletta, 2006) and look at particular classes of elastic coefficients c_{ijkh}.

7.5.1 *Monoclinic materials*

We now consider a class of monoclinic materials with $x_1 O x_2$ as a plane of elastic symmetry. As (Chirita and Ciarletta, 2006) observe this includes various systems in elasticity more general than isotropic. The class is those for which

$$c_{\alpha 333} = 0 \quad \text{and} \quad c_{3\alpha\beta\gamma} = 0. \tag{7.57}$$

Next, introduce the (Chirita and Ciarletta, 2006) measure $I(z)$ by

$$I(z) = \int_{D_z} (c_{\alpha 3\beta 3} u_\alpha u_\beta + \delta c_{3333} u_3^2) dA, \tag{7.58}$$

where $\delta > 0$ is a constant at our disposal. The following analysis simply follows that of (Chirita and Ciarletta, 2006), *mutatis mutandis*. By differentiation,

$$\begin{aligned} I'' =& 2\int_{D_z} (c_{\alpha 3\beta 3} u_{\alpha,3} u_{\beta,3} + \delta c_{3333} u_{3,3}^2) dA \\ &+ 2\int_{D_z} (c_{\alpha 3\beta 3} u_\alpha u_{\beta,33} + \delta c_{3333} u_3 u_{3,33}^2) dA. \end{aligned} \tag{7.59}$$

Now, from equations $(7.47)_1$,

$$c_{\alpha 3\beta 3} u_{\beta,33} = -c_{\alpha\beta\mu\lambda} u_{\mu,\lambda\beta} - c_{\alpha\beta 33} u_{3,3\beta} - c_{\alpha 33\lambda} u_{3,3\lambda} - a_{\alpha j} \theta_{,j} \,,$$

and

$$c_{3333} u_{3,33} = -c_{3\alpha 3\beta} u_{3,\beta\alpha} - c_{3\alpha 3\lambda} u_{\lambda,3\alpha} - c_{33\beta\alpha} u_{\beta,3\alpha} - a_{3j} \theta_{,j} \,.$$

We now substitute the last two equations in (7.59) and then integrate by parts in x_α and rearrange the c_{ijkh} terms as in (Chirita and Ciarletta, 2006) to find

$$I'' = I_1 + I_2 - 2\int_{D_z} a_{\alpha j} \theta_{,j} u_\alpha dA - 2\delta \int_{D_z} a_{3j} u_3 \theta_{,j} dA, \tag{7.60}$$

where I_1 and I_2 are given by

$$I_1 = 2\int_{D_z} (\delta c_{3333} u_{3,3}^2 + f_{\alpha\beta} u_{3,3} u_{\alpha,\beta} + c_{\alpha\beta\lambda\mu} u_{\alpha,\beta} u_{\lambda,\mu}) dA,$$

and

$$I_2 = 2\int_{D_z} (c_{\alpha 3\beta 3} u_{\alpha,3} u_{\beta,3} + \delta f_{\alpha\beta} u_{3,\alpha} u_{\beta,3} + \delta c_{\alpha 3\beta 3} u_{3,\alpha} u_{3,\beta}) dA,$$

where $f_{\alpha\beta}$ is given by

$$f_{\alpha\beta} = c_{3\alpha 3\beta} + c_{\alpha\beta 33} \,.$$

The idea is now to derive conditions such that

$$I_1 \geq c_1 \int_{D_z} u_{\alpha,\beta} u_{\alpha,\beta} dA \quad \text{and} \quad I_2 \geq c_2 \int_{D_z} u_{3,\alpha} u_{3,\alpha} dA,$$

for positive constants c_1 and c_2. To do this one employs the arithmetic-geomteric mean inequality on the middle terms of both I_1 and I_2. Thus, for positive constants μ and ϵ to be selected we have

$$\begin{aligned} 2f_{\alpha\beta}u_{3,3}u_{\alpha,\beta} \geq& - \mu u_{3,3}^2 - \frac{1}{\mu}(f_{\alpha\beta}u_{\alpha,\beta})^2 \\ \geq& - \mu u_{3,3}^2 - \frac{1}{\mu}(f_{\alpha\beta}f_{\alpha\beta})u_{\gamma,\zeta}u_{\gamma,\zeta}\,, \end{aligned} \tag{7.61}$$

where the Cauchy-Schwarz inequality has also been used, and

$$\begin{aligned} 2\delta f_{\alpha\beta}u_{3,\alpha}u_{\beta,3} \geq& - \frac{\delta}{\epsilon}f_{\alpha\beta}u_{\beta,3}f_{\alpha\zeta}u_{\zeta,3} - \epsilon\delta u_{3,\alpha}u_{3,\alpha} \\ \geq& - \frac{\delta}{\epsilon}f_{\mu\lambda}f_{\mu\lambda}u_{\alpha,3}u_{\alpha,3} - \epsilon\delta u_{3,\alpha}u_{3,\alpha}\,. \end{aligned} \tag{7.62}$$

We next need inequality (3.12) of (Chirita and Ciarletta, 2006), namely,

$$\int_{D_z} c_{\alpha\beta\lambda\mu}u_{\alpha,\beta}u_{\lambda,\mu}dA \geq \gamma_0 \int_{D_z} u_{\alpha,\beta}u_{\alpha,\beta}dA. \tag{7.63}$$

Upon using (7.61) and (7.63) in the expression for I_1, we obtain

$$I_1 \geq (2\delta c_{3333} - \mu)\int_{D_z} u_{3,3}^2 dA + \left(2\gamma_0 - \frac{f_{\mu\lambda}f_{\mu\lambda}}{\mu}\right)\int_{D_z} u_{\alpha,\beta}u_{\alpha,\beta}dA\,. \tag{7.64}$$

Similarly, we employ (7.62) in the expression for I_2, together with inequality (7.52), to deduce

$$I_2 \geq \left(2\hat{k}_m - \frac{\delta}{\epsilon}f_{\mu\lambda}f_{\mu\lambda}\right)\int_{D_z} u_{\alpha,3}u_{\alpha,3}dA + (2\hat{k}_m - \epsilon)\delta\int_{D_z} u_{3,\alpha}u_{3,\alpha}dA\,. \tag{7.65}$$

The coefficients of the integrals on the right of (7.64) and (7.65) are required to be positive. This means choosing δ so that

$$\delta > \frac{f_{\mu\lambda}f_{\mu\lambda}}{4c_{3333}\gamma_0} \qquad \text{and} \qquad \delta < \frac{4\hat{k}_m^2}{f_{\mu\lambda}f_{\mu\lambda}}\,.$$

This is possible, as (Chirita and Ciarletta, 2006) observe if we are in the class of strongly elliptic materials such that

$$f_{\alpha\beta}f_{\alpha\beta} < 4\hat{k}_m\sqrt{\gamma_0 c_{3333}}\,.$$

Next, we use (7.64) and (7.65) in equation (7.60) recalling the restrictions on δ to deduce that for constants $c_1 > 0, c_2 > 0$,

$$\begin{aligned} I'' \geq& c_1\int_{D_z} u_{\alpha,\beta}u_{\alpha,\beta}dA + c_2\int_{D_z} u_{3,\alpha}u_{3,\alpha}dA \\ &- 2\int_{D_z} a_{\alpha j}u_\alpha\theta_{,j}dA - 2\delta\int_{D_z} a_{3j}u_3\theta_{,j}dA. \end{aligned} \tag{7.66}$$

Let $|a| = \max_{i,j} |a_{ij}|$, then with the aid of the arithmetic-geometric mean inequality we derive from (7.66), for other constants ϵ and $\delta > 0$,

$$\begin{aligned} I'' \geq & c_1 \int_{D_z} u_{\alpha,\beta} u_{\alpha,\beta} dA + c_2 \int_{D_z} u_{3,\alpha} u_{3,\alpha} dA - \frac{a}{\epsilon} \int_{D_z} \theta_{,i}\theta_{,i} dA \\ & - \epsilon a \int_{D_z} u_\alpha u_\alpha dA - \delta\gamma a \int_{D_z} u_3^2 dA - \frac{\delta a}{\gamma} \int_{D_z} \theta_{,i}\theta_{,i} dA \\ \geq & \Big(\frac{c_1\lambda_1 - \epsilon a}{\hat{k}_M}\Big) \int_{D_z} c_{\alpha 3\beta 3} u_\alpha u_\beta dA + \Big(\frac{c_2\lambda_1 - \delta\gamma a}{\delta c_{3333}}\Big) \int_{D_z} \delta c_{3333} u_3^2 dA \\ & - \Big(\frac{a}{\epsilon} + \frac{\delta a}{\gamma}\Big) \int_{D_z} \theta_{,i}\theta_{,i} dA \end{aligned} \tag{7.67}$$

where λ_1 is the constant for Poincaré's inequality on D_z.

Next, we use (7.55) and for a constant $\omega > 0$ we form

$$\begin{aligned} I'' + \omega F'' \geq & \gamma_1 \int_{D_z} c_{\alpha 3\beta 3} u_\alpha u_\beta dA + \gamma_2 \delta \int_{D_z} c_{3333} u_3^2 dA \\ & + \gamma_3 \int_{D_z} \theta_{,i}\theta_{,i} dA \end{aligned} \tag{7.68}$$

where

$$\gamma_1 = \frac{c_1\lambda_1 - \epsilon a}{\hat{k}_M}, \qquad \gamma_2 = \frac{c_2\lambda_1 - \delta\gamma a}{\delta c_{3333}}, \qquad \gamma_3 = 2\omega k_0 - \frac{a}{\epsilon} - \frac{\delta a}{\gamma}.$$

We pick ϵ, γ small enough that

$$\frac{c_1\lambda_1}{a} > \epsilon \qquad \text{and} \qquad \frac{c_2\lambda_1}{\delta a} > \gamma$$

and then pick ω large enough that

$$\omega > \frac{a}{2k_0\epsilon} + \frac{\delta a}{2k_0\gamma}.$$

This ensures $\gamma_1, \gamma_2, \gamma_3 > 0$. Next, use the inequality

$$\int_{D_z} \theta_{,i}\theta_{,i} dA \geq \frac{2k_0\lambda_1}{k_{33}} \int_{D_z} k_{33}\theta^2 dA = \frac{2k_0\lambda_1}{k_{33}} F.$$

Put

$$\gamma_4^2 = \min\Big\{\gamma_1, \gamma_2, \frac{2k_0\lambda_1\gamma_3}{k_{33}\omega}\Big\}.$$

Then, from inequality (7.68) we find

$$I'' + \omega F'' \geq \gamma_4^2 (I + \omega F). \tag{7.69}$$

Let us define $H(z)$ by

$$H(z) = I + \omega F,$$

then after multiplying by an integrating factor, inequality (7.69) may be written as

$$\frac{d}{dz}\Big[(H' + \gamma_4 H)e^{-\gamma_4 z}\Big] \geq 0. \tag{7.70}$$

We require

$$\lim_{z\to\infty} H(z)e^{-\gamma_4 z} = 0 \qquad \text{and} \qquad \lim_{z\to\infty} H'(z)e^{-\gamma_4 z} = 0.$$

Then integrating (7.70) from z to ∞ we obtain

$$H' + \gamma_4 H \leq 0.$$

Upon integration from 0 to z this yields

$$H(z) \leq e^{-\gamma_4 z} H(0). \tag{7.71}$$

If we recollect the data (7.50) then inequality (7.71) yields the spatial decay estimate

$$\begin{aligned} &\int_{D_z} (c_{\alpha 3\beta 3} u_\alpha u_\beta + \delta c_{3333} u_3^2) dA + \omega \int_{D_z} k_{33}\theta^2 dA \\ &\leq e^{-\gamma_4 z}\left[\int_{D_z} (c_{\alpha 3\beta 3} g_\alpha g_\beta + \delta c_{3333} g_3^2) dA + \omega \int_{D_z} k_{33}\theta_0^2 dA\right]. \end{aligned} \tag{7.72}$$

(Chirita and Ciarletta, 2006) show how one may specialize the above result to the case of isotropic materials. They further analyse the decay rate for particular metals.

7.5.2 Triclinic materials

We now again follow (Chirita and Ciarletta, 2006) and consider a class of triclinic materials for which

$$c_{\alpha 333} \neq 0 \qquad \text{or} \qquad c_{\alpha\beta\gamma 3} \neq 0. \tag{7.73}$$

While (Chirita and Ciarletta, 2006) restrict attention to isothermal stationary elasticity we consider the equivalent thermal problem defined by equations (7.47).

Again, the analysis herein parallels the analysis of (Chirita and Ciarletta, 2006) for the isothermal problem. We start with the function $J(z)$ given by

$$J(z) = \int_{D_z} c_{i3k3} u_i u_k dA. \tag{7.74}$$

Upon differentiation,

$$J'' = 2\int_{D_z} c_{i3k3} u_{i,3} u_{k,3} dA + 2\int_{D_z} c_{i3k3} u_i u_{k,33} dA. \tag{7.75}$$

Now use the differential equation $(7.47)_1$ to see that

$$c_{i3k3}u_{k,33} = -c_{i\alpha k\beta}u_{k,\beta\alpha} - c_{i\alpha k3}u_{k,3\alpha} - c_{i3k\beta}u_{k,3\beta} - a_{ij}\theta_{,j}\,.$$

We use this expression in (7.75) and integrate by parts to find

$$\begin{aligned} J'' =& 2\int_{D_z} c_{i3k3}u_{i,3}u_{k,3}dA + 2\int_{D_z} c_{i\alpha k\beta}u_{k,\beta}u_{i,\alpha}dA \\ &+ 2\int_{D_z} g_{ik\alpha}u_{i,\alpha}u_{k,3}dA - 2\int_{D_z} a_{ij}u_i\theta_{,j}dA\,, \end{aligned} \tag{7.76}$$

where

$$g_{ik\alpha} = c_{i\alpha k3} + c_{i3k\alpha}\,.$$

We next use the inequalities

$$c_{i3k3}u_{i,3}u_{k,3} \geq k_m u_{i,3}u_{i,3}$$

and

$$\int_{D_z} c_{i\alpha k\beta}u_{k,\beta}u_{i,\alpha}dA \geq \gamma_0 \int_{D_z} u_{i,\alpha}u_{i,\alpha}dA$$

in (7.76) and further use the arithmetic-geometric mean inequality on the $g_{ik\alpha}$ and a_{ij} terms, for arbitrary $\mu, \epsilon > 0$. With $m^2 = g_{rk\beta}g_{rk\beta}$ we may derive from (7.76),

$$\begin{aligned} J'' \geq& (2k_m - \epsilon)\int_{D_z} u_{i,3}u_{i,3}dA + (2\gamma_0 - m^2\epsilon)\int_{D_z} u_{i,\alpha}u_{i,\alpha}dA \\ &- \frac{a}{\mu}\int_{D_z} u_iu_idA - a\mu\int_{D_z}\theta_{,i}\theta_{,i}dA\,. \end{aligned} \tag{7.77}$$

We require the coefficients of the first two terms on the right of (7.77) to be positive, so $\epsilon < 2k_m$ and $\epsilon < 2\gamma_0/m^2$, which is valid provided we restrict attention to the class of coefficients for which

$$4\gamma_0 k_m^2 > m^2.$$

Recall inequality (7.55), then form $J'' + \omega F''$ to find

$$\begin{aligned} J'' + \omega F'' \geq& (2k_m - \epsilon)\int_{D_z} u_{i,3}u_{i,3}dA + \delta_1\int_{D_z} u_iu_idA \\ &+ \delta_2\int_{D_z}\theta_{,i}\theta_{,i}dA\,, \end{aligned} \tag{7.78}$$

where

$$\delta_1 = (2\gamma_0 - m^2\epsilon)\lambda_1 - \frac{a}{\mu}\,, \qquad \delta_2 = 2\omega k_0 - a\mu,$$

and where also Poincaré's inequality has been employed. We now pick μ such that $\delta_1 > 0$ and then select ω so large that $\delta_2 > 0$. Hence, from inequality (7.78) we may see that

$$J'' + \omega F'' \geq \frac{\delta_1}{k_M} \int_{D_z} c_{i3k3} u_i u_k dA + \delta_2 \int_{D_z} \theta_{,i}\theta_{,i} dA\,. \tag{7.79}$$

With the aid of Poincaré's inequality the last term may be bounded below as

$$\int_{D_z} \theta_{,i}\theta_{,i} dA \geq \frac{\lambda_1}{\omega k_{33}}\,\omega \int_{D_z} k_{33}\theta^2 dA\,.$$

Then, setting

$$\delta_3^2 = \min\Big\{\frac{\delta_1}{k_M}, \frac{\delta_2\lambda_1}{\omega k_{33}}\Big\}$$

and putting $G = J + \omega F$, we obtain from (7.79)

$$G'' \geq \delta_3^2 G. \tag{7.80}$$

This inequality may be integrated as (7.69) to obtain

$$G(z) \leq e^{-\delta_3 z} G(0). \tag{7.81}$$

Upon insertion of the data terms in $G(0)$ into (7.81), this inequality yields a spatial decay estimate in the measure $G(z)$.

(Chirita and Ciarletta, 2006) also derive some estimates for cross-sectional measures involving the stress and displacement. They specifically analyse transversely isotropic materials under the strong ellipticity condition, and especially investigate how sharp their decay estimates may be. They also specifically apply their results to the material which constitutes a cortical bone. One may be able to include thermal effects in this case, but we do not do this here.

Other spatial decay estimates in continuum mechanics which illustrate the importance of this area overall have appeared frequently, for example, (Ames et al., 2001), (Ames et al., 1993), (Ames and Payne, 1998), (Amick, 1977; Amick, 1978), (Aouadi, 2009), (Chirita, 1995a; Chirita, 1995b; Chirita, 1997; Chirita, 2007), (Chirita and Ciarletta, 1999; Chirita and Ciarletta, 2003b; Chirita and Ciarletta, 2003a; Chirita and Ciarletta, 2006), (Chirita and Danescu, 2000), (Chirita and Quintanilla, 1996), (Chirita et al., 2001), (Fabrizio and Morro, 2003), pp. 366–373, (Flavin and Rionero, 1995), chapter 7, (Horgan and Knowles, 1983), (Horgan, 1989; Horgan, 1996), (Horgan and Payne, 1997),(Horgan and Quintanilla, 2005), (Ignaczak, 1998; Ignaczak, 2000; Ignaczak, 2002), (Iovane and Passarella, 2004a; Iovane and Passarella, 2004b), (Knops and Payne, 2005), (Lin and Lin, 2008), (Lin and Payne, 2004a; Lin and Payne, 2004b), (Mielke, 1992), (Nappa, 1998), (Payne and Song, 1996; Payne and Song, 1997b; Payne and Song, 1997a; Payne and Song, 2002; Payne and Song, 2004a; Payne and Song, 2005; Payne and Song, 2006; Payne and Song, 2007b; Payne and Song, 2007a; Payne and Song, 2008), (Quintanilla, 1996; Quintanilla, 1999; Quintanilla, 2001a).

8
Thermal convection in nanofluids

8.1 Heat transfer enhancement in nanofluids

Nanofluids consist of a suspension of very small metallic like particles suspended in a carrier fluid. Typically these fluids are manufactured by using a suspension of copper, Cu, copper oxide, CuO, or aluminium oxide, Al_2O_3, in water or ethylene glycol, cf. (Vadasz et al., 2005), (Kwak and Kim, 2005), (Wong and Kurma, 2008), or by creating a suspension of carbon nanotubes in an appropriate oil, cf. (Vadasz et al., 2005).

The use of nanofluids in heat transfer devices is very appealing and they appear to have highly desirable properties for greatly increasing heat transfer by comparison with ordinary fluids. It is well known that the thermal conductivities of metals like copper, or oxides such as CuO,Al_2O_3, are much greater than those of a typical carrier fluid. The resulting suspension is believed to have a greatly increased thermal conductivity due to the presence of metallic like particles, cf. (Kwak and Kim, 2005), (Hwang et al., 2007), (Masoumi et al., 2009), (Xuan et al., 2003), (Wong and Kurma, 2008), (Xuan and Roetzel, 2000), (Kim et al., 2007), (Putra et al., 2003), and this may have a pronounced effect on heat transfer.

The very interesting article of (Vadasz et al., 2005), questions the basis on which the increased thermal conductivity of a nanofluid suspension relies. They point out that experimental measurements on the thermal conductivity usually employ theory which is based on the classical law of Fourier heat conduction. However, they assert that key factors underlying thermal properties of nanofluids are ballistic as opposed to diffusive. They argue that thermal wave effects should be taken into account when interpreting experimental results. In particular, (Vadasz et al., 2005), propose six possible reasons for the increased effective thermal conductivity of a nanofluid and these may be summarized as,

a) hyperbolic or phase - lagging thermal wave effects;

b) thermal resonance because of the combination of hyperbolic thermal waves combined with an amplified periodic signal which may arise from a mobile phone or a short wavelength radio wave;

c) particle driven, or thermally driven, fluid convection;

d) convection due to electro-phoresis;

e) hyperbolic thermal convection;

f) any combination of a) - e).

(Vadasz et al., 2005) analyse in detail the experimental methods used to determine the effective thermal conductivity of a nanofluid suspension. They use a Cattaneo theory to analyse heat transfer in a slab and in this way are able to make a direct comparison with the experiments by employing both the Cattaneo and Fourier theories of heat flow. Their results are highly interesting and they conclude that they "*cannot confirm the validity of either one of the models as the correct one.*" However, they also deduce that, "*the apparent thermal conductivity evaluated via the Fourier conduction constitutive relationship could indeed produce results that show substantial apparent enhancement of the effective thermal conductivity of the nano-fluid suspension if the actual conduction process is governed by a hyperbolic thermal conduction process.*" They do conclude that further investigation is necessary before a definite conclusion may be reached as to why an anomalous thermal conductivity enhancement is observed.

In view of the very interesting results and conclusions of (Vadasz et al., 2005) we believe it is worth presenting some recent work on models to describe the behaviour of a nanofluid. We also review and extend recent work on hyperbolic thermal convection in a fluid or fluid-saturated porous medium, thereby directly incorporating possibility e) provided by (Vadasz et al., 2005).

(Vadasz, 2006) and (Buongiorno, 2006) investigate theroretically heat transfer in a nanofluid. (Tzou, 2008) also produces a model and investigates in some detail instability of thermal convection in his model. (Savino and Paterna, 2008) likewise produce and analyse a model which allows the fluid to be compressible. (Kuznetsov and Nield, 2010b; Kuznetsov and Nield, 2010a) develop a theory and analysis for thermal convection in a nanofluid which saturates a porous medium. The implication is likely that the pore size is considerably larger than the nanoparticle size.

8.2 The Tzou model

(Tzou, 2008) develops an interesting model to describe thermal convection in a nanofluid suspension. He uses an incompressible fluid with a Boussinesq approximation and his theory involves differential equations for the velocity in the suspension, v_i, the pressure, p, the temperature, T, and the concentration of nanoparticles, $\phi(\mathbf{x},t)$. In keeping with the style of this book we now describe the model of (Tzou, 2008) but use notation consistent with elsewhere in this volume. There is the equation of continuity of mass,

$$v_{i,i}=0, \tag{8.1}$$

the momentum balance equation,

$$\rho_0(v_{i,t}+v_jv_{i,j})=-p_{,i}+\mu\Delta v_i-\rho gk_i, \tag{8.2}$$

the energy balance law,

$$\rho_0c_F(T_{,t}+v_iT_{,i})=-q_{i,i}+h_SJ^S_{i,i}\,, \tag{8.3}$$

and an equation describing the conservation of nanoparticles,

$$\rho_S(\phi_{,t}+v_i\phi_{,i})=-J^S_{i,i}\,. \tag{8.4}$$

In these equations F or S refer to (bulk) fluid or (nanoparticle) solid components, ρ is the density at a point $\mathbf{x}$, ρ_0 is the constant fluid density at a reference temperature T_0, μ is the dynamic viscosity of the bulk fluid, g is gravity, $\mathbf{k}=(0,0,1)$, c_F or c_S denote the specific heat (at constant pressure) of the fluid or solid particles, q_i is the heat flux, h_S is the enthalpy of the solid, ρ_S is the solid density, and $\mathbf{J}^S$ is a flux vector which is associated to the nanoparticle density. In fact, (Tzou, 2008) assumes

$$J^S_i=-\rho_SD_B\phi_{,i}-\rho_S\frac{D_T}{T_B}T_{,i}\,, \tag{8.5}$$

where T_B is a constant (bulk fluid temperature) and Tzou treats D_B and D_T as constants. ((Tzou, 2008) does remark that D_B and D_T may depend on temperature and particle concentration and we return to this below.)

The theory of (Tzou, 2008) writes the heat flux as

$$q_i=-kT_{,i}+h_SJ^S_i\,, \tag{8.6}$$

where k is the thermal conductivity of the bulk fluid.

To simplify equation (8.2) (Tzou, 2008) adopts a Boussinesq approximation and writes

$$\begin{aligned}\rho=&\phi\rho_S+(1-\phi)\rho_F\\ =&\phi\rho_S+\rho_0(1-\phi)\big[1-\alpha(T-T_0)\big],\end{aligned} \tag{8.7}$$

where α is the thermal expansion coefficient of the fluid. He further writes

$$h_{S,i}=c_ST_{,i}\,, \tag{8.8}$$

cf. (Lighthill, 1963), p. 8, for a treatment of enthalpy in a fluid. Then the energy balance equation (8.3) may be rewritten with the aid of (8.6) as

$$\begin{aligned}\rho_0 c_F(T_{,t}+v_iT_{,i}) &= k\Delta T-(h_SJ_i^S)_{,i}+h_SJ_{i,i}^S\\&=k\Delta T-h_{S,i}J_i^S\\&=k\Delta T+h_{S,i}\rho_SD_B\phi_{,i}+h_{S,i}\rho_S\frac{D_B}{T_B}T_{,i}\\&=k\Delta T+\rho_Sc_SD_B\phi_{,i}T_{,i}+\frac{\rho_Sc_SD_T}{T_B}T_{,i}T_{,i}\,,\end{aligned}\tag{8.9}$$

where equation (8.8) has been employed.

Using the above relations one may rewrite equations (8.1) - (8.4) explicitly as

$$\begin{aligned}&v_{i,i}=0,\\&v_{i,t}+v_jv_{i,j}=-\frac{1}{\rho_0}p_{,i}+\frac{\mu}{\rho_0}\Delta v_i\\&\qquad\qquad -gk_i\Big\{\phi\frac{\rho_S}{\rho_0}+(1-\phi)\big(1-\alpha[T-T_0]\big)\Big\}\,,\\&T_{,t}+v_iT_{,i}=\kappa\Delta T+k_1\phi_{,i}T_{,i}+k_2T_{,i}T_{,i}\,,\\&\phi_{,t}+v_i\phi_{,i}=D_B\Delta\phi+\frac{D_T}{T_B}\Delta T,\end{aligned}\tag{8.10}$$

where the constants k_1 and k_2 have form

$$k_1=\frac{\rho_Sc_SD_B}{\rho_0c_F}\,,\qquad k_2=\frac{\rho_SD_Tc_S}{\rho_0c_FT_B}\,,$$

and κ is the thermal diffusivity of the fluid given by $\kappa=k/\rho_0c_F$.

(Tzou, 2008) remarks that the $D_B\Delta\phi$ term in equation $(8.10)_4$ represents diffusion of nanofluid particles due to Brownian motion whereas the $(D_T/T_B)\Delta T$ term is thermophoresis due to particle movement along a temperature gradient. It is worth remarking that the latter term is essentially a Soret effect. As (Tzou, 2008) remarks, the k_1 and k_2 terms in $(8.10)_3$ represent nonlinear contributions due to particle/temperature interactions. Let us observe that equations (8.10) are a coupled system of six nonlinear partial differential equations for the six variables v_i, p, T and ϕ.

Tzou non-dimensionalizes his equations and determines a basic solution for non-convective fluid motion in a layer $z\in(0,d),(x,y)\in\mathbb{R}^2$, with boundary conditions of no slip and the temperature and nanoparticle concentration prescribed on the boundaries $z=0$ and $z=d$. In particular, he has

$$\phi=\phi_L,\ T=T_L,\quad \text{at}\quad z=0;\qquad \phi=\phi_U,\ T=T_U\quad\text{at}\quad z=d,\tag{8.11}$$

for ϕ_L,T_L,ϕ_U,T_U constants. He seeks a basic solution for which $\bar{v}_i\equiv 0$, $\bar{\phi}=\bar{\phi}(z)$ and $\bar{T}=\bar{T}(z)$. Once these are known the pressure $\bar{p}$ follows from equation $(8.10)_2$. To determine the basic solution Tzou observes that equation $(8.10)_4$ integrates to yield

$$\bar{\phi} + \frac{D_T}{T_B D_B}\bar{T} = c_1 z + c_2 \tag{8.12}$$

where the constants c_1 and c_2 are determined by use of the boundary conditions. Inserting $\bar{\phi}(z)$ as given by (8.12) into equation $(8.10)_3$ yields the following equation for $\bar{T}$

$$\kappa\bar{T}_{zz} + k_1 c_1 \bar{T}_z = 0. \tag{8.13}$$

The basic temperature is found by integrating equation (8.13) twice (with an integrating factor) and this yields $\bar{T}(z)$ as a nonlinear function of z (involving z and exponentials of z). Once $\bar{T}$ is known, $\bar{\phi}$ follows from (8.12).

It is interesting to note that $\bar{T}, \bar{\phi}$ are nonlinear functions of z which is very different from the classical Bénard problem.

(Tzou, 2008) then introduces perturbations u_i, π, θ, ϕ to $\bar{v}_i, \bar{p}, \bar{T}, \bar{\phi}$ and derives *linearized* equations for the perturbation variables. His goal is to determine an instability threshold where thermal convection will begin. In fact, (Tzou, 2008) derives the following non-dimensional linearized system of equations for u_i, π, θ, ϕ,

$$\begin{aligned}
u_{i,i} &= 0,\\
u_{i,t} &= -\pi_{,i} + Pr\Delta u_i - k_i RaPr(\phi_U - 1)\theta\\
&\quad - (\phi_L - \phi_U)\big\{H[R_\rho - 1 + \alpha(T_U - T_L)]\big\}k_i\phi\\
&\quad + RaPrk_i(\bar{T}\phi + \bar{\phi}\theta),\\
\theta_{,t} &= u_3 + \Delta\theta - \frac{1}{Le}\phi_{,z} - \left(\frac{1+2R_N}{Le}\right)\theta_{,z}\,,\\
\phi_{,t} &= u_3 + N_{BT}\Delta\phi + N_{TT}\Delta\theta,
\end{aligned} \tag{8.14}$$

where Ra is a Rayleigh number and $Pr, R_N, R_\rho, H, Le, N_{BT}, N_{TT}$ are other non-dimensional parameters defined in (Tzou, 2008).

Tzou analyses equations (8.14) and finds an Ra, a^2 (wavenumber) boundary. However, he is interested in an overall qualitative behaviour and makes several approximations. His approach is standard and seeks solutions of the form

$$u_i = u_i(\mathbf{x})e^{\sigma t}, \quad \pi = \pi(\mathbf{x})e^{\sigma t}, \quad \theta = \theta(\mathbf{x})e^{\sigma t}, \quad \phi = \phi(\mathbf{x})e^{\sigma t},$$

where these represent Fourier modes in what is really an infinite Fourier series for each of u_i, π, θ and ϕ. (Tzou, 2008) looks for stationary convection, i.e. where $\sigma = 0$. A more complete analysis should allow σ to be complex. Oscillatory convection might well be the dominant mechanism. With odd derivatives, as are present in $(8.14)_3$, oscillatory convection cannot be ruled out. Indeed, in the next section, section 8.3, we find oscillatory convection is present for another theory pertaining to nanofluid behaviour. (Tzou, 2008) also allows $\phi_U, \phi_L \to 0$ so $Le \to \infty$ so that he can seek solutions like

$\Theta(z) = \sum_{m=1}^{\infty} A_m \sin m\pi z$. A more general numerical procedure would not require this. In this way Tzou uses a weighted residual method to find his Ra, a^2 boundary. Tzou also concentrates on the lowest mode $m = 1$. With a system as complicated as (8.14) care must be taken. Critical instability thresholds in other linearized hydrodynamic stability problems are found with $m > 1$, cf. (Webber, 2007; Webber, 2008), (Chen, 1993). Nevertheless, (Tzou, 2008) demonstrates that the nanofluid model leads to a dramatic lowering of the critical Rayleigh number threshold as compared to what one finds with a standard linearly viscous fluid when no nanoparticles are present. This does indicate convection occurs more easily in a nanofluid and heat transfer is, therefore, facilitated.

8.2.1 Coefficient dependence on nanoparticles

Even though Tzou treats the coefficients D_B and D_T in equation (8.5) as constants he does point out that they really have the forms

$$D_B = \alpha_1 T, \qquad D_T = \alpha_2 \phi, \tag{8.15}$$

where specific forms for the coefficients α_1, α_2 are given in (Tzou, 2008), equation (5). In the penultimate paragraph of (Tzou, 2008) he does write that, ... "*it remains worthwhile to reinstate the volume - fraction dependent thermal properties in the analysis and reexamine the drastic reductions of Ra_c obtained in this work.*"

If one adopts expressions (8.15) then from equation $(8.10)_3$ one finds $\bar{T}, \bar{\phi}$ satisfy the equation

$$\kappa \bar{T}_{zz} + \tilde{k}_1 \bar{T}\bar{T}_{,z}\bar{\phi}_{,z} + \tilde{k}_2 \bar{\phi}(\bar{T}_{,z})^2 = 0, \tag{8.16}$$

where

$$\tilde{k}_1 = \frac{\rho_s c_s \alpha_1}{\rho_0 c_F}, \qquad \tilde{k}_2 = \frac{\rho_s c_s \alpha_2}{\rho_0 c_F T_B}.$$

Let us observe that

$$\tilde{k}_1 = \tilde{k}_2 \frac{\alpha_1 T_B}{\alpha_2}. \tag{8.17}$$

On the other hand from equation $(8.10)_4$,

$$\alpha_1 \frac{d}{dz}\Big(\bar{T}\frac{d\bar{\phi}}{dz}\Big) + \frac{\alpha_2}{T_B}\frac{d}{dz}\Big(\bar{\phi}\frac{d\bar{T}}{dz}\Big) = 0. \tag{8.18}$$

If the normal component of the flux $\mathbf{J}^s$ is zero on the boundaries $z = 0, d$, i.e. $J_i^s n_i = 0$ there, then equation (8.18) integrates to find

$$\frac{\alpha_1 T_B}{\alpha_2} \bar{T}\bar{\phi}_{,z} + \bar{\phi}\bar{T}_{,z} = 0. \tag{8.19}$$

Thus,

$$\frac{\tilde{k}_2\alpha_1 T_B}{\alpha_2}\bar{T}\bar{\phi}_{,z}+\tilde{k}_2\bar{\phi}\bar{T}_{,z}=0$$

and using (8.17)

$$\tilde{k}_1\bar{T}\bar{\phi}_{,z}+\tilde{k}_2\bar{\phi}\bar{T}_{,z}=0. \tag{8.20}$$

Upon insertion of equation (8.20) into equation (8.16) we find

$$\kappa\bar{T}_{zz}=0$$

which leads to $\bar{T}(z)$ being a linear function of z, as in the standard Bénard problem. The function $\bar{\phi}(z)$ may then be found from (8.19). One may then use $\bar{T}$ and $\bar{\phi}$ and derive equations for u_i,π,θ,ϕ to perform a linearized instability analysis. By using a numerical method such as the Chebyshev tau one, cf. (Dongarra et al., 1996), one may then analyse whether oscillatory convection occurs by using the full, exact system of linear equations. It will also be possible to look for possible interchange of modes as parameters are varied to see if the lowest fundamental mode is indeed the one responsible for convective overturning, according to linear theory.

8.3 Convection with Cattaneo theories

The theories of viscous fluid motion coupled with heat transfer via a Cattaneo - like law were introduced in section 3.1. In view of that fact that (Vadasz et al., 2005) pointed out, see section 8.1, that hyperbolic thermal convection may be a mechanism which induces an effective increased thermal conductivity in a nanofluid, we deem it useful to include an account of such thermal convection using a linear viscous fluid coupled with suitable Cattaneo laws for heat transfer. We commence with the work which historically initiated this area.

8.3.1 Cattaneo - Fox law

The (Cattaneo, 1948) law was introduced in thermal convection in fluid mechanics by (Straughan and Franchi, 1984), and later work followed by (Lebon and Cloot, 1984), (Franchi and Straughan, 1994b) and (Dauby et al., 2002). These writers all used a Jaumann - like derivative to modify the rate of change of the heat flux.

(Straughan and Franchi, 1984) used an invariant form of derivative suggested by (Fox, 1969b) in the Cattaneo law and concentrated on two free surfaces. They showed oscillatory convection was possible for large enough Prandtl number provided the non-dimensional form of the relaxation time τ exceeded 0.0338. However, they did not investigate when stationary convection or oscillatory convection is preferred. (Straughan, 2009b) investigates thermal convection by employing the equations of (Straughan and Franchi, 1984), but he uses boundary conditions appropriate to two fixed surfaces. We now describe and elaborate this work.

Assume an incompressible Newtonian fluid is contained in the layer between the planes $z = 0, d$. The relevant equations for non-isothermal flow employing a Cattaneo - Fox law are given by (Straughan and Franchi, 1984) (see also section 3.1) as

$$\begin{aligned}
&\frac{\partial v_i}{\partial t} + v_j \frac{\partial v_i}{\partial x_j} = -\frac{1}{\rho}\frac{\partial p}{\partial x_i} + \nu \Delta v_i + g_i[1 - \alpha(T - T_R)],\\
&\frac{\partial v_i}{\partial x_i} = 0,\\
&\frac{\partial T}{\partial t} + v_i \frac{\partial T}{\partial x_i} = -\frac{\partial Q_i}{\partial x_i},\\
&\tau\Big(\frac{\partial Q_i}{\partial t} + v_j \frac{\partial Q_i}{\partial x_j} - \epsilon_{ijk}\omega_j Q_k\Big) = -Q_i - \kappa T_{,i}
\end{aligned} \tag{8.21}$$

where standard notation is employed. Here v_i, T, Q_i and p are the velocity, temperature, heat flux and pressure, $\boldsymbol{\omega} = \mathrm{curl}\,\mathbf{v}/2$, $\mathbf{g} = (0, 0, -g)$ the gravity vector, T_R is a reference temperature, and τ is a constant with the dimensions of time.

The boundary conditions are

$$v_i = 0,\ z = 0, d, \qquad T = T_L,\ z = 0, \quad T = T_U,\ z = d, \tag{8.22}$$

with $T_L > T_U$, both constants. We are interested in the instability of the conduction solution

$$\bar{v}_i = 0, \qquad \bar{T} = -\beta z + T_L, \qquad \bar{q}_i = (0, 0, \kappa\beta) \tag{8.23}$$

where β is the temperature gradient,

$$\beta = \frac{T_L - T_U}{d}.$$

Introduce perturbations (u_i, θ, π, q_i) to $(\bar{v}_i, \bar{T}, \bar{p}, \bar{q}_i)$ and then from equations (8.21) one derives the *linearized* perturbation equations as

$$\begin{aligned}
&\frac{\partial u_i}{\partial t} = -\frac{1}{\rho}\frac{\partial \pi}{\partial x_i} + \nu \Delta u_i + g\alpha\theta k_i,\\
&\frac{\partial u_i}{\partial x_i} = 0,\\
&\frac{\partial \theta}{\partial t} = \beta w - \frac{\partial q_i}{\partial x_i},\\
&\tau \frac{\partial q_i}{\partial t} = \frac{1}{2}\tau\kappa\beta\Big(\frac{\partial u_i}{\partial z} - \frac{\partial w}{\partial x_i}\Big) - q_i - \kappa\theta_{,i},
\end{aligned} \tag{8.24}$$

where $\mathbf{k} = (0, 0, 1)$, $w = u_3$. These equations are written in non-dimensional form using the length, time, pressure, heat flux, and temperature scales

$$d, \quad \frac{d^2}{\nu}, \quad \frac{\nu U}{d}, \quad \frac{\kappa T^\sharp}{d}, T^\sharp = U\sqrt{\frac{\beta\nu}{\alpha\kappa g}}.$$

The Rayleigh, Prandtl and Cattaneo numbers are written as (cf. (Straughan and Franchi, 1984), (Straughan, 2009b)),

$$Ra = R^2 = \frac{g\alpha\beta d^4}{\nu\kappa}, \qquad Pr = \frac{\nu}{\kappa}, \qquad C = \frac{\tau\kappa}{2d^2},$$

and in these papers an instability analysis is performed in terms of these quantities. (Papanicolaou et al., 2011) introduce another non-dimensional parameter, Sg, by

$$Sg = \frac{\tau\nu}{d^2}.$$

(Papanicolaou et al., 2011) argue that Sg is the more pertinent parameter when considering thermal convection in viscous fluids, and we employ this quantity here. Equations (8.24) are rewritten in the non-dimensional form

$$\begin{aligned} u_{i,t} &= -\pi_{,i} + R\theta k_i + \Delta u_i\,, \\ u_{i,i} &= 0, \\ Pr\theta_{,t} &= Rw - q_{i,i}\,, \\ Sg\,q_{i,t} &= \frac{SgR}{2Pr}(u_{i,z} - w_{,i}) - q_i - \theta_{,i}\,. \end{aligned} \tag{8.25}$$

To study instability and hence find a convection threshold we take curl curl of equation $(8.25)_1$ and retain the w component. Introducing the variable $\xi = q_{i,i}$ we then reduce system (8.25) to solving

$$\begin{aligned} \Delta^2 w + R\Delta^*\theta &= \sigma\Delta w, \\ Pr\sigma\theta &= Rw - \xi, \\ \sigma Sg\xi &= -\frac{SgR}{2Pr}\Delta w - \xi - \Delta\theta, \end{aligned} \tag{8.26}$$

where $\Delta^* = \partial^2/\partial x^2 + \partial^2\partial y^2$, and a time dependence like $e^{\sigma t}$ has been introduced. Supposing w, θ, ξ satisfy a plane tiling form $f(x, y)$ with $\Delta^* f = -a^2 f$, where a is a wavenumber, cf. (Chandrasekhar, 1981), p. 43-52, (Straughan, 2004), p. 51, then we solve equations (8.26) numerically in the form

$$\begin{aligned} \Delta W - \chi &= 0, \\ \Delta\chi - Ra^2\Theta &= \sigma\chi, \\ \Delta\Theta + \Xi + \frac{SgR}{2Pr}\chi &= -Sg\,\Xi\,\sigma, \\ RW - \Xi &= \sigma Pr\Theta, \end{aligned} \tag{8.27}$$

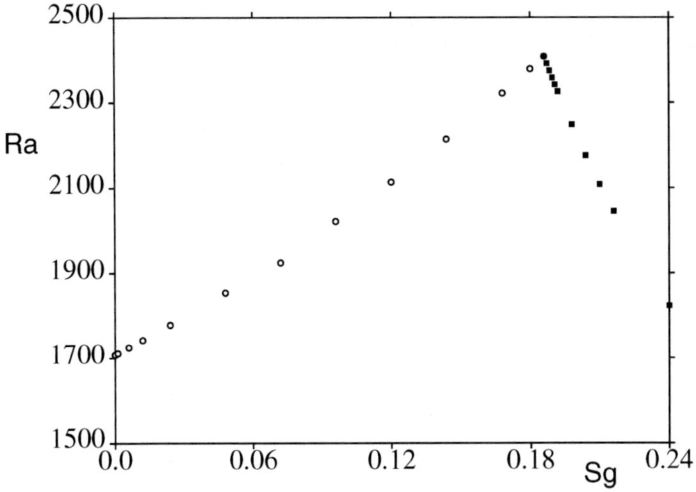

Figure 8.1. Critical values of Ra vs. Sg, Cattaneo-Fox model.

where W, Θ and Ξ are the $z-$parts of w, θ, ξ (e.g. $w = W(z)f(x, y)$), and χ is defined as ΔW. The boundary conditions for two fixed surfaces follow from equations (8.22) and take the form

$$W = DW = \Theta = 0, \qquad z = 0, 1, \tag{8.28}$$

where $D = d/dz$.

The Chebyshev tau D^2 numerical method is used to solve (8.27) subject to the boundary conditions (8.28) in (Straughan, 2009b).

(Straughan, 2009b), table 1, gives instability values with Prandtl number equal to 6. We present Ra and a^2 at criticality for values of Sg varying from 0 to 1.2 in table 8.1. We also include figures of the instability thresholds, see figure 8.1.

It is seen that the stationary convection behaviour witnessed asymptotically by (Straughan and Franchi, 1984) for small Sg, persists for two fixed surfaces. As Sg increases (Sg small) the critical Rayleigh number Ra likewise increases. However, at $Sg = Sg_T = 0.18602568$, we witness a striking transition. For $Sg > Sg_T$, a Hopf bifurcation occurs and convection switches from stationary convection to one where oscillatory convection is dominant. The critical Rayleigh number then begins to rapidly decrease as seen in figure 8.1. Also, the wave number increases and this means the transition is accompanied by a switch from a larger to a narrower convection cell, see table 8.1. Mathematically, the transition is manifest by the lowest critical Rayleigh number value switching from one eigenvalue $\sigma^{(1)}$ to another $\sigma^{(2)}$.

The dramatic reduction of Ra accompanied by the switch to a narrower convection cell is interesting. In view of the remarks of (Vadasz et al., 2005),

Table 8.1. Critical values of Ra and a against Sg.

Sg	a	Ra	σ_1
0	3.12	1707.765	0
1.2×10^{-3}	3.12	1711.180	0
6×10^{-3}	3.11	1724.935	0
1.2×10^{-2}	3.09	1742.393	0
2.4×10^{-2}	3.07	1778.194	0
4.8×10^{-2}	3.03	1853.544	0
7.2×10^{-2}	2.98	1934.276	0
9.6×10^{-2}	2.94	2020.868	0
0.12	2.89	2113.893	0
0.144	2.84	2213.969	0
0.168	2.80	2321.775	0
0.174	2.78	2350.027	0
0.18	2.77	2378.814	0
0.1812	2.77	2384.638	0
0.1824	2.77	2390.490	0
0.1836	2.76	2396.367	0
0.1848	2.76	2402.252	0
0.18602568	2.760	2408.291	0
0.18602568	4.994	2408.291	±3.932125
0.1872	4.99	2391.625	±3.9375
0.1884	4.99	2374.830	±3.9492
0.1896	4.99	2358.267	±3.9604
0.1908	4.99	2341.932	±3.9711
0.1920	4.99	2325.822	±3.9814
0.1980	4.99	2248.468	±4.0266
0.204	4.98	2176.066	±4.0484
0.21	4.98	2108.164	±4.0775
0.216	4.97	2044.356	±4.0866
0.24	4.96	1823.474	±4.1194
0.36	4.92	1183.489	±3.8889
0.48	4.90	876.008	±3.5737
0.6	4.89	695.403	±3.3032
1.2	4.87	342.568	±2.4797

see section 8.1, it will be interesting to observe if this effect is measurable in thermal convection in a nanofluid suspension.

8.3.2 Cattaneo - Christov law

In the previous section we employed a Cattaneo - Fox model. However, such models which use Jaumann derivatives for the heat flux can lead to instability when heating from above. (Christov, 2009) has recently proposed a Lie derivative form of invariant time derivative for the heat flux when dealing with a Cattaneo type theory for a fluid. This derivative has been employed in thermal convection studies by (Papanicolaou et al., 2009; Papanicolaou et al., 2011) and by (Straughan, 2010d; Straughan, 2010c). We now report on the work of (Straughan, 2010d) and investigate Christov's theory in the context of thermal convection of a layer of linearly viscous, incompressible fluid heated from below. We point out that the paper of (Papanicolaou et al., 2011) is an appealing one which studies thermal convection in a two-dimensional domain, solving the two-dimensional eigenvalue problem rather than by assuming a normal mode form in the x, y directions.

The basic equations are given in section 3.1.2, but we repeat them here for clarity. They consist of the balance of mass, balance of linear momentum, balance of energy, together with an equation relating the heat flux to the temperature gradient, namely the (Christov, 2009) reformulation of the (Cattaneo, 1948) equation. The system of equations is then

$$\frac{\partial v_i}{\partial t} + v_j \frac{\partial v_i}{\partial x_j} = -\frac{1}{\rho}\frac{\partial p}{\partial x_i} + \alpha g k_i T + \nu \Delta v_i \,, \tag{8.29}$$

$$\frac{\partial v_i}{\partial x_i} = 0, \tag{8.30}$$

$$\rho c_p \Big(\frac{\partial T}{\partial t} + v_i \frac{\partial T}{\partial x_i} \Big) = -\frac{\partial Q_i}{\partial x_i} \,, \tag{8.31}$$

$$\tau \Big(\frac{\partial Q_i}{\partial t} + v_j \frac{\partial Q_i}{\partial x_j} - Q_j \frac{\partial v_i}{\partial x_j} \Big) = -Q_i - \kappa \frac{\partial T}{\partial x_i} \,, \tag{8.32}$$

in which v_i, p, T, Q_i are the velocity, pressure, temperature and heat flux. The quantity ρ is the (constant) density, ν is the kinematic viscosity, g is the gravity, $\mathbf{k} = (0, 0, 1)$, α is the thermal expansion coefficient, and c_p is the specific heat at constant pressure. The Boussinesq approximation has been employed in the body force term. In Christov's equation (8.32), κ is the thermal conductivity and τ is a (constant) relaxation coefficient.

Since we are studying thermal convection the fluid occupies the horizontal layer $(x, y) \in \mathbb{R}^2$, $z \in (0, d)$ and equations (8.29) - (8.32) hold in the domain $\mathbb{R}^2 \times (0, d) \times \{t > 0\}$. The relevant boundary conditions are those of no-slip and temperatures prescribed, so

$$\begin{aligned} &v_i = 0 \qquad \text{on} \quad z = 0, d, \\ &T = T_L, \quad z = 0, \qquad T = T_U, \quad z = d, \end{aligned} \tag{8.33}$$

where T_L, T_U are constants with $T_L > T_U$.

The conduction (steady) solution is

$$\bar{v}_i \equiv 0, \qquad \bar{T} = -\beta z + T_L\,, \qquad \bar{\mathbf{Q}} = (0, 0, \kappa\beta), \tag{8.34}$$

with β being the temperature gradient, $\beta = (T_L - T_U)/d$.

To determine a threshold for thermal convection we investigate the instability of the steady solution (8.34) and so introduce perturbations (u_i, θ, π, q_i) such that $v_i = \bar{v}_i + u_i$, $T = \bar{T} + \theta$, $p = \bar{p} + \pi$, $Q_i = \bar{Q}_i + q_i$. Equations for (u_i, θ, π, q_i) are derived from (8.29) - (8.32) and are then non-dimensionalized with the length, time and velocity scales $L, \mathcal{T}$ and U given by $L = d$, $\mathcal{T} = d^2/\nu$, $U = \nu/d$. The temperature, pressure, and heat flux scales $T^\sharp, P, Q^*$ are

$$T^\sharp = U\sqrt{\frac{\beta\nu}{\alpha g k}}\,, \qquad P = \frac{\rho\nu U}{d}\,, \qquad Q^* = \frac{kT^\sharp}{d}\,,$$

and we put $k = \kappa/\rho c_p$. The Prandtl number, Pr, the Rayleigh number, $Ra = R^2$, and the non-dimensional number Sg are similar to those in section 8.3.1, namely

$$Pr = \frac{\rho\nu U}{d}\,, \qquad Sg = \frac{\tau\nu}{d^2}\,, \qquad R = \sqrt{\frac{\alpha g d^4\beta}{\nu k}}\,.$$

Then, the fully nonlinear, non-dimensional equations for (u_i, θ, π, q_i) are, cf. (Straughan, 2010d)

$$\begin{aligned}
&u_{i,t} + u_j u_{i,j} = -\pi_{,i} + Rk_i\theta + \Delta u_i,\\
&u_{i,i} = 0,\\
&Pr(\theta_t + u_i\theta_{,i}) = Rw - q_{i,i},\\
&Sg(q_{i,t} + u_j q_{i,j} - q_j u_{i,j}) = -q_i + \frac{SgR}{Pr}u_{i,z} - \theta_{,i}\,.
\end{aligned} \tag{8.35}$$

An analysis of the instability of the steady solution (8.34) discards the nonlinear terms in equations (8.35). Then an exponential time dependence is proposed, i.e.

$$\begin{aligned}
u_i(\mathbf{x}, t) &= e^{\sigma t}u_i(\mathbf{x}), \quad \theta(\mathbf{x}, t) = e^{\sigma t}\theta(\mathbf{x}),\\
q_i(\mathbf{x}, t) &= e^{\sigma t}q_i(\mathbf{x}), \quad \pi(\mathbf{x}, t) = e^{\sigma t}\pi(\mathbf{x}).
\end{aligned}$$

This leads to the linearized equations

$$\begin{aligned}
&\sigma u_i = -\pi_{,i} + Rk_i\theta + \Delta u_i,\\
&u_{i,i} = 0,\\
&\sigma Pr\theta = Rw - q_{i,i},\\
&\sigma Sg q_i = -q_i + \frac{SgR}{Pr}u_{i,z} - \theta_{,i}\,.
\end{aligned} \tag{8.36}$$

Next eliminate the pressure and then define $Q = q_{i,i}$ so that from (8.36) one obtains the equations

$$
\begin{aligned}
\sigma\Delta w &= R\Delta^*\theta + \Delta^2 w \\
\sigma Pr\theta &= Rw - Q \\
\sigma SgQ &= -Q - \Delta\theta ,
\end{aligned} \tag{8.37}
$$

with $\Delta^* = \partial^2/\partial x^2 + \partial^2/\partial y^2$ being the horizontal Laplacian.

For stationary convection, $\sigma = 0$, and then one derives from equations (8.37) the single equation for w,

$$\Delta^3 w = R^2\Delta^* w. \tag{8.38}$$

For two free surfaces this yields the classical instability threshold

$$Ra = R^2 = \frac{27\pi^4}{4}, \qquad a_c^2 = \frac{\pi^2}{2},$$

whereas for two fixed surfaces

$$Ra = R^2 = 1707.762, \qquad a_c = 3.117,$$

cf. Chandrasekhar (Chandrasekhar, 1981).

(Straughan, 2010d) solves equations (8.37) numerically for the case of two rigid surfaces $z = 0, d$, without assuming $\sigma \in \mathbb{R}$. In this way he investigates the Cattaneo effect upon oscillatory convection. He does, however, also analyse the case of two free surfaces. The solutions are supposed to have a spatial dependence in (x, y) commensurate with a plane tiling periodicity, cf. Chandrasekhar (Chandrasekhar, 1981), Straughan (Straughan, 2004), p. 51, so that on $z = 0, 1$,

$$w = \theta = w_{zz} = 0.$$

The Laplace operator is equivalent to $\Delta = D^2 - a^2$, where $D = \partial/\partial z$ and a is a wavenumber arising from a spatial dependence like $\Delta^* f = -a^2 f$, cf. Straughan (Straughan, 2004), p. 51.

(Straughan, 2010d) follows Chandrasekhar's method and puts $\sigma = i\sigma_1$ and eliminates Q and θ to find a single equation for W. By taking $W(z) = \sin n\pi z$, he shows that the critical wavenumber for oscillatory convection is

$$a_c^2 = \pi^2\Big(1 + \frac{A_1}{Sg}\Big)^{1/2}, \tag{8.39}$$

where $A_1 = (1 + Pr)/\pi^2 Pr$, while the corresponding critical value of R^2 for oscillatory convection is

$$R^2 = \frac{[SgPr(\pi^2 + a_c^2) + 1 + Pr](\pi^2 + a_c^2)}{Sg^2 a_c^2}. \tag{8.40}$$

(Straughan, 2010d) observes that for fixed Prandtl number, as $Sg \to 0$,

$$a_c^2 \sim \pi^2\sqrt{\frac{A_1}{Sg}}.$$

Likewise, as $Sg \to 0$,

$$R^2 \sim \frac{(1+Pr)}{Sg^2}.$$

Thus, for Sg small R^2 is very large and one finds stationary convection is the dominant mechanism with $Ra = 27\pi^4/4$. However, equation (8.40) may be rearranged as

$$R^2 = \frac{Pr\pi^2\Big[Sg\Big\{1+\Big(1+\frac{A_1}{Sg}\Big)^{1/2}\Big\}+A_1\Big]\Big[1+\Big(1+\frac{A_1}{Sg}\Big)^{1/2}\Big]}{Sg^2\Big(1+\frac{A_1}{Sg}\Big)^{1/2}}.$$

One then sees that as Sg increases one finds eventually R^2 is less than $27\pi^4/4$ and a_c^2 jumps from $\pi^2/2$ to the value given by (8.39), an a^2 value such that $a^2 \geq \pi^2$. (Straughan, 2010d) concludes that there is a transition value of $Sg = Sg_T$ such that once Sg exceeds this threshold stationary convection is not the observed mechanism and oscillatory convection prevails. We here compute $Sg_T = 0.2669184$ for the two fixed surface situation, with Pr=6.

(Straughan, 2010d) solves equations (8.37) numerically by a D^2 Chebyshev tau method. Further output is computed here.

Figure 8.2 shows critical values for the solution of equations (8.37) when two fixed surfaces are employed. The eigenvalue σ is written as $\sigma = \sigma_r + i\sigma_1$, and all values are critical values for instability, i.e. they represent

$$\min_{a^2} R^2(a^2) \qquad \text{when} \qquad \sigma_r = 0,$$

i.e. the linear instability threshold. The Prandtl number has value 6.

From figure 8.2 one observes that for values of Sg below a transition value $Sg_T = 0.2669184$, stationary convection is the mechanism by which thermal convection starts. The wavenumber $a = 3.12$ in this region, as seen in table 8.2. Once Sg increases beyond Sg_T there is a bifurcation and the dominant eigenvalue changes. Convection is then by oscillatory convection, $\sigma_1 \neq 0$, with a different, and larger wavenumber, as seen in figure 8.2 and table 8.2. (Straughan, 2010d) notes that this implies that the convection cells become narrower. As Sg increases further the convection cells continue to become narrower and the Rayleigh number decreases.

(Straughan, 2010d) concludes that the (Christov, 2009) model coupled with the Cattaneo one (see section 3.1.2) leads to a very interesting effect in thermal convection. For very small Sg convection is by stationary convection only and the convection cells have a fixed aspect ratio. As Sg increases a threshold is reached and convection then switches to oscillatory convection (Hopf bifurcation) with narrower cells. Further increase in Sg leads to further narrowing of the convection cells and lowering of the critical Rayleigh number which means thermal convection occurs more easily.

Table 8.2. Critical values of Ra and a against Sg.

Sg	a	Ra	σ_1
$0 - 0.2669184$	3.12	1707.765	0
0.2669184	4.874	1707.765	± 2.306
0.2676	4.872	1703.071	± 2.309
0.2688	4.872	1694.871	± 2.319
0.2700	4.872	1686.750	± 2.329
0.2712	4.872	1678.706	± 2.338
0.2724	4.872	1670.738	± 2.347
0.2736	4.870	1662.845	± 2.353
0.2748	4.870	1655.025	± 2.362
0.276	4.870	1647.279	± 2.371
0.288	4.866	1573.612	± 2.439
0.300	4.862	1506.229	± 2.492
0.312	4.860	1444.362	± 2.534
0.324	4.856	1387.362	± 2.564
0.336	4.854	1334.679	± 2.587
0.348	4.850	1285.840	± 2.603
0.360	4.848	1240.441	± 2.615

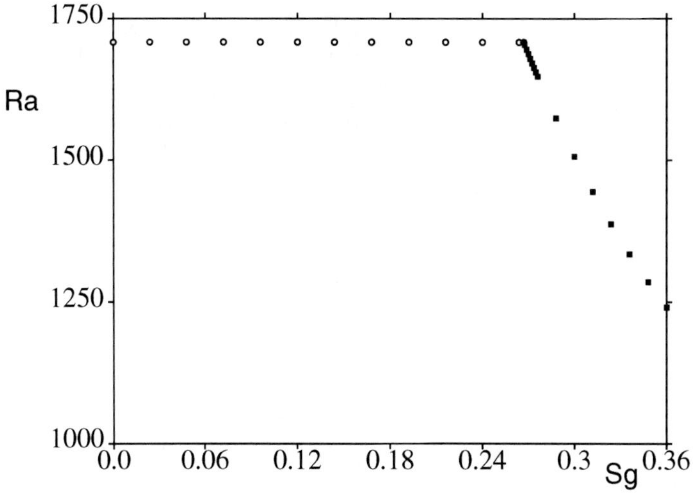

Figure 8.2. Critical values of Ra vs. Sg, Cattaneo-Christov model.

He deduces that the properly invariant heat flux law of (Christov, 2009) leads to an important effect in the field of thermal convection.

8.3.3 Cattaneo theories and porous materials

(Straughan, 2010c) has investigated analogous thermal convection problems to those of sections 8.3.1 and 8.3.2 in a porous medium of Darcy type. Apart from the fact that such analysis might prove valuable in a practical convection situation in a star or planet, the Darcy equations are lower order than Navier-Stokes and so one is able to proceed in greater detail analytically without having to resort to numerical solution of the relevant equations.

For thermal convection in a porous medium the basic equations for balances of linear momentum, mass, and energy, are given by (Straughan, 2010c) as

$$\begin{aligned} \frac{\partial v_i}{\partial t} &= -\frac{1}{\rho}\frac{\partial p}{\partial x_i} + \alpha g k_i T - \frac{\mu}{\rho K}\, v_i\,, \\ \frac{\partial v_i}{\partial x_i} &= 0, \\ \frac{1}{M}\frac{\partial T}{\partial t} + v_i\frac{\partial T}{\partial x_i} &= -\frac{\partial Q_i}{\partial x_i}\,. \end{aligned} \tag{8.41}$$

Here v_i, p, T are the velocity, pressure and temperature fields, ρ, α, g, μ and K are density, thermal expansion coefficient, gravity, dynamic viscosity and permeability, respectively. The quantity Q_i is the heat flux vector, $\mathbf{k} = (0,0,1)$, $(8.41)_1$ represents Darcy's law, and $M = (\rho_0 c_p)_f/(\rho_0 c)_m$, where $(\rho_0 c)_m = \phi(\rho_0 c_p)_f + (1-\phi)(\rho_0 c)_s$, ϕ being the porosity, and the f and s denote fluid and solid, respectively.

One may write separate energy balance and heat flux equations for the solid and fluid parts of the porous medium and then combine them to arrive at either the Cattaneo - Fox equation in a porous medium,

$$\tau\frac{\partial Q_i}{\partial t} + \tau_f\Big(v_j\frac{\partial Q_i}{\partial x_j} - \frac{1}{2}Q_j\frac{\partial v_i}{\partial x_j} + \frac{1}{2}Q_j\frac{\partial v_j}{\partial x_i}\Big) = -Q_i - \kappa\frac{\partial T}{\partial x_i}\,, \tag{8.42}$$

or the analogous Cattaneo-Christov equation

$$\tau\frac{\partial Q_i}{\partial t} + \tau_f\Big(v_j\frac{\partial Q_i}{\partial x_j} - Q_j\frac{\partial v_i}{\partial x_j}\Big) = -Q_i - \kappa\frac{\partial T}{\partial x_i}\,. \tag{8.43}$$

Details of the derivation may be found in (Straughan, 2010c) and we note $\tau = \phi\tau_f + (1-\phi)\tau_s$, τ_f and τ_s being relaxation times for the fluid and solid, respectively.

The saturated porous medium is assumed to occupy the horizontal layer $\{(x,y)\in\mathbb{R}^2,\ z\in(0,d)\}$ with the partial differential equations holding in the domain $\mathbb{R}^2\times(0,d)\times\{t>0\}$. The appropriate boundary conditions for Darcy's law are

$$\begin{aligned} &w \equiv v_3 = 0 \qquad \text{on} \quad z = 0, d, \\ &T = T_L, \quad z = 0, \qquad T = T_U, \quad z = d, \end{aligned} \tag{8.44}$$

$T_L > T_U$. The steady solution for either heat flux law is

$$\bar{v}_i \equiv 0, \qquad \bar{T} = -\beta z + T_L\,, \qquad \bar{\mathbf{Q}} = (0, 0, \kappa\beta), \tag{8.45}$$

where β is the temperature gradient, $\beta = (T_L - T_U)/d$.

In terms of the Darcy number, Da, the Rayleigh number, $Ra = R^2$, and the number Sg, introduced as

$$Da = \frac{K}{d^2}\,, \qquad Sg = \frac{\tau\mu}{\rho d^2}\,, \qquad R = \sqrt{\frac{\alpha g d^2 \beta K \rho}{\mu\kappa}}\,,$$

the linearized, non-dimensional equations which arise for perturbations (u_i, θ, π, q_i) to the steady solution $(\bar{v}_i, \bar{T}, \bar{p}, \bar{Q}_i)$ are, for Cattaneo-Fox,

$$\begin{aligned} u_{i,t} &= -\pi_{,i} + Rk_i\theta - u_i, \\ u_{i,i} &= 0, \\ \frac{Pr}{MDa}\theta_t &= Rw - q_{i,i}, \\ \frac{Sg}{Da}q_{i,t} &= \frac{SgR}{2Pr}\hat{\tau}(u_{i,z} - w_{,i}) - q_i - \theta_{,i}\,, \end{aligned} \tag{8.46}$$

or Cattaneo-Christov,

$$\begin{aligned} u_{i,t} &= -\pi_{,i} + Rk_i\theta - u_i, \\ u_{i,i} &= 0, \\ \frac{Pr}{MDa}\theta_t &= Rw - q_{i,i}, \\ \frac{Sg}{Da}q_{i,t} &= \frac{SgR}{Pr}\hat{\tau}u_{i,z} - q_i - \theta_{,i}\,, \end{aligned} \tag{8.47}$$

where $\hat{\tau} = \tau_f/\tau$.

(Straughan, 2010c) shows that Cattaneo - Fox theory for stationary convection leads to the equation

$$\Delta^2 w = -R^2\Delta^*\Big(\frac{Sg\hat{\tau}}{2Pr}\,\Delta w + w\Big). \tag{8.48}$$

He shows further that this then yields

$$R^2 = \frac{2Pr\Lambda^2}{a^2(2Pr - Sg\hat{\tau}\Lambda)} \tag{8.49}$$

where $\Lambda = n^2\pi^2 + a^2$. He notes that equation (8.49) leads to some interesting possibilities including R^2 switching to negative values, which does not contradict physics since one replaces R^2 by Ra and interprets it as heating from above. (Straughan, 2010c) concentrates on the case $n = 1$ in equation (8.49). He shows that the critical value of a^2, a_c^2, at which R^2 achieves a minimum is when

$$a_c^2 = \frac{\pi^2(2Pr - Sg\hat{\tau}\pi^2)}{(2Pr + \pi^2 Sg\hat{\tau})}\,.$$

This yields a critical Rayleigh number of the form

$$Ra = R^2 = \frac{8Pr\pi^2}{(2Pr - Sg\hat{\tau}\pi^2)}. \tag{8.50}$$

For oscillatory convection he finds

$$R^2 = \frac{\dfrac{\Lambda^2 Da}{Sg} + \dfrac{\Lambda}{M}\left(\dfrac{Pr}{Da} + \dfrac{Pr}{Sg}\right)}{a^2\left(\dfrac{Sg}{Da} + \dfrac{\Lambda Sg\hat{\tau}}{2Pr}\right)}. \tag{8.51}$$

He takes $\hat{\tau} = 1$ and restricts attention to the case of $M = 1$. It is necessary to analyse equation (8.51). To this end it is useful to note that one expects Sg to be small and Da to be likewise small. For example, for water a typical value of $Pr = 6$, whereas for sand K takes a value in the range 2×10^{-7} to 1.8×10^{-6} cm^2, see Nield & Bejan (Nield and Bejan, 2006). For a 3cm layer $d = 3$cm and this yields a value of $Da = d^2/K$ in the range 2×10^{-6} to 2×10^{-5}. Thus, for practical values one finds the critical value of R^2 by allowing $a^2 \to \infty$, and so

$$Ra_{osc} = \frac{2DaPr}{Sg^2}. \tag{8.52}$$

(Straughan, 2010c) deduces that with the Cattaneo-Fox model in porous convection one finds the Rayleigh number threshold is the smaller of (8.50) and (8.52). He shows that for $Sg < Sg_T$, $Ra = 8\pi^2 Pr/(2Pr - Sg\pi^2)$, whereas for $Sg > Sg_T$, $Ra = 2DaPr/Sg^2$. The transition Cattaneo number is given by

$$Sg_T = \frac{2Pr\sqrt{Da}}{\sqrt{Da}\pi^2 + \sqrt{8Pr}\,\pi}. \tag{8.53}$$

For Sg small one finds stationary convection, but once Sg exceeds Sg_T the convection mechanism switches to one of oscillatory convection and the cell structure breaks down.

(Straughan, 2010c) also considers the Cattaneo-Christov theory. He shows that stationary convection reduces to

$$a_c = \pi, \qquad R^2_{stat} = 4\pi^2. \tag{8.54}$$

For the oscillatory case he derives

$$a_c^2 = \pi\sqrt{\pi^2 + (Pr/DaM)(Sg/Da + 1)} \tag{8.55}$$

and with $M = 1$,

$$R^2_{osc} = \frac{2Da^2\pi^2}{Sg^2} + \frac{2\pi Da^2}{Sg^2}\sqrt{\pi^2 + \frac{Pr}{Da}\left(1 + \frac{Sg}{Da}\right)} + \frac{DaPr}{Sg^2} + \frac{Pr}{Sg}. \tag{8.56}$$

(Straughan, 2010c) deduces from equations (8.54) and (8.56) that when $Sg < Sg_T$ one finds $R^2 = 4\pi^2$ with $a^2 = \pi^2$. However, if $Sg > Sg_T$,

R^2 has the value given by (8.56) with a^2 given by (8.55). Thus, at Sg_T the convection changes from stationary convection to oscillatory convection and the wave number increases discontinuously which means the convection cells switch to a narrower hexagonal shape. The transition depends on Pr and Da and further details are given in (Straughan, 2010c) .

8.4 Green - Naghdi model

We now describe work of (Straughan, 2010b) who adapts the Green-Naghdi theory explained in section 3.5 to be applicable to a nanofluid. (Straughan, 2010b) points out that this theory accounts for a non-Newtonian behaviour of a nanofluid suspension.

The work of (Straughan, 2010b) modifies the entropy flux vectors of (Green and Naghdi, 1996). He replaces H by F and T by S and then equations (3.73) of section 3.5 are modified to

$$p_i^F = -\frac{k_F}{\theta_F}\frac{\partial\theta_F}{\partial x_i}, \qquad p_i^S = -\frac{k_S}{\theta_S}\frac{\partial\theta_S}{\partial x_i}, \tag{8.57}$$

while the equations for the intrinsic entropy supply functions ξ_F and ξ_S become

$$\begin{aligned} \rho\xi_F\theta_F &= \frac{k_F}{\theta_F}\frac{\partial\theta_F}{\partial x_i}\frac{\partial\theta_F}{\partial x_i} + 2\mu D_{ij}D_{ij} + \phi, \\ \rho\xi_S\theta_S &= \frac{k_S}{\theta_S}\frac{\partial\theta_S}{\partial x_i}\frac{\partial\theta_S}{\partial x_i} + 4\mu_1 D_{ij}P_{ij} + \frac{2\mu_1^2}{\mu}P_{ij}P_{ij} - \phi, \end{aligned} \tag{8.58}$$

where k_F and k_S are thermal conductivities. While (Green and Naghdi, 1996), see section 3.5 take ϕ constant (Straughan, 2010b) assumes

$$\phi = -h(\theta_F - \theta_S) \tag{8.59}$$

for $h > 0$ constant.

(Straughan, 2010b) assumes the body force term is gravity with ρ depending on θ_F, θ_S so that

$$\rho b_i = -gk_i\rho_0\big[1 - \alpha_F(\theta_F - \theta_F^0) + \alpha_S(\theta_S - \theta_S^0)\big] \tag{8.60}$$

where g is gravity, $\mathbf{k} = (0,0,1)$, α_F, α_S are the thermal expansion coefficients of the fluid and solid, respectively, and θ_F^0, θ_S^0 are reference (constant) temperatures, ρ_0 being a constant. He also considers another relation which questions whether the solid particles really do contribute to the buoyancy in equation (8.60). This is,

$$\rho b_i = -gk_i\big[\rho_S\phi + (1-\phi)\rho_F^0(\theta_F - \theta_F^0)\big] \tag{8.61}$$

where ϕ is the volume fraction of particles in the suspension.

The system of equations considered by (Straughan, 2010b) consists of the momentum and conservation of mass equations together with the entropy balance equations and these are,

$$\dot{v}_i - \frac{\mu_1}{\mu}\frac{d}{dt}\Delta v_i = -\frac{1}{\rho_0}p_{,i} + \alpha_F g k_i \theta_F - \alpha_S g k_i \theta_S + \nu\Delta v_i - 2\nu_1\Delta^2 v_i\,, \tag{8.62}$$

$$v_{i,i} = 0, \tag{8.63}$$

$$\rho c_F \dot{\theta}_F = \rho s_F \theta_F + 2\mu D_{ij}D_{ij} + k_F\Delta\theta_F - h(\theta_F - \theta_S), \tag{8.64}$$

$$\rho c_S \dot{\theta}_S = \rho s_S \theta_S + k_S\Delta\theta_S + 4\mu_1 D_{ij}P_{ij} + \frac{2\mu_1^2}{\mu}P_{ij}P_{ij} + h(\theta_F - \theta_S), \tag{8.65}$$

where $\nu = \mu/\rho_0$, $\nu_1 = \mu_1/\rho_0$, and p absorbs the constant terms which arise from (8.60).

The objective of the article of (Straughan, 2010b) is to present a theory for nanofluid behaviour which allows the suspension to exhibit non-Newtonian characteristics, but he also wishes to investigate thermal convection. Thus, suppose the fluid occupies the horizontal layer $\{(x,y)\in\mathbb{R}^2\}\times\{z\in(0,d)\}$ with gravity acting downward. The basic equations (8.62) - (8.65) thus hold on the domain $\{(x,y)\in\mathbb{R}^2\}\times\{z\in(0,d)\}\times\{t>0\}$. The boundaries are assumed fixed with the temperatures maintained at constants T_L, T_U, $T_L > T_U$, so that

$$\theta_F = \theta_S = T_L, \quad z = 0, \qquad \theta_F = \theta_S = T_U, \quad z = d. \tag{8.66}$$

The steady (conduction) solution whose instability is investigated is then

$$\bar{v}_i = 0, \qquad \bar{\theta}_F = \bar{\theta}_S = -\beta z + T_L\,, \tag{8.67}$$

where β is the temperature gradient, $\beta = (T_L - T_U)/d$, and the steady pressure $\bar{p}(z)$ is then determined from equation (8.62).

Instability is analysed by letting $(u_i, \pi, \theta_F, \theta_S)$ be perturbations to the basic state $(\bar{v}_i, \bar{p}, \bar{\theta}_F, \bar{\theta}_S)$, so that $v_i = \bar{v}_i + u_i$, $p = \bar{p} + \pi$, $\theta_F = \bar{\theta}_F + \theta_F$, $\theta_S = \bar{\theta}_S + \theta_S$.

Employing the non-dimensional variables

$$\begin{aligned}
&x_i = x_i^* d, \qquad t = t^*\frac{d^2}{\nu}, \qquad U = \frac{\nu}{d}, \qquad P = \frac{\nu^2\rho_0}{d}, \\
&H = \frac{hd^2}{\rho_0 c_F \kappa_F}, \qquad \kappa_F = \frac{k_F}{\rho_0 c_F}, \qquad \kappa_S = \frac{k_S}{\rho_0 c_S}, \qquad \frac{\kappa_F}{\kappa_S} = \frac{k_F c_S}{k_S c_F}, \\
&\hat{\mu} = \frac{\mu_1}{\mu d^2}, \qquad a_1 = 2\sqrt{Pr}\frac{\alpha_F g}{d c_F\sqrt{\beta}}, \qquad b_1 = \frac{4\mu_1 U\sqrt{\kappa_F \alpha_F g}}{\rho_0 c_S d^2 \kappa_S\sqrt{\beta\nu}}, \\
&b_2 = \frac{2\mu_1^2 U\sqrt{\kappa_F\alpha_F g}}{\mu d^4 \kappa_S\sqrt{\beta\nu}}, \qquad Pr = \frac{\nu_F}{\kappa_F},
\end{aligned}$$

with the Rayleigh number $Ra = R^2$ defined by

$$R^2 = \frac{\alpha_F g d^4\beta}{\nu\kappa_F}, \tag{8.68}$$

(Straughan, 2010b) derives the full system of non-dimensional equations for the perturbations, where $*$s are discarded,

$$\begin{aligned}
u_{i,t} + u_j u_{i,j} - \hat{\mu}\Delta u_{i,t} - \hat{\mu} u_j \Delta u_{i,j} &= -\pi_{,i} + Rk_i\theta_F - \frac{\alpha_S}{\alpha_F} Rk_i\theta_S \\
&\quad + \Delta u_i - 2\hat{\mu}\Delta^2 u_i, \\
u_{i,i} &= 0, \\
Pr(\theta_{F,t} + u_j\theta_{F,j}) &= Rw + \Delta\theta_F - H(\theta_F - \theta_S) + a_1 d_{ij} d_{ij}\,, \\
\frac{\kappa_F}{\kappa_S} Pr(\theta_{S,t} + u_j\theta_{S,j}) &= \frac{\kappa_F}{\kappa_S} Rw + \Delta\theta_S + \frac{k_F}{k_S} H(\theta_F - \theta_S) \\
&\quad + b_1 d_{ij} p_{ij} + b_2 p_{ij} p_{ij}\,.
\end{aligned} \tag{8.69}$$

To investigate instability (Straughan, 2010b) linearizes equations (8.69) and then puts

$$u_i = e^{\sigma t} u_i(\mathbf{x}), \quad \pi = e^{\sigma t}\pi(\mathbf{x}), \quad \theta_F = e^{\sigma t}\theta_F(\mathbf{x}), \quad \theta_S = e^{\sigma t}\theta_S(\mathbf{x}).$$

After eliminating π from the linearized version one finds the system of equations

$$\begin{aligned}
\sigma(\hat{\mu}\Delta^2 w - \Delta w) &= -R\Delta^*\theta_F + \frac{\alpha_S}{\alpha_F} R\Delta^*\theta_S - \Delta^2 w + 2\hat{\mu}\Delta^3 w, \\
\sigma Pr\theta_F &= Rw + \Delta\theta_F - H(\theta_F - \theta_S), \\
\sigma\frac{\kappa_F}{\kappa_S} Pr\theta_S &= \frac{\kappa_F}{\kappa_S} Rw + \Delta\theta_S + \frac{k_F}{k_S} H(\theta_F - \theta_S),
\end{aligned} \tag{8.70}$$

where $\Delta^* = \partial^2/\partial x^2 + \partial^2/\partial y^2$ is the horizontal Laplacian.

The boundary conditions are those of two fixed surfaces at constant temperatures, so that

$$w = w_z = 0, \quad \theta_F = \theta_S = 0, \qquad \text{on} \qquad z = 0, d. \tag{8.71}$$

However, (Straughan, 2010b) observes that due to the higher order terms in the Green-Naghdi momentum equation we require an extra boundary condition. He follows (Green and Naghdi, 1996), equation (74). Introduce their tensor $M^1_{ij} = -p_1\delta_{ij} + \mu_1 N_{ij}$ where p_1 is an arbitrary scalar and $N_{ij} = \epsilon_{irs} u_{s,rj}$, i.e. $\partial/\partial x_j$ of $(\mathrm{curl}\,\mathbf{u})_i$. Then the couple at the walls m^1_i is $m^1_i = M^1_{ij} n_j$ where $\mathbf{n} = (0,0,1)$ at $z = 1$, $\mathbf{n} = (0,0,-1)$ at $z = 0$. The couple in the horizontal directions x, y is supposed zero. Thus,

$$\mathbf{e}^1 \cdot \mathbf{m}^1 = 0, \qquad \mathbf{e}^2 \cdot \mathbf{m}^1 = 0,$$

$\mathbf{e}^1, \mathbf{e}^2$ being standard basis vectors. By writing $\mathbf{u} = (u, v, w)$ one finds the equations

$$w_{yz} - v_{zz} = 0, \quad w_{xz} - u_{zz} = 0, \qquad \text{on } z = 0, 1. \tag{8.72}$$

But in the whole fluid domain $u_x + v_y + w_z = 0$, and (Straughan, 2010b) shows this together with (8.72) yields the boundary condition

$$w_{zzz} + \Delta^* w_z = 0. \tag{8.73}$$

One next puts $w = W(z)f(x,y)$, $\theta_F = \Theta_F(z)f(x,y)$, $\theta_S = \Theta_S(z)f(x,y)$, where f is a plane tiling function such that

$$\Delta^* f + a^2 f = 0,$$

a being a wavenumber.

Then, with $D = d/dz$, equations (8.70) reduce to

$$\begin{aligned}
&2\hat{\mu}(D^2 - a^2)^3 W - (D^2 - a^2)^2 W + Ra^2\Theta_F - \alpha Ra^2 \Theta_S \\
&\qquad = \sigma\left[\hat{\mu}(D^2 - a^2)^2 W - (D^2 - a^2)W\right], \\
&(D^2 - a^2)\Theta_F - H(\Theta_F - \Theta_S) + RW = \sigma Pr \Theta_F, \\
&(D^2 - a^2)\Theta_S + kH(\Theta_F - \Theta_S) + \kappa RW = \sigma\kappa Pr\Theta_S\,,
\end{aligned} \tag{8.74}$$

where $\alpha = \alpha_S/\alpha_F$, $k = k_F/k_S$, $\kappa = \kappa_F/\kappa_S$. The boundary conditions are

$$W = W_z = 0, \; W_{zzz} - a^2 W_z = 0, \; \Theta_F = 0, \; \Theta_S = 0, \quad \text{on } z = 0, 1. \tag{8.75}$$

The system (8.74) and (8.75) is solved numerically by a D^2 Chebyshev tau method in (Straughan, 2010b). Details of the numerical method are given there.

(Straughan, 2010b) reports numerical results for the thermal convection instability analysis just outlined. He employed four sets of nanofluid suspensions. These are those where the fluid was either water or ethylene glycol and where the particles are CuO or Al_2O_3. Values of $\alpha = \alpha_S/\alpha_F$, $k = k_F/k_S$, $\kappa = k_F c_S/k_S c_F$ and Pr were obtained from (Chandrasekhar, 1981), p. 66, (Dow, 2009), (Accuratus, 2009), and (EngineeringToolbox, 2009).

The numerical routine of (Straughan, 2010b) calculates the minimum value of a where instability will commence, i.e. that value of Ra for which $\sigma_r = 0$ where $\sigma = \sigma_r + i\sigma_1$. For *all* of the parameter values he investigated σ was found to be real at the instability transition. Thus, instability is by stationary convection and for the range of parameter values analysed, overstability is not witnessed.

(Straughan, 2010b) observes that the Rayleigh number (8.68) is based on the fluid properties only. Since it is well known that the thermal conductivity of a nanofluid may be considerably higher than that of the solvent, cf. (Kwak and Kim, 2005), (Hwang et al., 2007), (Masoumi et al., 2009), (Xuan et al., 2003), (Wong and Kurma, 2008), (Xuan and Roetzel, 2000), (Kim et al., 2007), (Putra et al., 2003), and in addition the recent work of (Kwak and Kim, 2005), and also (Masoumi et al., 2009), shows that the effective viscosity of the nanofluid may also be significantly different from that of the base fluid, he gives consideration to a Rayleigh number which might more accurately reflect the effective properties of the nanofluid suspension itself. In fact, (Straughan, 2010b) considers two other definitions of Rayleigh number, one which explicitly depends on the particle concentration ϕ and another which accounts for the viscosity correction given by (Hwang et al., 2007). These are denoted by $Ra_N^{(J)}$, where J denotes the percentage by volume of particles, and by Ra_1.

(Straughan, 2010b) presents numerical values for the H_2O - CuO case, noting that the other combinations yield similar outcomes. He finds that if one uses the nanofluid Rayleigh numbers $Ra_N^{(4)}$, $Ra_N^{(2)}$, or Ra_1 then for small $\hat{\mu}$ values we may certainly obtain a large reduction in the Rayleigh number as compared to that for a classical Newtonian fluid with no particles, i.e. $Ra \approx 1707$. This is completely in line with the findings of (Tzou, 2008), although we stress that the Green-Naghdi model incorporates non-Newtonian effects in the nanofluid due to the presence of very small metallic oxide particles. The reduction of the Rayleigh number, very significantly in some cases, e.g. $Ra_N^{(2)}$ when $H = 0.1, \hat{\mu} = 10^{-4}$, drops from 1707 to 277 which means convective motion occurs much more easily in the nanofluid suspension. This, in turn, means heat transfer occurs more readily and this agrees with the perceived use of a nanofluid in a heat transfer device.

9
Other applications

In this chapter we examine further applications where thermal wave propagation is likely to be important, or we include applications of the ideas used in deriving the thermal wave models. This chapter is split into four sub-headings. The first two deal with applications of the various heat propagation theories presented earlier in this book. The second two sub-chapters deal with applications of the mathematical ideas which were used to develop a theory of finite speed heat propagation, but are applied to other problems in real life. These classes of problem are in traffic flow and in biological / medical contexts. It is worth observing that in twelve of the fourteen subsections the key idea involves essentially the model of Cattaneo. Thus, one should not underestimate the influence the paper of (Cattaneo, 1948) has in the field reported in this book.

9.1 Applications in continuum mechanics

9.1.1 Nanoscale heat transport

Modern technology is employing and inventing devices which are increasingly smaller. As (Pilgrim et al., 2004) write, ... "*there is a growing demand for greater understanding of thermal transport in nanoscale devices.*" When discussing finite speed heat transport and mathematical models for this phenomenon (Pilgrim et al., 2004) also write, ... the "*hyperbolic description will become increasingly important as device dimensions move even further into the deep sub-micron regime.*"

In fact, (Pilgrim et al., 2004) analyse two possible mechanisms of heat transport at the nanoscale such as will be required in semiconductor devices. They apply their results explicitly to samples of gallium arsenide, GaAs. They investigate a microscopic Monte Carlo Model, and they analyse a Cattaneo model with a heat source. The second of these is of interest here.

If one begins with the Cattaneo system with a heat source $g(\mathbf{x}, t)$ then the energy balance and Cattaneo equations may be written as, cf. equations (1.45), section 1.2,

$$\begin{aligned} \rho c \frac{\partial T}{\partial t} &= -\frac{\partial q_i}{\partial x_i} + g, \\ \tau \frac{\partial q_i}{\partial t} + q_i &= -\kappa \frac{\partial T}{\partial x_i} \,. \end{aligned}$$

In keeping with (Pilgrim et al., 2004), $\kappa = \kappa(T)$, so the thermal conductivity depends on temperature, T. The heat flux q_i may be eliminated and then one derives the equation

$$\tau \rho c \frac{\partial^2 T}{\partial t^2} + \rho c \frac{\partial T}{\partial t} = \frac{\partial}{\partial x_i}\Big(\kappa \frac{\partial T}{\partial x_i}\Big) + g + \tau \frac{\partial g}{\partial t} \,. \tag{9.1}$$

In equation (9.1) ρ and c are the density and specific heat of the solid under consideration and τ is the relaxation time. To identify this equation with equation (6) of (Pilgrim et al., 2004) we put $\tau\rho c = \kappa/\sigma^2$ and $k/\sigma^2 = \tau$ so that k is the thermal diffusivity of the solid. (Pilgrim et al., 2004) identify σ with the thermal wavespeed, and they estimate $\sigma \approx 5000$ m s^{-1} for GaAs with τ the order of 10^{-12}s. They take Laplace transforms of equation (9.1) to solve for the temperature in the transform space, using s as the $t-$Laplace transform variable. They find numerical (and analytical) solutions of the inverse transform for an appropriate choice of $\kappa(T)$. (Pilgrim et al., 2004) analyse their results for two samples of GaAs, these being pieces measuring $500\times500\times100$ microns (micron=μm=10^{-6}m) and $0.05\times0.05\times0.01$ microns, respectively. A heat pulse is input into the sample and its transient behaviour studied according to their mathematical model. For the larger sample they find the thermal behaviour to be very close to that obtained by using an analogous parabolic heat transfer equation. However, for the smaller sample they find a rapid temperature increase between 0.01 and 1 picoseconds (picosecond $= 10^{-12}s$) after input and this temperature reaches a maximum which decays (cools) to a steady state. They conclude that in the latter situation the hyperbolic solution is leading to different behaviour.

9.1.2 Heat transport in nanowires

A very recent area of research relevant to the present section was reported by (Hochbaum et al., 2008) and by (Boukai et al., 2008), and involves the possible production of electricity from from heat which is normally lost to the environment. As (Majumdar and Yang, 2008) report, approximately 90 per cent of the world's power is created by using fossil fuel, but this operation works at only 30 to 40 per cent efficiency, leading to a massive loss of heat. (Majumdar and Yang, 2008) grew from a silicon wafer a forest of silicon nanowires some 20-300 nm in diameter, see (Hochbaum et al., 2008). They have been able to increase dramatically the ZT value of the nanomaterial by a factor of 60, to 0.6, and this is yielding the exciting possibility of generating electricity from waste heat. The number ZT is given by ZT= $S^2T/\hat{\rho}\kappa$, where S is the thermoelectric power, T is the absolute temperature, $\hat{\rho}$ is the electrical resistivity, and κ is the thermal conductivity. Normally such high ZT values are found only in materials composed of much rarer elements than silicon. (Boukai et al., 2008) have reported similar findings for silicon nanowires which have cross - sectional areas 10 nm $\times$ 20 nm and 20 nm $\times$ 20 nm. They indicate that the improved thermoelectric efficiency is due to phonon effects, and the same results may be expected from other types of semiconductor nanomaterials.

In connection with the above results, very interesting theoretical studies of temperature wave propagation along possible models for nanowires have begun by (Jou and Sellitto, 2009),(Jou et al., 2009), (Jou et al., 2010b), (Jou et al., 2011), (Cimmelli et al., 2010a), and by (Sellitto et al., 2011). (Jou et al., 2010b) investigate the interesting phenomenon of heat slip in a Guyer-Krumhansl system appropriate to heat transport in nanomaterials. Furthermore, a recent interesting study by (Jou et al., 2011) shows that the roughness of the wall can play a major role on the speed of a heat wave in an nanowire. (Majumdar and Yang, 2008) report that the roughness of the nanowires is an important factor in the thermoelectric efficiency.

9.1.3 Heat transport in thin films

There are various examples where heat transport in a thin film of material is believed to be ballistic rather than diffusive. We give a brief exposition of two.

(Lor and Chu, 1999) observe that since high temperature superconductors have been discovered several electronic devices employ a thin film of such a superconductor which is deposited on a substrate such as a layer of a metallic oxide or sapphire. In fact, (Lor and Chu, 1999) employ a Cattaneo theory to model heat transfer in such a scenario and present numerical results for a thin film of the high - temperature superconductor Yttrium barium copper oxide, a crystalline chemical compound with the formula $YBa_2Cu_3O_7$, written by (Lor and Chu, 1999) as YBaCuO. This thin film is deposited on a substrate of magnesium oxide, MgO, Strontium titanate, $SrTiO_3$, Lanthanum aluminate, $LaAlO_3$ or sapphire. Interestingly, (Lor and Chu, 1999) estimate the relaxation time τ in a Cattaneo theory for the superconductor YBaCuO to have values of 300 picoseconds (ps=10^{-12}s) when the temperature T is $4°K$, 0.6 ps when $T = 50°K$ and 0.4 ps when $T = 77°K$.

(Lor and Chu, 1999) develop a mathematical model for heat transport through a two layer system, $x \in (0, x_1) \equiv X_1$, the superconducting thin film, $x \in (x_1, x_2) \equiv X_2$, the substrate. A heat flux is input at $x = 0$ at time $t = t_p$, so that $q = q_0(1 - H(t - t_p))$, where q_0 is a constant and H is the Heaviside function. The boundary temperature at $x = x_2$ is fixed at a value T_0. The equations of (Lor and Chu, 1999) for each region X_1, X_2 are both of Cattaneo type and if the temperature and heat flux in each layer are denoted by T_1, q_1, and T_2, q_2, then the equations are, cf. equations (1.45), section 1.2,

$$\begin{aligned} \frac{\partial T_1}{\partial t} &= -\frac{1}{k_1 \alpha_1} \frac{\partial q_1}{\partial x}, \\ \tau_1 \frac{\partial q_1}{\partial t} + 2q_1 &= -k_1 \frac{\partial T_1}{\partial x}, \end{aligned} \qquad (9.2)$$

in $X_1, t > 0$, and

$$\begin{aligned} \frac{\partial T_2}{\partial t} &= -\frac{1}{k_2 \alpha_2} \frac{\partial q_2}{\partial x}, \\ \tau_2 \frac{\partial q_2}{\partial t} + 2q_2 &= -k_2 \frac{\partial T_2}{\partial x}, \end{aligned} \qquad (9.3)$$

in $X_2, t > 0$, for suitable constants $k_1, \alpha_1, \tau_1, k_2, \alpha_2, \tau_2$.

An important issue for (Lor and Chu, 1999) is what is the correct condition at the interface between the domains X_1 and X_2, i.e. at $x = x_1$. They take either

$$q_1 = q_2 \quad \text{and} \quad T_1 = T_2 \quad \text{at } x = x_1,$$

for what they call a perfect contact interface, or

$$q_1 = q_2 = \kappa(T_1^4 - T_2^4), \quad \text{at } x = x_1, \qquad (9.4)$$

for an interface with thermal resistance. The coefficient κ is a constant.

(Lor and Chu, 1999) take as initial condition T constant with $q = 0$ throughout their thin - layer - substrate configuration and they solve the resulting boundary - initial value problem numerically by a finite difference Godunov method. A variety of numerical results are presented for the superconductor YBaCuO for each of the substrates MgO, $LaAlO_3$, $SrTiO_3$ and sapphire. Comparison is made with an analogous parabolic model and the perfect contact and thermal resistance interface conditions are seen to lead to very different results. In particular, the thermal resistance condition is seen to have an effect of prolonging the hyperbolic thermal wave.

(Niu and Dai, 2009) discuss another situation of heat transfer through a layered system, the layers being metallic in nature and with more or less equal thicknesses. They particularly apply their model to a gold layer connected to a chromium one. (Niu and Dai, 2009) contains a useful brief review of such topics over approximately the last twenty years. They note that one has to consider temperatures and heat fluxes for the electrons in the layers, and for the rest of the lattice. They denote the electron and lattice temperatures in a layer by T_e and T_ℓ, respectively, with corresponding heat fluxes $\mathbf{q}_e$ and $\mathbf{q}_\ell$. When no deformation of the metal is accounted for they note that a successful model for heat transport in a metal film has been derived by (Chen and Beraun, 2001; Chen and Beraun, 2003), (Chen et al., 2002b) and (Chen et al., 2002a). This model is a generalization of the Cattaneo one, cf. equations (1.45), section 1.2, and suggests employing the equations

$$\begin{aligned} c_e \frac{\partial T_e}{\partial t} &= -\nabla \mathbf{q}_e - G(T_e - T_\ell) + S, \\ \tau_e \frac{\partial \mathbf{q}_e}{\partial t} + \mathbf{q}_e &= -k_e \nabla T_e, \end{aligned} \tag{9.5}$$

and

$$\begin{aligned} c_\ell \frac{\partial T_\ell}{\partial t} &= -\nabla \mathbf{q}_\ell + G(T_e - T_\ell), \\ \tau_\ell \frac{\partial \mathbf{q}_\ell}{\partial t} + \mathbf{q}_\ell &= -k_\ell \nabla T_\ell \,. \end{aligned} \tag{9.6}$$

In these equations c_e, c_ℓ, k_e, k_ℓ are constants, τ_e is the electron relaxation time of free electrons in the metal, τ_ℓ is the lattice relaxation time in phonon collisions, S is a heat source due to laser heating, and the G terms represent electron - lattice interactions.

(Niu and Dai, 2009) generalize the above equations to allow for deformation within each film of a two layer structure. They specifically study heat transport through a two-film scenario in two-space dimensions, x, y, the films, denoted by L_1 and L_2, being contained by the boundaries $x = 0, x = L_x/2$, and $x = L_x/2, x = L_x$, the horizontal extent of the two layer film being $y \in (0, L_y)$. (Niu and Dai, 2009) write down equations of form (9.5) and (9.6) for each film L_1 and L_2, adding a term to the lattice equations $(9.6)_1$ of the form $\zeta^{(m)}(\partial/\partial t)(u^{(m)}_{\alpha,\alpha})$ where $u^{(m)}_\alpha$ denotes the components of displacement, $\alpha = 1, 2$, in each layer, $m = 1, 2$, and $u^{(m)}_{\alpha,\alpha} = \sum_{\alpha=1}^2 \partial u^{(m)}_\alpha / \partial x_\alpha$. They couple their system of four equations in each layer to momentum equations for $u^{(m)}_\alpha$, cf. (Tzou et al., 2002). These equations correspond to those of linear elastodynamics with a nonlinear temperature - temperature gradient contribution which is referred to as a hot electron - blast effect.

(Niu and Dai, 2009) focusses on developing an efficient numerical method for solving their equations in the two layer domain. Explicit calculations are performed for parameters which correspond to a Gold - Chromium layer situation. Extensive numerical results are presented.

(Ignaczak, 2009) also considers an interesting model for heat transfer in thin metal films.

Use of temperature waves to measure the thermal diffusivity of a low dielectric constant thin film is analysed by (Morikawa and Hashimoto, 2005). Further employment of temperature waves in the analysis of relaxation transitions and thermal diffusivity in polymers is due to (Hashimoto et al., 1997) and to (Polikarpov and Slutsker, 1997).

9.1.4 Reactor fuel rods

(Espinosa-Paredes and Espinosa-Martinez, 2009) develop a model for heat transport in a nuclear fuel rod in a light water reactor. They explain that an understanding of the heat transfer mechanism to the coolant in the reactor is absolutely essential. They deal with a situation where they have cylindrical fuel rods surrounded by a small gap which is filled with an inert gas and this is then surrounded by a concentric cylindrical layer of cladding.

They introduce a series of assumptions and then write their mathematical model. This is based on Cattaneo's equations, (1.45), section 1.2, in a cylindrical geometry. Thus, they have the equations

$$\begin{aligned} \rho c_p \frac{\partial T}{\partial t} &= \frac{1}{r}\frac{\partial}{\partial r}(r q_r) + s(t), \\ \tau \frac{\partial q_r}{\partial t} + q_r &= -k \frac{\partial T}{\partial r}, \end{aligned} \tag{9.7}$$

holding in $r \in (r_0, r_{c\ell})$ for $t > 0$. Here T is the temperature in the fuel rod, q_r is the heat flux in the radial direction, r is the radial coordinate, r_0 being the centre of the cylinder, and $s(t)$ is a heat source. The boundary conditions of (Espinosa-Paredes and Espinosa-Martinez, 2009) are

$$\begin{aligned} -k\frac{\partial T}{\partial r} &= H_\infty (T - T_m), \qquad r = r_{c\ell} \\ \frac{\partial T}{\partial r} &= 0, \qquad r = r_0. \end{aligned}$$

Here H_∞ is a heat transfer coefficient and T_m is a temperature outside the cladding. (Espinosa-Paredes and Espinosa-Martinez, 2009) give a precise form for $s(t)$ which is based on the reactor power and they give details of how they calculate this explicitly. For initial conditions, it is assumed the temperature is known initially, i.e.

$$T(r, 0) = T_0(r).$$

(Espinosa-Paredes and Espinosa-Martinez, 2009) employ a control volume numerical technique to solve equations (9.7) subject to their boundary and initial conditions. Many detailed numerical results are presented choosing appropriate parameter values from the nuclear reactor industry. They deduce that employing a Cattaneo theory shows the heat fluxes on the surface of the cladding will be substantially different over a long period of time, and they deduce that a propagative heat mechanism can be important.

(Gabaraev et al., 2003) note that accidental coolant loss can lead to significant temperature rise in a nuclear reactor core. This can be followed by a change from film to nucleate boiling due to the passage of a travelling temperature wave. These writers develop and analyse a mathematical model for this situation.

9.1.5 Phase changes

The idea of a material changing phase is an important one in real life. We can think of water changing to ice, an example of a fluid to solid transition, water changing to steam, an example of a fluid to vapour transition, or one may think of a fluid to gas transition. However, there are many other phase transitions of interest in modern technology, such as those in thermoelastic solids, or those in shape memory alloys. While traditionally many of the processes associated with phase transition have been regarded as parabolic this has recently been a rich area for studying analogous processes by theories which allow for finite speed of propagation, i.e. are more "hyperbolic" - like. A good review of this area is given in the introduction of (Galenko and Jou, 2005) and a similar useful review of some of the mathematical literature on the subject is given in the introduction of (Jiang, 2009).

The theory of phase tansitions is interesting in that it has employed Cattaneo-like approaches to study finite speed of solute transport, as opposed to classical parabolic diffusion, finite speed of the "phase field" itself, and finite speed of heat propagation. One way to investigate a phase transition is to employ a moving front approach. This has some connection with acceleration waves and shock waves in that one studies a moving discontinuity surface. This is appealing because one can incorporate thermodynamics naturally. A lucid description of this may be found in (Berezovski and Maugin, 2005) where they treat a moving phase transition front in thermoelasticity. They solve their equations numerically and make detailed predictions on the phase change.

Another approach to phase change is to assume that the transition is not abrupt but takes place with a finite interfacial thickness. This has led to the idea of a phase field in the context of a phase transition. However, before introducing this concept we point out that in the field of phase transitions the idea of a solute concentration moving with a finite speed of propagation has been proposed for some time, see e.g. (Galenko and Danilov, 1997), (Sobolev, 1997), and the references therein. For example, (Galenko and Danilov, 1997) study a solidifcation process and introduce conditions at a liquid-solid interface. In the liquid and solid phases they write that the liquid and solid temperatures T_L, T_S and the concentration of solute in the liquid, C_L, satisfy the equations,

$$\begin{aligned}\frac{\partial T_L}{\partial t} &= a_L \Delta T_L\,,\\ \frac{\partial T_S}{\partial t} &= a_S \Delta T_S\,,\\ \tau\frac{\partial^2 C_L}{\partial t^2} + \frac{\partial C_L}{\partial t} &= D\Delta C_L\,.\end{aligned} \tag{9.8}$$

Hence, while the heat transport is parabolic they are using a Cattaneo - like theory for solute transport. (Galenko and Jou, 2005) allow also the possibility of finite speed heat transport (and, in fact, finite speed phase field transport) and consider estimates for the relaxation times τ_T for heat, τ_D for solute concentration. They give values which suggest $10^{-13}s < \tau_T < 10^{-11}s$ for metallic systems with $10^{-11}s < \tau_D < 10^{-7}s$ in a binary alloy system. While these values are for entirely different materials it does suggest "hyperbolic" solute transport cannot be entirely neglected in phase transition.

(Auriault et al., 2007) consider diffusion in a composite material. They introduce the solute balance equation

$$\frac{\partial c}{\partial t} = -J_{i,i}\,,$$

c being a solute concentration and J_i the solute flux. To incorporate finite speed of propagation they suggest for $\mathbf{J}$ the equation

$$A\frac{\partial J_i}{\partial t} + J_i = -Dc_{,i}\,.$$

Together these equations lead to the equation for c

$$A\frac{\partial^2 c}{\partial t^2} + \frac{\partial c}{\partial t} = \frac{\partial}{\partial x_i}\Big(D\frac{\partial c}{\partial x_i}\Big)\,. \tag{9.9}$$

They estimate the diffusion coefficient and consider where their equation may be valid. They also analyse memory effects in composite materials. Their calculations do not support the use of equation (9.9) for solute transport in general. The propagation values suggested by (Galenko and Jou, 2005) do contrast with the work of (Auriault et al., 2007).

An interesting approach to hyperbolic diffusion is given by (Malysiak et al., 2007). They analyse hyperbolic diffusion in the context of what they call a ball and chain problem. Their mathematical results are compared to electrophysiological data and they estimate the diffusion coefficient for their model employing the diffusion coeffcient for a single aminoacid in water.

For a diffuse interface transition a phase field Φ is introduced, cf. (Galenko, 2001), (Galenko and Jou, 2005). This quantity has a fixed numerical value in a particular phase, e.g. $\Phi = -1$ for an unstable liquid phase with $\Phi = +1$ in the liquid phase, (Galenko and Jou, 2005). Between these extremes lies the diffuse interface region where Φ changes smoothly, albeit steeply, from -1 to 1. As (Galenko and Jou, 2005) remark one may find numerical solutions and then locate an interface where $\Phi = 0$. (Galenko, 2001) writes a nonlinear parabolic equation for Φ but couples this to a Cattaneo-like equation for the solute flux. In this way he obtains an equation for the solute concentration which has second derivatives in both time and space. He places this within the context of other models. (Galenko and Jou, 2005) develop a model where the energy density, e, solute concentration, X, and the phase field, Φ, satisfy equations of the form

$$\begin{aligned} \tau_T \frac{\partial^2 e}{\partial t^2} + \frac{\partial e}{\partial t} &= -\frac{\partial}{\partial x_i}\left[M_{ee} \frac{\partial}{\partial x_i}\Big(\frac{\partial \eta}{\partial e} + \epsilon_e^2 \Delta e \Big) \right], \\ \tau_D \frac{\partial^2 X}{\partial t^2} + \frac{\partial X}{\partial t} &= -\frac{\partial}{\partial x_i}\left[M_{XX} \frac{\partial}{\partial x_i}\Big(\frac{\partial \eta}{\partial X} + \epsilon_X^2 \Delta X \Big) \right], \\ \tau_\Phi \frac{\partial^2 \Phi}{\partial t^2} + \frac{\partial \Phi}{\partial t} &= M_\Phi \Big(\frac{\partial \eta}{\partial \Phi} + \epsilon_\Phi^2 \Delta \Phi \Big). \end{aligned} \tag{9.10}$$

Here η is the entropy, $M_{ee}, M_{XX}, M_\Phi, \epsilon_e^2, \epsilon_X^2, \epsilon_\Phi^2$ are appropriate functions, and $\tau_T, \tau_D, \tau_\Phi$ are relaxation times for temperature, solute concentration, and the phase field Φ. The phase field itself satisfies a hyperbolic-like equation $(9.10)_3$.

Finite speed of heat transport has been also introduced into phase transitions, see e.g. (Colli and Recupero, 2002), (Bonetti, 2002b; Bonetti, 2002a), (Jiang, 2009), and the references therein. Interestingly, (Liu et al., 2009) suggest that one is better using the theory of (Coleman et al., 1982) (which generalizes that of Cattaneo, see page 13 of this book) rather than simply the model of (Cattaneo, 1948).

Certainly, the field of phase transitions where solute concentration, phase field, and heat transport are described by finite speed of propagation (hyperbolic-like) models is an area of much research. Various combinations of these models such as hyperbolic phase field - parabolic heat transport, hyperbolic phase field - hyperbolic heat transport, have been studied intensely recently. Studies of existence, attractors, uniqueness, long time behaviour of solutions, stability and related issues may be found in e.g. (Gatti et al., 2005), (Bonetti et al., 2007), (Grasselli et al., 2007), (Wu et al., 2007), (Jiang, 2009), (Miranville and Quintanilla, 2009), and the many references therein. An interesting application of a hyperbolic solute equation is to drug release in the human body. Here, the experimental results necessitate a hyperbolic model, cf. (Ferreira and de Oliveira, 2010).

Another interesting development of phase transitions is to incorporate finite speed of heat transport by means of Green-Naghdi type III thermodynamics. This has been done by (Miranville and Quintanilla, 2010), who investigate existence and uniqueness of solutions and dissipative properties, as well as spatial decay in a cylinder.

9.2 Stellar, planetary heat propagation

9.2.1 Cryovolcanism on Enceladus

I find the article by (Bargmann et al., 2008a) to be a beautiful and extremely interesting paper dealing with a possible explanation for some of the findings on Enceladus, which is a moon of the planet Saturn. (Bargmann et al., 2008a) give precise details of Enceladus and its interesting properties which make it an object of intense study in our solar system. They note that observations from the Cassini spacecraft have shown the occurrence of active volcanoes on Enceladus. In fact, since Enceladus has a mean surface temperature of 77°K they refer to the phenomenon of volcanism as cryovolcanism. (Bargmann et al., 2008a) also note that volcanic eruptions have been observed on four bodies in the solar system, the Earth, Jupiter's moon Io, Neptune's moon Triton, and Saturn's moon Enceladus.

(Bargmann et al., 2008a) observe that volcanoes on Enceladus erupt water rather than magma, and no evidence has been found of ammonia and/or methane which is also found in the eruptions of other cryovolcanoes on other icy moons. They are particularly interested in trying to explain a warm spot which is centred on Enceladus' south pole. Since the eruptions on Enceladus occur under very cold conditions (Bargmann et al., 2008a) argue that this is a very good place to model heat transfer using the Green-Naghdi model of thermoelasticity of type III, see section 2.4. To me personally, their arguments are very convincing. They argue that type III thermoelasticity is the best because it incorporates both classical thermoelasticity and thermoelasticity of type II. They also argue that the model should include energy dissipation due to the experiences of volcanoes on Earth and this is further argument for type III theory.

(Bargmann et al., 2008a) use a finite element method to simulate the temperature field in the vicinity of a volcanic vent on Enceladus. The surface temperature is computed using a radiation balance accounting for solar radiation, black-body radiation from the ice surface, and geothermal heating from below. They derive their appropriate form of equations for type III thermoelasticity, and these are

$$\rho \ddot{u}_i = (E_{ijkh} e_{kh})_{,j} - 3\big[K\omega(T)(T - T_0)\big]_{,i} + b_i \,,$$

and

$$\rho c(T)\dot{T} = (k_1 \alpha_{,i})_{,i} + (k_2 T_{,i})_{,i} + \rho r - 3T_0 \omega(T) K \dot{e}_{ii} \,,$$

where u_i, T are the elastic displacements and temperature field, e_{ij} being the strain. (Bargmann et al., 2008a) carefully estimate the temperature dependent functions $k(T), \omega(T)$ and $c(T)$ for the situation appropriate to Enceladus, and they pay particular attention to the forms for the thermal coeffcients k_1 and k_2. For k_2 they use known fits for ice whereas for

k_1, which is intrinsic to type III thermoelasticity, they produce various plausible arguments but vary this coefficient to see the effect of variation.

The numerical results of (Bargmann et al., 2008a) are carefully presented together with their conclusions. They particularly note that observations on the surface of Enceladus show "stripes" which are probably caused by cryomagma being distributed during the volcanic eruptions. Their computations with type III theory thermoelasticity do allow them to predict the occurrence of such stripes. In fact, they write, ... "*Therefore, with all due caution arising from our simplified model and the data uncertainties, it seems that non-classical heat transport in ice at cryogenic temperatures may play a role in explaining the observed temperature distribution in the vicinity of the volcanically active troughs in Enceladus' south polar region.*" The paper of (Bargmann et al., 2008a) is a particularly appealing use of Green-Naghdi type III thermoelasticity to a problem of real interest.

9.2.2 Thermohaline convection

(Herrera and Falcón, 1995) discuss the possibility of heat transported as a wave being responsible for convective overturning in certain stars. They draw an analogy with thermohaline instability whereby a layer of warm salty water can overlie a layer of cooler fresh water, but when the salty layer cools it becomes less dense and tends to fall under gravity creating a convective overturning instability. (Herrera and Falcón, 1995) discuss applying a Cattaneo model of heat transport to this situation. They argue that in a close binary system star a helium rich outer layer may form and this may lead to an instability not dissimilar to a thermohaline one due to helium burning which creates a carbon enriched outer layer which has higher molecular weight than the stellar material below.

(Herrera and Falcón, 1995) also suggest similar instability mechanisms may be present in neutron stars, in radio pulsars, and in the collapsed core of a supernovae progenitor. (Falcón, 2001) continues this investigation, in particular, looking at cooling in white dwarfs and neutron stars. He argues that the superfluid interior of a neutron star promotes heat wave propagation and he employs a Cattaneo theory to estimate the cooling time. He shows that Cattaneo theory increases the cooling time, although he stresses that numerical values of the relaxation time are currently uncertain.

(Straughan, 2011) develops and analyses linear instability of a model for thermohaline instability in a porous layer employing a Cattaneo-Christov equation, cf. section 3.1.2, for the heat flux. He derives the following equations for balance of linear momentum, mass, salt concentration, energy, and a Cattaneo-Christov equation, for a fluid saturated porous medium,

$$
\begin{aligned}
&p_{,i} = -\frac{\mu}{K} v_i + \rho_0 \alpha g T k_i - \rho_0 \alpha_S g C k_i \, , \\
&v_{i,i} = 0, \\
&\phi C_{,t} + v_i C_{,i} = \phi k_C \Delta C, \\
&\frac{1}{M} T_{,t} + v_i T_{,i} = -Q_{i,i} \, , \\
&\tau Q_{i,t} + \tau_f (v_j Q_{i,j} - Q_j v_{i,j}) = -Q_i - \kappa T_{,i} \, .
\end{aligned}
\tag{9.11}
$$

Here, v_i, T, C, Q_i, p are velocity, temperature, salt concentration, heat flux, and pressure, $\mu, K, \rho_0, \alpha, g, \alpha_S, \phi, k_C$ and κ are dynamic viscosity, permeability, a reference density, thermal expansion coefficient of the fluid, gravity, salt expansion coefficient, porosity, salt diffusivity, and thermal diffusivity. The coefficients M and τ are given by

$$
\begin{aligned}
&M = \frac{(\rho_0 c_p)_f}{\phi(\rho_0 c_p)_f + (1-\phi)(\rho_0 c)_S} \, , \\
&\tau = \phi \tau_f + (1-\phi)\tau_s
\end{aligned}
$$

where f, s denote fluid and solid parts in the saturated porous layer, and τ_f and τ_s are relaxation times for the fluid and solid, respectively.

Equations $(9.11)_{4,5}$ are derived by writing a Cattaneo system for the solid skeleton and a Cattaneo-Christov system for the saturating fluid and combining, cf. section 8.3.3.

(Straughan, 2011) employs the boundary conditions

$$
T = T_L, \quad C = C_L, \quad z = 0; \qquad T = T_U, \quad C = C_U, \quad z = d;
$$

where T_L, T_U, C_L, C_U are constants with $T_L > T_U, C_L > C_U$, so interest is in the situation of heating and simultaneously salting the porous layer from below. The steady solution is

$$
\begin{aligned}
&\bar{v}_i = 0, \qquad \bar{C} = -\beta_s z + C_L \, , \qquad \beta_s = \frac{C_L - C_U}{d} \, , \\
&\bar{T} = -\beta z + T_L \, , \qquad \beta = \frac{T_L - T_U}{d} \, , \qquad \bar{Q}_3 = \kappa\beta,
\end{aligned}
$$

with the steady pressure $\bar{p}(z)$ found from $(9.11)_1$.

Perturbations u_i, π, θ, q_i and φ to the steady values $\bar{v}_i, \bar{p}, \bar{T}, \bar{Q}_i, \bar{C}$ are introduced and from equations (9.11) one derives the nonlinear perturbation equations, see (Straughan, 2011). These are non-dimensionalized and linearized and then one finds the linearized perturbation equations to be

$$
\begin{aligned}
&\pi_{,i} = -u_i + R\theta k_i - R_s \varphi k_i, \\
&u_{i,i} = 0, \\
&\varepsilon Le \, \varphi_{,t} = R_s w + \Delta\varphi, \\
&\theta_{,t} = Rw - q_{i,i} \, , \\
&2MCa \, q_{i,t} = 2C_f R u_{i,z} - q_i - \theta_{,i} \, .
\end{aligned}
\tag{9.12}
$$

These are solved in (Straughan, 2011) with the boundary conditions

$$u_i n_i = 0, \quad \theta = \varphi = 0, \qquad \text{on} \quad z = 0, 1.$$

In equations (9.12), $R^2, R_s^2, \epsilon, Le, Ca$ and C_f are non-dimensional parameters,

$$R^2 = \frac{\alpha g \beta K d^2}{(\mu/\rho_0)\kappa}, \qquad R_s^2 = \frac{\alpha_s g \beta_s K d^2}{(\mu/\rho_0)\phi k_C}$$

$$\epsilon = \phi M, \qquad Le = \frac{Ud}{\phi k_C} \qquad C_f = \frac{\kappa \tau_f}{2d^2} \qquad Ca = \frac{\kappa \tau}{2d^2}\,.$$

In fact, R^2 is the Rayleigh number, R_s^2 is the salt Rayleigh number, Le is the Lewis number, C_f is the fluid Cattaneo number, and Ca is the Cattaneo number.

The instability surface in $R^2, R_s^2, \mathcal{C}(= 2MCa)$ space is determined in (Straughan, 2011). It is an interesting surface which is formed by eigenvalues changing places as the Cattaneo number increases. The instabilities may be stationary or oscillatory, depending on the parameter values, R_s^2 and Ca.

9.3 Traffic flow

Traffic flow on a one lane highway (road) is explained in (Whitham, 1974), pp. 68–80. Reviews of mathematical models for traffic flow under various conditions may be found in e.g. (Bellomo et al., 2002a), (Bellomo et al., 2002b), (Darbha et al., 2008). In this section we include a brief exposition of work of (Jordan, 2005b) who uses a mathematical technique not dissimilar to that explained in section 1.2 in connection with the Cattaneo equation to derive a hyperbolic - like model for traffic flow on a one lane highway.

Let $\rho(x,t)$ be the density of traffic at a point x on a single lane highway at time t and let $q(x,t)$ be the corresponding "flux" of cars across x at time t. Then, one may show ρ, q satisfy a conservation law, cf. (Whitham, 1974), pp. 68–80, of form

$$\frac{\partial \rho}{\partial t} + \frac{\partial q}{\partial x} = 0. \tag{9.13}$$

(Jordan, 2005a) begins by discussing a "constitutive equation" for the flux q in terms of ρ and ρ_x. He writes

$$q(x,t) = \rho(x,t)\left\{ v_m\left(1 - \frac{\rho(x,t)}{\rho_s}\right)\right\} - \nu \frac{\partial \rho}{\partial x}(x,t)\,. \tag{9.14}$$

In equation (9.14) ρ_s is a saturation value for the density of traffic ($0 < \rho < \rho_s$), v_m is the maximum speed of a vehicle when $\rho \to 0$, and $\nu > 0$ is a constant. However, (Jordan, 2005b) argues that the flux at time t ought not to depend directly on ρ, ρ_x at the same time. He argues that a driver will not react instantaneously to a change ahead and proposes instead that there be a delay in response time in equation (9.14). Thus, (Jordan, 2005b) suggests employing the constitutive equation

$$q(x,t+\tau) = \rho(x,t)\Big\{v_m\Big(1-\frac{\rho(x,t)}{\rho_s}\Big)\Big\} - \nu\frac{\partial\rho}{\partial x}(x,t)\,, \tag{9.15}$$

where $\tau > 0$ is a response time with $\tau << t_c$, the time t_c being a characteristic time for the traffic flow.

In order not to be dealing with a differential - delay equation (Jordan, 2005b) then expands q in (9.15) in a Taylor series to write

$$\tau\frac{\partial q}{\partial t} + q(x,t) = \rho(x,t)\Big\{v_m\Big(1-\frac{\rho(x,t)}{\rho_s}\Big)\Big\} - \nu\frac{\partial\rho}{\partial x}(x,t)\,, \tag{9.16}$$

where terms $O(\tau^2)$ have been neglected. By combining equations (9.13) and (9.16), (Jordan, 2005b) then arrives at the following partial differential equation for $\rho(x,t)$,

$$\tau\frac{\partial^2\rho}{\partial t^2} + \frac{\partial\rho}{\partial t} - \nu\frac{\partial^2\rho}{\partial x^2} + v_m\Big(1-\frac{2\rho}{\rho_s}\Big)\frac{\partial\rho}{\partial x} = 0. \tag{9.17}$$

(Jordan, 2005b) also notes that equation (9.17) was briefly introduced by (Lighthill and Whitham, 1955) in their influential paper.

(Jordan, 2005b) analyses equation (9.17) in detail. He first shows that one may derive an explicit solution in travelling wave form, $\rho(x,t) = f(\xi) = f(x-vt)$ for a constant $v > 0$, the propagation velocity. He shows that one may solve for f explicitly so that

$$f(\xi) = \frac{1}{2}\Big\{(\rho_1+\rho_2) + (\rho_2-\rho_1)\,\tanh\Big(\frac{2\xi}{\ell}\Big)\Big\}$$

where ρ_1 and ρ_2 are limits of f as $\xi \to \mp\infty$, and

$$\ell = \frac{4\nu\rho_s}{v_m(\rho_2-\rho_1)}\Big(1-\frac{v^2\tau}{\nu}\Big).$$

He interprets this as a diffusive soliton and shows this has a shock thickness given by ℓ. He shows that for this solution to be valid one must have

$$\tau < \tau^* = \frac{\nu}{[v_m(1-\tilde\rho)]^2}\,,$$

with $\tilde\rho = (\rho_1+\rho_2)/\rho_s$. This threshold τ^* is important in his interpretation for traffic flow.

(Jordan, 2005b) also develops a shock wave and an acceleration wave analysis for a solution to equation (9.17). A shock wave is defined to be a singular surface across which ρ has a finite jump whereas an acceleration wave is one where ρ is continuous but ρ_x has a finite jump. (Jordan, 2005b) shows that the shock amplitude satisfies a Bernoulli equation whereas the amplitude of the acceleration wave satisfies a linear equation. He solves for the amplitudes explicitly.

(Jordan, 2005b) provides a detailed explanation of the findings from his travelling wave, shock wave, and acceleration wave analysis in terms of traffic flow. For example, he shows that breakdown of the travelling wave solution corresponds to τ exceeding τ^*, i.e. the driver's reaction time is too slow and a collision will occur. He also indentifies another threshold time $\tau^\bullet \leq \tau$ which is based on the initial values of ρ. This threshold is also interpreted in terms of traffic flow and the potential for a collision to occur.

This section has considered a hyperbolic model for traffic flow and congestion (crowding) caused by vehicles. There are other interesting areas of crowding which may be modelled by suitable hyperbolic models. For example (Bellomo and Dogbé, 2008) discuss in detail strategies for examining crowds of people. They also discuss the modelling of the motion of swarms such as a swarm of bees, or a flock of birds.

9.4 Applications in biology

9.4.1 Population dynamics

(Mendez and Camacho, 1997) develop and investigate a model for a population, such as an animal population. They begin by observing that a reaction - diffusion equation has been used to model population growth, such as Fisher's equation where the growth term is a logistic one, i.e. if $n(x,t)$ denotes the population at time t and position x, then

$$\frac{\partial n}{\partial t} = D\frac{\partial^2 n}{\partial x^2} + \lambda n(1-n). \tag{9.18}$$

However, they oberserve that, ... "*it is well known that an animal's motion during a small time period has a tendency to proceed in the same direction as it did in the immediate period before*". Therefore, they infer that there is memory in the animal's motion whereas equation (9.18) does not contain any memory effect.

To incorporate memory (Mendez and Camacho, 1997) employ a random walk argument to derive a conservation law for n of form

$$\frac{\partial n}{\partial t} + \frac{\partial J}{\partial x} = 0, \tag{9.19}$$

where $J(x,t)$ is the flux. They derive a Cattaneo equation for J, namely,

$$\tau\frac{\partial J}{\partial t} + J = -D\frac{\partial n}{\partial x} = 0, \tag{9.20}$$

However, they argue that the population should have a source term (production of animals) added and so instead of (9.19) they write

$$\frac{\partial n}{\partial t} + \frac{\partial J}{\partial x} = F(n), \tag{9.21}$$

where F is the source. By combining equation (9.20) and (9.21), (Mendez and Camacho, 1997) derive the following equation for the population n,

$$\tau\frac{\partial^2 n}{\partial t^2}+\frac{\partial n}{\partial t}=D\frac{\partial^2 n}{\partial x^2}+F(n)+\tau\frac{\partial F}{\partial t}(n). \tag{9.22}$$

(Mendez and Camacho, 1997) also derive equation (9.22) by what they call a phenomenological derivation which uses a delay between $J(x,t+\tau)$ and $\partial n/\partial x(x,t)$. In addition they include a derivation by using extended thermodynamic arguments of (Jou et al., 2010a). (Mendez and Camacho, 1997) study travelling waves for their equation (9.22) and further analyse the flux term using arguments from non-equilibrium thermodyanamics.

9.4.2 Migration of a school of fish

In this section we describe a model for the behaviour of a large but highly organised school of fish. (A school of fish is a highly organised group of fish which synchronises their swimming, as opposed to a shoal which is a more loosely organised group.) The model was derived by (Niwa, 1998) in a very inspiring article. As (Niwa, 1998) points out, certain fish schools such as mackerel, can swim enormous distances but do so in a highly organised manner so that the school itself often behaves like a continuous body rather than a group of individuals. He argues that one of the reasons for this is that water temperature has a very strong effect on fish behaviour. Fish control their body temperature by moving from one place to another and (Niwa, 1998) quotes research where fish migration is performed in order to maximize comfort by sensing changes in water temperature. In order to account for this in a mathematical model, (Niwa, 1998) argues that fish must possess a memory mechanism for sensing temperature gradients. They need to be able to sense temperature gradients of the order of 0.01 to 0.1°C / 100m and so a memory record of the thermal environment is essential.

(Niwa, 1998) employs a non-trivial but very clever statistical argument to calculate the position of a school of fish taking into account the positions of individuals and incorporating thermal memory effects. If $P(\mathbf{x},t)$ is the probability distribution that a school of fish is at position $\mathbf{x}$ at time t, and $\mathbf{j}(\mathbf{x},t)$ is the associated probability current density then (Niwa, 1998) shows that P satisfies a conservation equation of form

$$\frac{\partial P}{\partial t}+\frac{\partial j_i}{\partial x_i}=0. \tag{9.23}$$

He further derives a Cattaneo - like equation for the function $\mathbf{j}$ of form

$$\tau\frac{\partial j_i}{\partial t}+j_i=-D\frac{\partial P}{\partial x_i}-\frac{\partial W(\mathbf{x})}{\partial x_i}P, \tag{9.24}$$

where D is a diffusion coefficient given explicitly in terms of an average of the velocity in the school in equation (61) of (Niwa, 1998). The function

$W(\mathbf{x})$ is also calculated explicitly and given by (Niwa, 1998) in his equation (69).

By combining equations (9.23) and (9.24), (Niwa, 1998) derives the following differential equation for the probability distribution of fish in the school

$$\tau \frac{\partial^2 P}{\partial t^2} + \frac{\partial P}{\partial t} = D\Delta P + \tau \frac{\partial}{\partial x_i}\Big(\frac{\partial W}{\partial x_i} P\Big). \tag{9.25}$$

(Niwa, 1998) includes further analysis of his model.

9.4.3 Spread of the Hantavirus

The Hantavirus, originally named after the Hanta river in Korea, is a virus transmitted by rodents through excreta or bites, and may be a cause of serious illness. In 1993 a new species, responsible for the Hantavirus cardiopulmonary syndrome (HPC), was discovered to be due to the Sin Nombre virus. Since both viruses can be fatal, their spread is of grave concern; see e.g. (Schmaljohn and Hjelle, 1997), (Mills and Childs, 1998), (Mills et al., 1999). Mathematical models to explain how they propagate include the recent proposals by (Abramson and Kenkre, 2002), (Abramson et al., 2003), (Allen et al., 2003), (Sauvage et al., 2003), and (Allen et al., 2006). We are particularly interested in the hyperbolic model developed by (Barbera et al., 2008) since it uses a wave - like process to describe virus transmission.

According to the last named writers, the model of (Abramson and Kenkre, 2002) involves equations in one space dimension for populations, $S(x,t), I(x,t)$, of susceptible and infected mice, respectively. For a total mouse population $M(x,t) = S + I$, these equations are

$$\begin{aligned} \frac{\partial S}{\partial t} + \frac{\partial J^{(S)}}{\partial x} &= bM - cS - \frac{SM}{K} - aSI, \\ \frac{\partial I}{\partial t} + \frac{\partial J^{(I)}}{\partial x} &= -cI - \frac{IM}{K} + aSI, \end{aligned} \tag{9.26}$$

where a, b, c, respectively, are the rate of infection, the birth rate, and the death rate for the mice. The function $K(x,t)$ represents the capacity of the medium to maintaining the population of mice, e.g. accessibility of food, shelter, etc., while $J^{(S)}$ and $J^{(I)}$ are fluxes for the susceptible and infected populations. Substitution in equations (9.26) of the constitutive equations

$$J^{(S)} = -D\frac{\partial S}{\partial x}, \qquad J^{(I)} = -D\frac{\partial I}{\partial x}, \tag{9.27}$$

where D is the diffusion coefficient for the mice, leads to the parabolic reaction - diffusion system

$$\begin{aligned}\frac{\partial S}{\partial t} &= D\frac{\partial^2 S}{\partial x^2} + bM - cS - \frac{SM}{K} - aSI,\\ \frac{\partial I}{\partial t} &= D\frac{\partial^2 I}{\partial x^2} - cI - \frac{IM}{K} + aSI.\end{aligned}$$

(Barbera et al., 2008) proposed replacing equations (9.27) by using arguments of extended thermodynamics to obtain a hyperbolic system. Instead of the variables S and I, consider the total flux $J = J^{(S)} + J^{(I)}$, and combine equations (9.27) into the single equation

$$J = -D\frac{\partial M}{\partial x}. \tag{9.28}$$

Equations (9.26) now may be equivalently written in terms of M and I, as

$$\begin{aligned}\frac{\partial M}{\partial t} + \frac{\partial J}{\partial x} &= \left(b - c - \frac{M}{K}\right) \equiv h(M),\\ \frac{\partial I}{\partial t} + \frac{\partial J^{(I)}}{\partial x} &= \left[\left(a - \frac{1}{K}\right)M - aI - c\right]I \equiv g(M,I),\end{aligned} \tag{9.29}$$

where h and g are the indicated functions. (Barbera et al., 2008) now appeal to extended thermodynamics to establish equations for J and $J^{(I)}$. The same equations may be derived under the assumption that due to the reaction slight time delays occur in J and $J^{(I)}$. Equations (9.27) and (9.28) are replaced by

$$\begin{aligned}J(x, t + \tau_1) &= -D\frac{\partial M}{\partial x}(x,t),\\ J^{(I)}(x, t + \tau_2) &= -D\frac{\partial I}{\partial x}(x,t).\end{aligned} \tag{9.30}$$

We approximate the left hand sides by a first order Taylor series expansion (cf. section 1.2, equation (1.44)) in which terms of order $O(\tau_1^2), O(\tau_2^2)$ are discarded. Equations (9.30) become reduced to the system of equations

$$\begin{aligned}\frac{\partial J}{\partial t} + \frac{1}{\tau_1}J &= -\frac{D}{\tau_1}\frac{\partial M}{\partial x},\\ \frac{\partial J^{(I)}}{\partial t} + \frac{1}{\tau_2}J^{(I)} &= -\frac{D}{\tau_2}\frac{\partial I}{\partial x}.\end{aligned} \tag{9.31}$$

On substituting $\gamma' = D/\tau_1$ and $\mu' = D/\tau_2$ in equations (9.29) and (9.31), we recover equations (4) and (8) of (Barbera et al., 2008) which, these writers observe, is a hyperbolic system with characteristic speeds

$$\lambda = \pm\sqrt{\gamma'}, \qquad \lambda = \pm\sqrt{\mu'}.$$

Moreover, the system possesses four equilibrium (steady state) solutions,

$$P_1 \equiv (0,0,0,0), \qquad P_2 \equiv (K(b-c),0,0,0),$$
$$P_3 \equiv (K(b-c), K(b-c) - \frac{b}{a}, 0, 0), \qquad P_4 \equiv (0, -\frac{c}{a}, 0, 0).$$

But P_4 is meaningless, while P_3 requires $K > K_c = b/a(b-c)$. A linearized instability analysis performed around the steady states demonstrates that the threshold K_c dictates stability. In fact, (Barbera et al., 2008) establish that

$$K < K_c \implies P_2 \text{ stable, } P_3 \text{ meaningless,}$$
$$K > K_c \implies P_2 \text{ unstable, } P_3 \text{ stable.}$$

These writers for the same system also investigate travelling waves, analyse their stability, and construct numerical solutions to examine the evolution of solutions from state P_2 to P_3, and from P_1 to P_3. Detailed numerical results are presented for various carrying capacities $K(x,t)$, and confirm that this particular hyperbolic model for the propagation of the Hantavirus deserves exploration.

9.4.4 Chemotaxis

The term chemotaxis frequently occurs in biology and refers to the phenomenon of chemically directed movement. Consider, for example, a species with a densely crowded population. In a diffusion process, the population will spread outward in space. By contrast, chemotaxis (or more properly, chemoattraction), is the opposite effect in which the species is attracted towards a high chemical concentration. For illustration (Murray, 2003a), p. 405, cites the example of the species in which female members exude a chemical to attract males.

A much studied chemotaxis process in mathematical biology is that of the formation of amoebae into a slime mold. A mathematical theory, developed by (Keller and Segel, 1970), is described in (Lin and Segel, 1974) p. 22, and also (Murray, 2003a), section 11.4. (Keller and Segel, 1971a) formulated other models for chemotaxis, and subsequently (Keller and Segel, 1971b) studied the effects of travelling waves.

The basic biological process of the slime mold amoebae in (Lin and Segel, 1974), p. 22. explains how amoebae feed on bacteria and when their food supply is abundant they propagate by division into two. Whenever, however, the food supply becomes scarce, an interphase period begins in which the amoebae move in a weak and random manner. In this phase the amoebae are effectively spread evenly over their environment. After some hours the amoebae begin to group together in a striking manner and form a many celled slug which can contain approximately 200,000 cells. This slug sometime later stops moving around and erects a stalk which contains spores.

These spores eventually re-emerge as amoebae and the life cycle of the slime mold amoebae begins afresh.

To describe the above process in terms of a mathematical model, (Keller and Segel, 1970) use a coupled system of reaction - diffusion partial differential equations into which is incorporated the chemotaxis effect. The general procedure behind diffusion models in biology is aptly explained in (Murray, 2003a), chapter 11. Basically, the idea is that one writes equations for functions c_i, $i = 1, \ldots, N$, say, which represent a species population(s), and possibly chemicals they feed on, absorb, or emit. The general form of a diffusion equation in a domain Ω (in biology the spatial domain Ω is usually a subset of $\mathbb{R}^2$ or $\mathbb{R}^1$), is

$$\frac{\partial}{\partial t} \int_\Omega c_i(\mathbf{x}, t)\, dx = - \int_\Gamma \mathbf{J}^{(i)} \cdot \mathbf{n}\, dA + \int_\Omega f_i\, dx. \tag{9.32}$$

Here Γ is the boundary of Ω, $\mathbf{n}$ is the unit outward normal to Γ, $\mathbf{J}^{(i)}$ is a flux vector associated with the $i-$th equation, and f_i is a source term. By using the divergence theorem to write (9.32) as an equation over Ω and reducing to point form, this equation assumes the form

$$\frac{\partial c_i}{\partial t} + \nabla \cdot \mathbf{J}^{(i)} = f_i, \qquad i = 1, \ldots, N, \tag{9.33}$$

holding typically in $\Omega \times (0, \infty)$. Such a procedure of deriving diffusion equations is familiar in continuum mechanics. In the same spirit as is necessary in continuum mechanics, in mathematical biology one needs to specify the domain Ω, the conditions on Γ, and the fluxes $\mathbf{J}^{(i)}$ and source terms f_i. This is where an intimate knowledge of the biological process is essential. (Murray, 2003a), page xi, writes, ... "*mathematical biology is ... the most exciting modern application of mathematics ... the use of esoteric mathematics arrogantly applied to biological problems by mathematicians who know little about the real biology, together with unsubstantiated claims as to how important such theories are, does little to promote the interdisciplinary involvement which is so essential.*"

Most research based on the (Keller and Segel, 1970) theory would appear to have focussed on use of a simplified theory which uses equations for the concentrations of cell density of Dictyostelium discoideum and the chemoattractant it secretes, namely cAMP. For this scenario (Keller and Segel, 1970) produce the following coupled system of partial differential equations

$$\begin{aligned} \frac{\partial a}{\partial t} &= -\nabla(D_1 \nabla \rho) + \nabla(D_2 \nabla a), \\ \frac{\partial \rho}{\partial t} &= D_\rho \Delta \rho - k(\rho)\rho + a f(\rho). \end{aligned} \tag{9.34}$$

(Keller and Segel, 1970) adopt the form

$$k(\rho) = \frac{\eta_0 k_2 K}{1 + K\rho},$$

for η_0, K constants. (Murray, 2003a), p. 407, uses D_2 constant and gives various forms for the chemotaxis coefficient D_1, such as

$$D_1 = \chi_0 a, \qquad D_1 = \frac{\chi_0 a}{\rho}, \qquad \text{or} \qquad D_1 = \frac{\chi_0 K a}{(K+\rho)^2},$$

where χ_0 and K are positive constants.

Thresholds for decay of a solution to the steady state are derived from the full nonlinear equations (9.34) by (Payne and Straughan, 2009).

In this book our interest centres on a Cattaneo - like modification to a chemotaxis system like (9.34) as derived by (Dolak and Hillen, 2003). In fact, we write equations (9.34) as

$$\begin{aligned} \frac{\partial a}{\partial t} &= -\nabla \mathbf{J}, \\ \mathbf{J} &= -D\nabla a + a(1-a)\nabla \rho, \\ \frac{\partial \rho}{\partial t} &= \Delta\rho + \alpha a - \rho, \end{aligned} \tag{9.35}$$

where $\mathbf{J}$ is a flux and to identify with (9.34) one takes $D_\rho \equiv 1, K(\rho) \equiv 1, f(\rho) \equiv \alpha, D_2 \equiv D$ and $D_1 = a(1-a)$. For α, D constant this corresponds to a "parabolic" version of the (Dolak and Hillen, 2003) system.

However, (Dolak and Hillen, 2003) are interested in a modification of (9.35) in which $\mathbf{J}$ is replaced by $\tau\mathbf{J}_t + \mathbf{J}$ in equations (9.35). This may be accounted for by employing a delay argument such as that leading to equation (1.44), for example. Thus, the (non-dimensional) system of chemotaxis equations studied by (Dolak and Hillen, 2003) has form

$$\begin{aligned} \frac{\partial a}{\partial t} &= -\nabla \mathbf{J}, \\ \tau\frac{\partial \mathbf{J}}{\partial t} + \mathbf{J} &= -D\nabla a + a(1-a)\nabla \rho, \\ \frac{\partial \rho}{\partial t} &= \Delta\rho + \alpha a - \rho. \end{aligned} \tag{9.36}$$

This is thus a sort of hyperbolic in a and parabolic in ρ system of equations. (Dolak and Hillen, 2003) adopt flux conditions such that

$$\frac{\partial \rho}{\partial n} = 0, \qquad \frac{\partial \mathbf{J}}{\partial n} = 0$$

on the boundary of their domain. They take realistic values for the coefficients τ, D and α and perform numerical simulations on a square spatial domain. Their results are very interesting and show clearly how cell aggregation may be achieved.

(Dolak and Hillen, 2003) also employ a Cattaneo-like modification to a system for the bacterium Salmonella typhimurium. Full details of models for Salmonella typhimurium are contained in (Murray, 2003b), pp. 281–306, where the equations studied are parabolic - like. If u denotes the (non-dimensional) cell density and S the concentration of aspartate which the cells produce then the Cattaneo - chemotaxis system governing u and S derived by (Dolak and Hillen, 2003) is

$$\begin{aligned} \frac{\partial u}{\partial t} &= -\nabla \mathbf{J} + \rho u\Big(1 - \frac{u}{c}\Big), \\ \tau \frac{\partial \mathbf{J}}{\partial t} + \mathbf{J} &= -D\nabla u + \frac{\alpha u}{(1+\beta S)^2}\, \nabla S, \\ \frac{\partial S}{\partial t} &= \Delta S + \frac{cu}{1+\gamma u} - S. \end{aligned} \tag{9.37}$$

(Dolak and Hillen, 2003) estimate values for the parametrs using data and solve (9.37) numerically on a square spatial domain employing zero-flux boundary conditions on S and $\mathbf{J}$. Their numerical simulations clearly display the ring patterns formed during aggregation of cells.

Finite time blow up results for hyperbolic chemotaxis models are given by (Hillen and Levine, 2003), and (Wang and Hillen, 2008) study shock formation in a chemotaxis model.

9.4.5 Radiofrequency heating

(Mitra et al., 1995) report experiments on heat transfer in processed meat and conclude that a hyperbolic model of heat transfer is appropriate. (Saidane et al., 2005) also consider wave - like heat transfer in biological tissues. (López Molina et al., 2008) and (Tung et al., 2009) specifically develop models for incorporating finite speed of propagation heat transfer in medical procedures.

(López Molina et al., 2008) treat a (Cattaneo, 1948) model coupled to the energy equation with a heat source, namely

$$\begin{aligned} \tau \frac{\partial q_i}{\partial t} + q_i &= -kT_{,i}\,, \\ \rho c \frac{\partial T}{\partial t} &= -q_{i,i} + S(\mathbf{x},t), \end{aligned} \tag{9.38}$$

where T, q_i, ρ, c, τ, k are temperature, heat flux, density, specific heat, relaxation time, thermal conductivity, and $S(\mathbf{x},t)$ is a heat source. They put $\kappa = k/\rho c$ and combine equations (9.38) into a single equation for T. The heat source they use is $S(r,t) = Pr_0 H(t)/4\pi r^4$, where r is the radial coordinate in a spherical coordinate system, P and r_0 are constants and $H(t)$ is the Heaviside function. By considering only variation in r the equation they employ is

$$\begin{aligned}\tau\frac{\partial^2 T}{\partial t^2}(r,t)+\frac{\partial T}{\partial t}(r,t) =&\kappa\Big[\frac{\partial^2 T}{\partial r^2}(r,t)+\frac{2}{r}\frac{\partial T}{\partial r}(r,t)\Big]\\ &+\frac{P\kappa r_0}{4\pi k r^4}\left[H(t)+\tau\delta(t)\right]\end{aligned} \tag{9.39}$$

where $\delta(t)$ is the Dirac delta function. (López Molina et al., 2008) solve the problem of heat flow according to equation (9.39) outside a sphere of radius r_0 together with the initial conditions

$$T(r,0)=T_0\,, \qquad \frac{\partial T}{\partial t}(r,0)=0 \qquad r>r_0. \tag{9.40}$$

The boundary conditions they use are

$$\lim_{r\to\infty} T(r,t)=T_0\,, \quad \forall t>0, \tag{9.41}$$

and they assume the thermal conductivity of the electrode (inside $r \le r_0$) is much greater than that of the surrounding tissue which leads them to propose

$$\tau T_{tt}(r_0,t)+T_t(r_0,t)=\frac{3k}{\rho c_0 r_0}\frac{\partial T}{\partial r}(r_0,t). \tag{9.42}$$

(López Molina et al., 2008) employ a Laplace transform method to solve (9.39)-(9.42) analytically. They evaluate their solutions using realistic values of an electrode used in radiofrequency heating in cardiac tissue and realistic values for the cardiac material. They compare their solutions with equivalent ones using a Fourier heat conduction model. The Cattaneo - like equation (9.39) leads to higher initial temperatures, but also to a cusp temperature profile with distance from the electrode. These temperature spikes travel through the cardiac tissue. (López Molina et al., 2008) point out that their spherical electrode is not a realistic shape and the electrical conductivity of the tissue should be considered as a function of temperature. As they wished to obtain an analytical solution such complexities had, of necessity, to be avoided.

(Tung et al., 2009) consider a similar model to that of (López Molina et al., 2008). They derive results for a one-space dimensional model applied to laser heating of the cornea. The laser beam falls on the entire surface of a biological tissue and heat propagation in a direction orthogonal to this surface is studied. A comparison is made with results of a similar model employing a Fourier heat transfer law. Notable differences are found, in particular, the hyperbolic model predicts a distinct temperature wave pulse in time as opposed to the smoothly decaying solution obtained from the Fourier model.

(Tung et al., 2009) also consider radiofrequency heating of a biological tissue. Their model is essentially that of (López Molina et al., 2008). They observe that radiofrequency heating is a surgical procedure employed in areas such as elimination of cardiac arrhythmias, destroying tumours, heating of the cornea, or treatment of gastroesophageal reflux disease. Again, they find a distinct travelling temperature pulse.

9.4.6 Skin burns

A one space dimensional model for skin burnt by "flash" heating was developed by (Torvi and Dale, 1994). They allowed for three skin layers, the epidermis, dermis, and sub-cutaneous layers. By flash heating we envisage skin burnt by flash fires which arise through combustible chemicals igniting rapidly, gas leaks, or such as petrochemical fires. The work of (Torvi and Dale, 1994) used a finite element method on a parabolic equation for the skin temperature T. J. Liu and his co-workers suggested employing a Cattaneo-like model for skin burns, see (Liu et al., 1995), (Liu et al., 1999), (Liu, 2000). (Liu, 2000) shows how the temperature equation may be derived, in general, from an energy balance, but essentially the equation he uses is based on a Cattaneo model with a chosen heat source, $S(\mathbf{x},t)$. In fact, with q_i being heat flux, τ relaxation time, ρ, c, k density, specific heat, and thermal conductivity of skin, the model of (Liu et al., 1995), (Liu et al., 1999), (Liu, 2000) relies on the equations

$$\begin{aligned} \tau\frac{\partial q_i}{\partial t} + q_i &= -kT_{,i}\,, \\ \rho c\frac{\partial T}{\partial t} &= -\frac{\partial q_i}{\partial x_i} + S, \end{aligned} \tag{9.43}$$

where S has the specific form

$$S(\mathbf{x},t) = W_b C_b(T_b - T) + Q_m + Q_r. \tag{9.44}$$

In equation (9.44) W_b, C_b and T_b are blood perfusion rate, specific heat of blood, and the blood temperature in the blood vessels enclosed in the affected skin. Thus, the model of Liu and his co-workers does take account of the fact that many blood vessels are passing through the skin layers. The terms Q_m and Q_r are a metabolic rate of tissue and a spatial heating term, respectively. (Liu et al., 1999) and (Liu, 2000) put $Q_r = 0$ in order to study only surface burning of skin. They suppose there is a steady state skin temperature T_i before burning takes place and they work with the variable $\theta = T - T_i$.

For constant Q_m, (Liu, 2000) shows equations (9.43) and (9.44) lead to the equation for the difference temperature θ in the skin

$$\frac{\tau}{\kappa}\frac{\partial^2\theta}{\partial t^2} + \Big(\frac{\rho c + \tau W_b C_b}{k}\Big)\frac{\partial\theta}{\partial t} = \frac{\partial^2\theta}{\partial x^2} + \Big(\frac{\partial}{\partial x}\ln k\Big)\frac{\partial\theta}{\partial x} - \frac{W_b C_b}{k}\,\theta\,, \tag{9.45}$$

where $\kappa = k/\rho c$.

(Liu, 2000) considers skin burning by solving (9.45) numerically with θ given at $x = 0$, $\theta = \theta^*$, and $\theta = 0$ deep into the body, i.e. at $x = L$ for some L large. As initial conditions he chooses θ and θ_t equal to 0 so that heating is by the function θ^* only. (Liu, 2000) considers four specific heating functions $\theta^* = \theta(0,t)$. These are $\theta^* = 5°$C, which corresponds to heating the skin by a hot plate, $\theta^* = 5°$C for $0 \leq t \leq 5$s, then $\theta^* = 0°$C thereafter. He writes that this function will be appropriate for such as eye surgery with laser heating, or flash burns due to heating over a short time, or possibly burns through an atomic explosion. He also considers linear and exponential heating, $\theta^* = (0.1°\text{C})t$ and $\theta^* = 5[1 - \exp(-0.1t)]°$C, which represents controlled skin heating, or sinusoidal heating, $\theta^* = 5\sin(0.2\pi t)°$C. The last form of heating might model heating due to repeated irradiation or laser heating.

(Liu, 2000) presents several numerical results for the various skin heating mechanisms proposed. He concludes that the Cattaneo-like model can have a big effect on the temperature in skin even a very long time after the heating period. He shows that the classical model based on Fourier's law leads to rapid temperature drop with skin depth. However, with the thermal wave model a much longer depth of penetration from the thermal wave is found.

Models which account specifically for the temperature of blood in the blood vessels in skin and the flow rate of blood therein have been proposed and analysed by (Dai et al., 2008) and by (Zhang, 2009). The model of (Dai et al., 2008) is a highly non-trivial extension of the model just described. Thus, (Dai et al., 2008) essentially employ equations (9.43) together with equation (9.44) for S with the internal heat source terms zero. They employ these equations in each of the three layers epidermis, dermis and sub-cutaneous with continuity of temperature and heat flux across the interfaces. However, (Dai et al., 2008) allow for rectangular vascular structures in the sub-cutaneous zone. They have a precise rule for defining how the blood vessels change cross sectional area, decreasing in area as the outer skin layer (epidermis) is approached. The blood temperature in the blood vessels (arteries and veins) is calculated using ordinary differential equations. The equations connecting heat transfer between the blood vessel and surrounding tissue have form

$$\frac{\partial T_b^m}{\partial n} = B(T_w^m - T_b^m)$$

where b, w denote blood and wall, and $m = 1, 2, 3$ depending on which blood vessel is chosen. On the outer skin surface (Dai et al., 2008) have the condition

$$-k\frac{\partial T}{\partial z} = h(T_a - T) + \epsilon\sigma(T_a^4 - T^4),$$

h, ϵ, σ being constants with T_a the ambient heating.

The model of (Dai et al., 2008) is solved numerically by a three-dimensional finite difference method which is unconditionally stable. Extensive numerical results are provided by (Dai et al., 2008). They conclude that the solution to their model exhibits time delay when compared to an analogous parabolic approach and their model also yields a lower tissue temperature.

(Zhang, 2009) employs a dual phase lag model to describe skin burns and writes equations for the blood flow. His model also includes the velocity of blood in the equations for blood flow. It is interesting to ask whether one might need an objective derivative when including the blood velocity and hence require something like a Cattaneo-Christov model, cf. section 3.1.2. (Zhang, 2009) includes an interesting account of phase lag times for blood and for biological (skin) tissue. He notes that the phase lag times significantly increase as the blood vessel diameter increases.

9.5 Exercises

Exercise 9.5.1 *Define equations* (9.5) *and* (9.6) *on* $\Omega \times \{t > 0\}$ *where* Ω *is a bounded domain in* $\mathbb{R}^3$ *with boundary* Γ. *Suppose* T_e *and* T_ℓ *are given on* $\Gamma \times \{t > 0\}$ *and* $T_e, T_\ell, \mathbf{q}_e$ *and* $\mathbf{q}_\ell$ *are given at* $t = 0$. *Show that the solution to this boundary - initial value problem is unique.*
Hint. Denote the boundary - initial value problem above by $\mathcal{P}$. *Let* $T_e^2, \mathbf{Q}_e^2, T_\ell^2, \mathbf{Q}_\ell^2$ *and* $T_e^1, \mathbf{Q}_e^1, T_\ell^1, \mathbf{Q}_\ell^1$ *be two solutions to* $\mathcal{P}$ *for the same boundary data* $T_B^e(\mathbf{x},t)$ *and* $T_B^\ell(\mathbf{x},t)$ *and for the same initial data* $T_0^e, T_0^\ell, \mathbf{Q}_0^e, \mathbf{Q}_0^\ell$. *Define the difference solution* $\theta_e = T_1^e - T_2^e, \theta_\ell = T_1^\ell - T_2^\ell$, $\mathbf{q}_e = \mathbf{Q}_1^e - \mathbf{Q}_2^e, \mathbf{q}_\ell = \mathbf{Q}_1^\ell - \mathbf{Q}_2^\ell$. *Show that if* $\|\cdot\|$ *denotes the norm on* $L^2(\Omega)$, *then*

$$\frac{d}{dt}\Big(\frac{c_e}{2}\|\theta_e\|^2+\frac{\tau_e}{2k_e}\|\mathbf{q}_e\|^2+\frac{c_\ell}{2}\|\theta_\ell\|^2+\frac{\tau_\ell}{2k_\ell}\|\mathbf{q}_\ell\|^2\Big)$$
$$=-\frac{1}{k_e}\|\mathbf{q}_e\|^2-\frac{1}{k_\ell}\|\mathbf{q}_\ell\|^2-G\|\theta_e-\theta_\ell\|^2.$$

Hence deduce uniqueness.

Exercise 9.5.2 *(See (Jordan, 2005b)). For equation* (9.17) *define an acceleration wave to be one for which* ρ *is continuous everywhere and* ρ_t, ρ_x *and higher derivatives are discontinuous across a surface* $\mathcal{S}$. *By taking the jumps of equations* (9.13) *and* (9.16) *show the wavespeeed* V *satisfies* $V^2 = \nu/\tau$. *Take the jump of equation* (9.17) *and show that* $a(t) = [\rho_x]$ *satisfies a linear ordinary differential equation in* t. *What can you deduce from this?*

Exercise 9.5.3 *Consider the global balance laws,*

$$\frac{d}{dt}\int_V c_\alpha dV = -\int_{\partial V} \mathbf{J}^\alpha\cdot\mathbf{n}\,dS + \int_V f_\alpha\,dV, \qquad \alpha = 1,\dots,N, \tag{9.46}$$

where c_α *are the concentrations of the* α*th biological species, there being* N *such species. Here* V *is a bounded domain in* $\mathbb{R}^m$, $m = 1,2$ *or 3, with boundary* ∂V. *Define the terms* $\mathbf{J}^\alpha, \mathbf{n}$ *and* f_α *and explain the meaning of the three terms in equation* (9.46). *Introduce a suitable set of constitutive equations for* $\mathbf{J}^\alpha$, *with a brief justification, and show how* (9.46) *may be reduced to the pointwise system of partial differential equations,*

$$\frac{\partial c_\alpha}{\partial t} = \frac{\partial}{\partial x_r}\left(D_{rs}\frac{\partial c_\alpha}{\partial x_s}\right) + f_\alpha(c_\beta, x_k, t), \qquad \alpha = 1, \ldots, N.$$

Exercise 9.5.4 *(See (Keller and Segel, 1971b).) To model the wavelike motion of a bacterium such as Escherichia coli in a substrate in a capillary tube (Keller and Segel, 1971b) developed a chemotaxis model and analysed a travelling wave. They introduce the following simplified mathematical model which consists of the system of partial differential equations,*

$$\begin{aligned}\frac{\partial a}{\partial t} =& \mu\frac{\partial^2 a}{\partial x^2} - \frac{\partial}{\partial x}\left(\chi\frac{\partial \rho}{\partial x}\right),\\ \frac{\partial \rho}{\partial t} =& D\frac{\partial^2 \rho}{\partial x^2} - k(\rho)a,\end{aligned} \tag{9.47}$$

where a *is the density of a biological species at* (x,t) *and* ρ *is the density of attractant. Explain the role of each of the terms in* (9.47).

Define what is meant by a travelling chemotactic wave for system (9.47). *Assume* a, ρ *satisfy the boundary conditions*

$$\begin{aligned} &a \to 0, |z| \to \infty, \qquad && a' \to 0, |z| \to \infty,\\ &\rho \to 0, z \to -\infty, && \chi\rho' \to 0, z \to \infty,\end{aligned}$$

where $z = x - ct$, c *being a wavespeed. Show that* (9.47) *reduce to the system*

$$\begin{aligned} -ca &= \mu a' - \chi\rho',\\ \rho' &= -\frac{D}{c}\rho'' + \frac{k(\rho)}{c}a\,.\end{aligned}$$

Now assume $D = 0$, *and* μ, k *are constant and take*

$$\chi = \chi_0\frac{a}{\rho}$$

for χ_0 *constant.*

Replace the boundary condition $\chi\rho' \to 0, z \to \infty$, *by the condition* $\rho \to 1, z \to \infty$, *and verify that*

$$\frac{a}{\rho} - \frac{\beta}{\alpha} = -\frac{\beta}{\alpha}\rho^\alpha,$$

where $\alpha = (\chi_0/\mu) - 1$ and $\beta = c^2/k\mu$. *(Find* $da/d\rho = a'/\rho'$ *and solve the equation for* a/ρ *which arises.)*

Finally, assume $\alpha > 0$ *and show that for a constant* K

$$\rho(z) = \frac{1}{(1 + Ke^{-cz/\mu})^\alpha}, \tag{9.48}$$

and

$$a = \frac{Kc^2\alpha}{k\mu} \frac{e^{-cz/\mu}}{(1+Ke^{-cz/\mu})^{1+\alpha}} \,. \tag{9.49}$$

Exercise 9.5.5 *Write system (9.47) in the form*

$$\begin{aligned} \frac{\partial a}{\partial t} &= -\frac{\partial J}{\partial x}, \\ J &= -\mu\frac{\partial a}{\partial x} + \chi\frac{\partial \rho}{\partial x}, \\ \frac{\partial \rho}{\partial t} &= D\frac{\partial^2 \rho}{\partial x^2} - k(\rho)a. \end{aligned} \tag{9.50}$$

Generalize this with a Cattaneo like substitution

$$\begin{aligned} \frac{\partial a}{\partial t} &= -\frac{\partial J}{\partial x}, \\ \tau\frac{\partial J}{\partial t} + J &= -\mu\frac{\partial a}{\partial x} + \chi\frac{\partial \rho}{\partial x}, \\ \frac{\partial \rho}{\partial t} &= D\frac{\partial^2 \rho}{\partial x^2} - k(\rho)a, \end{aligned} \tag{9.51}$$

and show this leads to the equations

$$\begin{aligned} \tau\frac{\partial^2 a}{\partial t^2} + \frac{\partial a}{\partial t} &= \mu\frac{\partial^2 a}{\partial x^2} - \frac{\partial}{\partial x}\left(\chi\frac{\partial \rho}{\partial x}\right), \\ \frac{\partial \rho}{\partial t} &= D\frac{\partial^2 \rho}{\partial x^2} - k(\rho)a. \end{aligned} \tag{9.52}$$

Develop a travelling chemotactic wave analysis for system (9.52). Assume a, ρ satisfy the boundary conditions

$$\begin{aligned} &a \to 0, |z| \to \infty, \qquad &a' \to 0, |z| \to \infty, \\ &\rho \to 0, z \to -\infty, \qquad &\chi\rho' \to 0, z \to \infty, \end{aligned}$$

where $z = x - ct$, c being a wavespeed. Show that (9.52) reduce to the system

$$\begin{aligned} \tau a' - ca &= \mu a' - \chi\rho', \\ \rho' &= -\frac{D}{c}\rho'' + \frac{k(\rho)}{c}\, a \,. \end{aligned}$$

Now assume $D = 0$, and μ, k are constant and take

$$\chi = \chi_0 \, \frac{a}{\rho}$$

for χ_0 constant.

Replace the boundary condition $\chi\rho' \to 0, z \to \infty$, by the condition $\rho \to 1, z \to \infty$, and verify that

$$\frac{a}{\rho} - \frac{\beta}{\alpha} = -\frac{\beta}{\alpha}\,\rho^{\alpha},$$

where now

$$\alpha = \frac{\chi_0}{(\mu - \tau)} - 1, \qquad \text{and} \qquad \beta = \frac{c^2}{k(\mu - \tau)}.$$

If $\mu > \tau$, *assume* $\alpha > 0$ *and show that* (9.48) *and* (9.49) *continue to hold.*

References

Abd-Alla, A. M. and Abo-Dahab, S. M. (2009). Time - harmonic sources in a generalized magneto - thermo - viscoelastic continuum with and without energy dissipation. *Applied Mathematical Modelling*, 33:2388–2402.

Abramson, G., Bishop, A. R., and Kenkre, V. M. (2001). Effects of transport memory and nonlinear damping in a generalized Fisher's equation. *Physical Review E*, 64:066615–1–066615–6.

Abramson, G. and Kenkre, V. M. (2002). Spatiotemporal patterns in the hantavirus infection. *Physical Review E*, 66:011912–1–5.

Abramson, G., Kenkre, V. M., Yates, T. L., and Parmenter, R. R. (2003). Travelling waves of infection in hantavirus epidemics. *Bull. Math. Biol.*, 65:519–534.

Accuratus (2009). http://www.accuratus.com/alumox.html.

Ackerman, C. C. and Overton, W. C. (1969). Second sound in Helium - 3. *Phys. Rev. Lett.*, 22:764–766.

Al-Qahtani, H. and Yilbas, B. S. (2010). The closed form solutions for Cattaneo and stress equations due to step input pulse heating. *Physica B*, 405:3869–3874.

Allen, L. J. S., Langlais, M., and Phillips, C. (2003). The dynamics of two viral infections in a single host population with applications to hantavirus. *Math. Biosciences*, 186:191–217.

Allen, L. J. S., McCormack, R. K., and Jonsson, C. B. (2006). Mathematical models for hantavirus infection in rodents. *Bull. Math. Biol.*, 68:511–524.

Alvarez-Ramirez, J., Fernandez-Anaya, G., Valdes-Parada, F. J., and Ochoa-Tapia, J. A. (2006). A high order extension for Cattaneo's diffusion equation. *Physica A - Statistical Mechanics and its Applications*, 368:345–354.

Alvarez-Ramirez, J., Valdes-Parada, F. J., and Ochoa-Tapia, J. A. (2008). A lattice - Boltzmann scheme for Cattaneo's diffusion equation. *Physica A - Statistical Mechanics and its Applications*, 387:1475–1484.

Alves, M. S., Muñoz Rivera, J. E., and Quintanilla, R. (2009). Exponential decay in a thermoelastic mixture of solids. *Int. J. Solids Structures*, 46:1659–1666.

Ames, K. A., Clark, G. W., Epperson, J. F., and Oppenheimer, S. F. (1998). A comparison of regularizations for an ill-posed problem. *Mathematics of Computation*, 67:1451–1471.

Ames, K. A. and Payne, L. E. (1991). Stabilizing solutions of the equations of dynamical linear thermoelasticity backward in time. *Stab. Appl. Anal. Continuous Media*, 1:243–260.

Ames, K. A. and Payne, L. E. (1994). Uniqueness and continuous dependence of solutions to a multidimensional thermoelastic contact problem. *J. Elasticity*, 34:139–148.

Ames, K. A. and Payne, L. E. (1995). Continuous dependence on initial-time geometry for a thermoelastic system with sign-indefinite elasticities. *J. Math. Anal. Appl.*, 189:693–714.

Ames, K. A. and Payne, L. E. (1998). Asymptotic behaviour for two regularizations of the Cauchy problem for the backward heat equation. *Math. Models Meth. Appl. Sci.*, 8:187–202.

Ames, K. A. and Payne, L. E. (1999). Continuous dependence on modelling for some well-posed perturbations of the backward heat equation. *J. Inequalities and Applications*, 3:51–64.

Ames, K. A., Payne, L. E., and Schaefer, P. W. (1993). Spatial decay estimates in time - dependent Stokes flow. *SIAM J. Math. Anal.*, 24:1395–1413.

Ames, K. A., Payne, L. E., and Schaefer, P. W. (2004a). Energy and pointwise bounds in some nonstandard parabolic problems. *Proc. Royal Soc. Edinburgh A*, 134:1–9.

Ames, K. A., Payne, L. E., and Schaefer, P. W. (2004b). On a nonstandard problem for heat conduction in a cylinder. *Applicable Analysis*, 83:125–133.

Ames, K. A., Payne, L. E., and Song, J. C. (2001). Spatial decay in pipe flow of a viscous fluid interfacing a porous medium. *Math. Models Meth. Appl. Sci.*, 11:1547–1562.

Ames, K. A. and Straughan, B. (1992). Continuous dependence results for initially prestressed thermoelastic bodies. *Int. J. Engng. Sci.*, 30:7–13.

Ames, K. A. and Straughan, B. (1997). *Non-standard and improperly posed problems*, volume 194 of *Mathematics in Science and Engineering*. Academic Press, San Diego.

Amick, C. J. (1977). Steady solution of the Navier-Stokes equations in unbounded channels and pipes. *Ann. Scuola Norm. Sup. Pisa*, 4:473–513.

Amick, C. J. (1978). Properties of steady Navier-Stokes solutions for certain unbounded channels and pipes. *Nonlinear Analysis*, 2:689–720.

Anile, A. M. and Romano, V. (2001). Numerical investigation of shock waves in a radiating gas described by a variable Eddington factor. In Straughan, B., Greve, R., Ehrentraut, H., and Wang, Y., editors, *Continuum Mechanics and Applications in Geophysics and the Environment*. Springer.

Aouadi, M. (2008). Some theorems in the generalized theory of thermomagnetoelectroelasticity under Green - Lindsay's model. *Acta Mechanica*, 200:25–43.

Aouadi, M. (2009). Spatial stability for the quasi-static problem in thermoelastic diffusion theory. *Acta Applicandae Mathematicae*, 106:307–323.

Atkin, R. J. and Fox, N. (1984). The dependence of thermal shock wave velocity on heat flux in helium II. *J. Phys. C: Solid State Physics*, 17:1191–1198.

Atkin, R. J. and Fox, N. (1985). Spherical thermal shock waves in helium II. *J. Phys. C: Solid State Physics*, 18:1585–1593.

Auriault, J. L., Lewandowska, J., and Royer, P. (2007). About non-Fickian diffusion. In *18th Congrès Francais de Mécanique*. Grenoble, 27-31 August.

Barbera, E., Curro, C., and Valenti, G. (2008). A hyperbolic reaction - diffusion model for the hantavirus infection. *Math. Meth. Appl. Sci.*, 31:481–499.

Bargmann, S., Greve, R., and Steinmann, P. (2008a). Simulation of cryovolcanism on Saturns moon Enceladus with the Green - Naghdi theory of thermoelasticity. *Bulletin of Glaciological Research*, 26:23–32.

Bargmann, S. and Steinmann, P. (2006). Theoretical and computational aspects of non-classical thermoelasticity. *Comput. Methods Appl. Mech. Engng.*, 196:516–527.

Bargmann, S. and Steinmann, P. (2008). Modeling and simulation of first and second sound in solids. *Int. J. Solids and Structures*, 45:6067–6073.

Bargmann, S., Steinmann, P., and Jordan, P. M. (2008b). On the propagation of second-sound in linear and nonlinear media: Results from Green - Naghdi theory. *Physics Letters A*, 372:4418–4424.

Batra, R. C. (1974). A thermodynamic theory of rigid heat conductors. *Arch. Rational Mech. Anal.*, 53:359–367.

Batra, R. C. (1975). On heat conduction and wave propagation in nonsimple rigid solids. *Letters in Applied and Engineering Sciences*, 3:97–107.

Bellomo, N., Delitala, M., and Coscia, V. (2002a). On the mathematical theory of vehicular traffic flow. I. Fluid dynamical and kinematic modelling. *Math. Models Meth. Appl. Sci.*, 12:1801–1843.

Bellomo, N. and Dogbé, C. (2008). On the modelling crowd dynamics from scaling to hyperbolic models. *Math. Models Meth. Appl. Sci.*, 18:1317–1345.

Bellomo, N., Marasco, A., and Romano, A. (2002b). From the modelling of driver's behaviour to hydrodynamic models and problems of traffic flow. *Nonlinear Analysis: Real World Applications*, 3:339–363.

Berezovski, A. and Maugin, G. A. (2005). Stress - induced phase - transition front propagation in thermoelastic solids. *European J. Mechanics A / Solids*, 24:1–21.

Biquard, P. (1936). Sur l'absorption des ondes ultra-sonores par les liquides. *Ann. Physique (Ser. 11)*, 6:195–304.

Bleustein, J. L. and Green, A. E. (1967). Dipolar fluids. *Int. J. Engng. Sci.*, 5:323–340.

Bonetti, E. (2002a). An asymptotic analysis of a diffusive model for shape memory alloys with the Cattaneo-Maxwell heat flux law. *Differential Integral Equations*, 60:527–566.

Bonetti, E. (2002b). Some results on the well-posedness of an integro-differential Frémond model for shape memory alloys. *Rend. Sem. Mat. Univ. Pol. Torino*, 60:115–128.

Bonetti, E., Colli, P., Fabrizio, M., and Gilardi, G. (2007). Global solution to a singular integro-differential system related to the entropy balance. *Nonlinear Analysis*, 66:1949–1979.

Bonfanti, G., Muñoz Rivera, J. E., and Naso, M. G. (2008). Global existence and exponential stability for a contact problem between two thermoelastic beams. *J. Math. Anal. Appl.*, 345:186–202.

Boukai, A. I., Bunimovich, Y., Tahir-Kheli, J., Yu, J. K., Goddard, W. A., and Heath, J. R. (2008). Silicon nanowires as efficient thermoelectric materials. *Nature*, 451:168–171.

Brusov, P., Brusov, P., Lawes, G., Lee, C., Matsubara, A., Ishikawa, O., and Majumdar, P. (2003). Novel sound phenomena in superfluid helium in aerogel and other impure superfluids. *Physics Letters A*, 310:311–321.

Bryner, J. (2009). Massive planet sees extreme heat waves. http://www.msnbc.msn.com/id/28898771/.

Buishvili, T., Kekutia, S., Tkeshelashvili, O., and Tkeshelashvili, L. (2002). Theory of sound propagation in superfluid - filled porous media. *Physics Letters A*, 300:672–686.

Buongiorno, J. (2006). Convective transport in nanofluids. *ASME J. Heat Transfer*, 128:240–250.

Burden, R. L. and Faires, J. D. (2001). *Numerical Analysis.* Brooks Cole, Pacific Grove, CA, seventh edition.

Cai, R., Gou, C., and Li, H. (2006). Algebraically explicit solutions of unsteady 3-D nonlinear non-Fourier (hyperbolic) heat conduction. *Int. J. Thermal Sciences*, 45:893–896.

Carrassi, M. and Morro, A. (1972). A modified Navier - Stokes equation, and its consequences on sound dispersion. *Nuovo Cimento B*, 9:321–343.

Casas, P. S. and Quintanilla, R. (2005a). Exponential decay in one-dimensional porous - thermo - elasticity. *Mech. Research Communications*, 32:652–658.

Casas, P. S. and Quintanilla, R. (2005b). Exponential stability in thermoelasticity with microtemperatures. *Int. J. Engng. Sci.*, 43:33–47.

Cattaneo, C. (1948). Sulla conduzione del calore. *Atti Sem. Mat. Fis. Modena*, 3:83–101.

Caviglia, G. and Morro, A. (2005). Harmonic waves in thermoviscoelastic solids. *Int. J. Engng. Sci.*, 43:1323–1336.

Caviglia, G., Morro, A., and Straughan, B. (1992). Thermoelasticity at cryogenic temperatures. *Int. J. Nonlinear Mech.*, 27:251–263.

Chadwick, P. and Currie, P. K. (1974). Intrinsically characterized acceleration waves in heat-conducting elastic materials. *Proc. Camb. Phil. Soc.*, 76:481–491.

Chadwick, P. and Currie, P. K. (1975). On the existence of transverse elastic acceleration waves. *Math. Proc. Camb. Phil. Soc.*, 77:405–413.

Chandrasekhar, S. (1981). *Hydrodynamic and Hydromagnetic Stability.* Dover, New York.

Chandrasekharaiah, D. S. (1986). Hyperbolic thermoelasticity: a review. *Appl. Mech. Rev.*, 51:705–729.

Chandrasekharaiah, D. S. (1998). Hyperbolic thermoelasticity: a review of recent literature. *Appl. Mech. Rev.*, 51:705–729.

Chen, F. (1993). Stability of Taylor-Dean flow in an annulus with arbitrary gap spacing. *Phys. Rev. E*, 48:1036–1045.

Chen, J. K. and Beraun, J. E. (2001). Numerical study of ultrashort laser pulse interactions with metal films. *Numer. Heat Transfer A*, 40:1–20.

Chen, J. K. and Beraun, J. E. (2003). Investigation of thermal response caused by pulsed laser heating. *Numer. Heat Transfer A*, 44:705–722.

Chen, J. K., Beraun, J. E., and Tham, C. L. (2002a). Comparison of one-dimensional and two-dimensional axisymmetric approaches to the thermomechanical response caused by ultrashort laser heating. *J. Optics A: Pure and Applied Optics*, 4:650–661.

Chen, J. K., Latham, W. P., and Beraun, J. E. (2002b). Axisymmetric modelling of femtosecond-pulse laser heating on metal films. *Numer. Heat Transfer B*, 42:1–17.

Chen, P. J. (1969a). On the growth and decay of one-dimensional temperature rate waves. *Arch. Rational Mech. Anal.*, 35:1–15.

Chen, P. J. (1969b). On the growth and decay of temperature rate waves of arbitrary form. *ZAMP*, 20:448–453.

Chen, P. J. (1973). Growth and decay of waves in solids. In Flügge, S. and Truesdell, C., editors, *Handbuch der Physik*, volume VIa/3, pages 303–402. Springer, Berlin.

Chen, P. J. and Gurtin, M. E. (1968). On a theory of heat conduction involving two temperatures. *ZAMP*, 19:614–627.

Chen, P. J., Gurtin, M. E., and Williams, W. O. (1969). On a thermodynamics of non-simple elastic materials with two temperatures. *ZAMP*, 20:107–112.

Chirita, S. (1995a). Saint-Venant's principle in linear thermoelasticity. *J. Thermal Stresses*, 18:485–496.

Chirita, S. (1995b). Spatial decay estimates for solutions describing harmonic vibrations in a thermoelastic cylinder. *J. Thermal Stresses*, 18:421–436.

Chirita, S. (1997). A Phragmen-Lindelof principle in dynamic linear thermoelasticity. *J. Thermal Stresses*, 20:505–516.

Chirita, S. (2006). On the strong ellipticity condition for transversely isotropic linearly elastic solids. *An. St. Univ. Iasi Matematica*, 52:245–250.

Chirita, S. (2007). Spatial behaviour in strongly elliptic anisotropic thermoelastic materials. *J. Thermal Stresses*, 30:859–873.

Chirita, S. (2009). On some non-standard problems in linear thermoelasticity. *J. Thermal Stresses*, 32:1256–1269.

Chirita, S. and Ciarletta, M. (1999). Time-weighted surface power function method for the study of spatial behaviour in dynamics of continua. *European J. Mech. A/Solids*, 18:915–933.

Chirita, S. and Ciarletta, M. (2003a). Some further growth and decay results in linear thermoelastodynamics. *J. Thermal Stresses*, 26:889–903.

Chirita, S. and Ciarletta, M. (2003b). Spatial behaviour of solutions in the plane Stokes flow. *J. Math. Anal. Appl.*, 277:571–588.

Chirita, S. and Ciarletta, M. (2006). Spatial estimates for the constrained anisotropic elastic cylinder. *J. Elasticity*, 85:189–213.

Chirita, S. and Ciarletta, M. (2008). Spatial evolution of harmonic vibrations in linear elasticity. *J. Mechanics of Materials and Structures*, 3:1675–1693.

Chirita, S. and Ciarletta, M. (2010a). Reciprocal and variational principles in linear thermoelasticity without energy dissipation. *Mech. Research Communications*, 37:271–275.

Chirita, S. and Ciarletta, M. (2010b). Spatial behaviour for some non-standard problems in linear thermoelasticity without energy dissipation. *J. Math. Anal. Appl.*, 367:58–68.

Chirita, S. and Ciarletta, M. (2010c). Spatial evolution in linear thermoelasticity. In Ciarletta, M., Fabrizio, M., Morro, A., and Rionero, S., editors, *New trends in fluid and solid models*, pages 37–47. World Scientific. Proceedings of the International Conference, Vietri sul Mare, 28 February - 1 March, 2008.

Chirita, S., Ciarletta, M., and Fabrizio, M. (2001). Some spatial decay estimates in time-dependent Stokes slow flows. *Appl. Anal.*, 77:211–231.

Chirita, S., Ciarletta, M., and Straughan, B. (2006). Structural stability in porous elasticity. *Proc. Roy. Soc. London A*, 462:2593–2605.

Chirita, S. and Danescu, A. (2000). Spatial decay and growth properties for the incremental thermoelasticity. *ZAMP*, 51:39–60.

Chirita, S. and Danescu, A. (2008). Strong ellipticity for tetragonal system in linearly elastic solids. *Int. J. Solids Structures*, 45:4850–4859.

Chirita, S., Danescu, A., and Ciarletta, M. (2007). On the strong ellipticity of the anisotropic linearly elastic materials. *J. Elasticity*, 87:1–27.

Chirita, S. and Ghiba, I. D. (2010a). Inhomogeneous plane waves in elastic materials with voids. *Wave Motion*, 47:333–342.

Chirita, S. and Ghiba, I. D. (2010b). Strong ellipticity and progressive waves in elastic materials with voids. *Proc. Roy. Soc. London A*, 466:439–458.

Chirita, S. and Quintanilla, R. (1996). Spatial decay estimates of Saint-Venant type in generalized thermoelasticity. *Int. J. Engng. Sci.*, 34:299–311.

Christov, C. I. (2008). On the evolution of localized wave packets governed by a dissipative wave equation. *Wave Motion*, 45:154–161.

Christov, C. I. (2009). On frame indifferent formulation of the Maxwell - Cattaneo model of finite - speed heat conduction. *Mechanics Research Communications*, 36:481–486.

Christov, C. I. and Jordan, P. M. (2005). Heat conduction paradox involving second sound propagation in moving media. *Physical Review Letters*, 94:154301–1–154301–4.

Christov, I. and Jordan, P. M. (2008). Shock bifurcation and emergence of diffusive solitons in a nonlinear wave equation with relaxation. *New Journal of Physics*, 10:043027–1–043027–13.

Christov, I. and Jordan, P. M. (2009). Shock and traveling wave phenomena on an externally damped, non-linear string. *Int. J. Nonlinear Mech.*, 44:511–519.

Christov, I. and Jordan, P. M. (2010). On the propagation of second - sound in nonlinear media: shock, acceleration and traveling wave results. *J. Thermal Stresses*, 33:1109–1135.

Christov, I., Jordan, P. M., and Christov, C. I. (2006). Nonlinear acoustic propagation in homentropic perfect gases: a numerical study. *Physics Letters A*, 353:273–280.

Christov, I., Jordan, P. M., and Christov, C. I. (2007). Modelling weakly nonlinear acoustic wave propagation. *Quart. Jl. Mech. Appl. Math.*, 60:473–495.

Ciancio, V. and Quintanilla, R. (2007). Thermodynamics of materials with internal variables in the context of the Green and Naghdi theories. *Balkan J. Geom. Applications*, 12:16–31.

Ciarletta, M. (1999). A theory of micropolar thermoelasticity without energy dissipation. *J. Thermal Stresses*, 22:581–594.

Ciarletta, M. and Iesan, D. (1993). *Non-classical elastic solids.* Longman, New York.

Ciarletta, M. and Scalia, A. (1993). Thermodynamic theory for porous piezoelectric materials. *Meccania*, 28:303–308.

Ciarletta, M. and Scarpetta, E. (1989). Some results in a generalized thermodynamic theory for elastic materials with voids. *Rev. Roum. Sci. Techn. Méc. Appl.*, 34:113–119.

Ciarletta, M. and Scarpetta, E. (1996). Some results on thermoelasticity for porous piezoelectric materials. *Mech. Research Communications*, 23:1–10.

Ciarletta, M. and Straughan, B. (2006). Poroacoustic acceleration waves. *Proc. Roy. Soc. London A*, 462:3493–3499.

Ciarletta, M. and Straughan, B. (2007a). Poroacoustic acceleration waves with second sound. *J. Sound Vibration*, 306:725–731.

Ciarletta, M. and Straughan, B. (2007b). Thermo-poroacoustic acceleration waves in elastic materials with voids. *J. Math. Anal. Appl.*, 333:142–150.

Ciarletta, M. and Straughan, B. (2010). Uniqueness and structural stability for the Cattaneo-Christov equations. *Mech. Res. Communications*, 37:445–447.

Ciarletta, M., Straughan, B., and Tibullo, V. (2010). Structural stability for a rigid body with thermal microstructure. *Int. J. Engng. Sci.*, 48:592–598.

Ciarletta, M., Straughan, B., and Zampoli, V. (2007). Thermo-poroacoustic acceleration waves in elastic materials with voids without energy dissipation. *Int. J. Engng. Sci.*, 45:736–743.

Cimmelli, V. A. and Frischmuth, K. (2007). Gradient generalization to the extended thermodynamic approach and diffusive - hyperbolic heat conduction. *Physica B*, 400:257–265.

Cimmelli, V. A., Jou, D., and Sellitto, A. (2010a). Propagation of temperature waves along core-shell nanowires. *J. Non-Equilibrium Thermodynamics*, 35:267–278.

Cimmelli, V. A., Sellitto, A., and Triani, V. (2010b). A generalized Coleman-Noll procedure for the exploitation of the entropy principle. *Proc. Roy. Soc. London A*, 466:911–925.

Cisotti, U. (1911). La ereditarietà lineare e i fenomeni dispersivi. *Nuovo Cimento*, 2:234–244.

Coleman, B. D., Fabrizio, M., and Owen, D. R. (1982). On the thermodynamics of second sound in dielectric crystals. *Arch. Rational Mech. Anal.*, 80:135–158.

Coleman, B. D., Fabrizio, M., and Owen, D. R. (1986). Thermodynamics and the constitutive relations for second sound in dielectric crystals. In Serrin, J., editor, *New perspectives in thermodynamics*. Springer.

Coleman, B. D. and Gurtin, M. E. (1965). Waves in materials with memory. III. Thermodynamic influences on the growth and decay of acceleration waves. *Arch. Rational Mech. Anal.*, 19:266–298.

Coleman, B. D. and Newman, D. C. (1988). Implications of a nonlinearity in the theory of second sound in solids. *Phys. Rev. B*, 37:1492–1498.

Coleman, B. D. and Noll, W. (1963). The thermodynamics of elastic materials with heat conduction and viscosity. *Arch. Rational Mech. Anal.*, 13:167–178.

Colli, P. and Recupero, V. (2002). Convergence to the Stefan problem of the phase relaxation problem with Cattaneo heat flux law. *J. Evol. Equ.*, 2:177–195.

Cramer, M. S., Watson, L. T., and Lowenkamp, B. B. (2001). Approximation models for nonlinear thermal waves. *Nonlinear Analysis: Real World Applications*, 2:221–247.

Curro, C., Valenti, G., Sugiyama, M., and Taniguchi, S. (2009). Propagation of an acceleration wave in layers of isotropic solids at finite temperatures. *Wave Motion*, 46:108–121.

Dafermos, C. M. (1968). On the existence and the asymptotic stability of solutions to the equations of linear thermoelasticity. *Arch. Rational Mech. Anal.*, 29:241–271.

Dafermos, C. M. (1985). Contemporary issues in the dynamic behavior of continuous media. LCDS Lecture Notes 85-1, Lefschetz Center for Dynamical Systems and Center for Control Sciences.

Dafermos, C. M. (2006). Continuous solutions for balance laws. *Ricerche di Matematica*, 55:79–91.

Dafermos, C. M. (2010). *Hyperbolic conservation laws in continuum physics*, volume 325 of *Grundleheren der mathematischen Wissenschaften*. Springer, 3rd edition.

Dafermos, C. M. and Hsiao, L. (1986). Development of singularities in solutions of the equations of nonlinear thermoelasticity. *Quart. Appl. Math.*, 44:463–474.

Dai, W. Z., Wang, H. J., Jordan, P. M., Mickens, R. E., and Bejan, A. (2008). A mathematical model for skin burn injury induced by radiation heating. *Int. J. Heat Mass Transfer*, 51:5497–5510.

Darbha, S., Rajagopal, K. R., and Tyagi, V. (2008). A review of mathematical models for the flow of traffic and some recent results. *Nonlinear Analysis, Theory, Methods and Applications*, 69:950–970.

Dauby, P. C., Nélis, M., and Lebon, G. (2002). Generalized Fourier equations and thermoconvective instabilities. *Revista Mexicana de Fisica*, 48:57–62.

De Cicco, S. and Diaco, M. (2002). A theory of thermoelastic materials with voids without energy dissipation. *J. Thermal Stresses*, 25:493–503.

Dharmawadane, P. M. N., Muñoz Rivera, J. E., and Kawashima, S. (2010). Decay property for second order hyperbolic systems of viscoelastic materials. *J. Math. Anal. Appl.*, 366:621–635.

Diebold, A. C. (2005). Subsurface imaging with scanning ultrasound holography. *Science*, 310:61–62.

Dolak, Y. and Hillen, T. (2003). Cattaneo models for chemosensitive movement. *J. Math. Biology*, 46:461–478.

Dongarra, J. J., Straughan, B., and Walker, D. W. (1996). Chebyshev tau - QZ algorithm methods for calculating spectra of hydrodynamic stability problems. *Appl. Numer. Math.*, 22:399–435.

Donnelly, R. J. (2009). The two-fluid theory and second sound in liquid helium. *Physics Today*, pages 34–39. October.

Dow (2009). http://www.dow.com/ethyleneglycol/about/properties.htm.

Dreher, M., Quintanilla, R., and Racke, R. (2009). Ill-posed problems in thermomechanics. *Appl. Math. Letters*, 22:1374–1379.

Dreyer, W. and Struchtrup, H. (1993). Heat pulse experiments revisited. *Continuum Mech. Thermodyn.*, 5:3–50.

Duhamel, P. (2001). A new finite integral transform pair for hyperbolic conduction problems in heterogeneous media. *Int. J. Heat Mass Trans.*, 44:3307–3320.

EngineeringToolbox (2009). http://www.engineeringtoolbox.com/water-thermal-properties-d_162.html.

Eremeyev, V. A. (2005). Acceleration waves in micropolar elastic media. *Doklady Physics*, 50:204–206.

Eringen, A. C. (1990). Theory of microstretch elastic solids. *Int. J. Engng. Sci.*, 28:1291–1301.

Eringen, A. C. (2004). Electromagnetic theory of microstretch elasticity and bone modeling. *Int. J. Engng. Sci.*, 42:231–242.

Espinosa-Paredes, G. and Espinosa-Martinez, E. G. (2009). Fuel rod model based on non-Fourier heat conduction equation. *Annals of Nuclear Energy*, 36:680–693.

Fabrizio, M. (1994). *Introduzione alla meccanica razionale e ai suoi metodi matematici.* Zanichelli, Bologna, second edition.

Fabrizio, M. (2011). Dario Graffi in a complex historical period. In *Mathematicians in Bologna 1861 – 1960.* Birkhauser.

Fabrizio, M., Franchi, F., and Straughan, B. (2008). On a model for thermo-poroacoustic waves. *Int. J. Engng. Sci.*, 46:790–798.

Fabrizio, M., Gentili, G., and Reynolds, D. W. (1998). On rigid linear heat conductors with memory. *Int. J. Engng. Sci.*, 36:765–782.

Fabrizio, M., Lazzari, B., and Muñoz Rivera, J. E. (2007). Asymptotic behaviour for a two-dimensional thermoelastic model. *Math. Meth. Appl. Sci.*, 30:549–566.

Fabrizio, M. and Morro, A. (2003). *Electromagnetism of continuous media.* Oxford University Press, Oxford.

Falcón, N. (2001). Compact star cooling by means of heat waves. In *RevMexAA (Serie de Conferencias).* See also http://adsabs.harvard.edu/full/2001RMxAC..11..41F.

Ferreira, J. A. and de Oliveira, P. (2010). Looking for the lost memory in diffusion - reaction equations. In Rannacher, R. and Sequeira, A., editors, *Advances in Mathematical Fluid Mechanics*, pages 229–251. Springer.

Fichera, G. (1992). Is the Fourier theory of heat propagation paradoxical? *Rend. Circolo Matem. Palermo*, 41:5–28.

Flavin, J. N. and Rionero, S. (1995). *Qualitative Estimates for Partial Differential Equations.* CRC Press, Boca Raton.

Fox, N. (1969a). Generalized thermoelasticity. *Int. J. Engng. Sci.*, 7:437–445.

Fox, N. (1969b). Low temperature effects and generalized thermoelasticity. *J. Inst. Maths. Applics.*, 5:373–386.

Franchi, F. (1985). Wave propagation in heat conducting dielectric solids with thermal relaxation and temperature dependent electric permittivity. *Riv. Mat. Univ. Parma*, 11:443–461.

Franchi, F. and Straughan, B. (1994a). Continuous dependence on the relaxation time and modelling, and unbounded growth, in theories of heat conduction with finite propagation speeds. *J. Math. Anal. Appl.*, 185:726–746.

Franchi, F. and Straughan, B. (1994b). Thermal convection at low temperature. *J. Non-Equilibrium Thermodynamics*, 19:368–374.

Fu, Y. B. and Scott, N. H. (1988). Acceleration wave propagation in an inhomogeneous heat conducting elastic rod of slowly varying cross section. *J. Thermal Stresses*, 15:253–264.

Fu, Y. B. and Scott, N. H. (1990). One-dimensional shock waves in simple materials with memory. *Proc. Roy. Soc. London A*, 428:547–571.

Fu, Y. B. and Scott, N. H. (1991). The transistion from acceleration wave to shock wave. *Int. J. Engng. Sci.*, 29:617–624.

Gabaraev, B. A., Kovalev, S. A., Molochnikov, Y. S., Soloviev, S. L., and Usatikov, S. V. (2003). Boiling curve in temperature wave region. *Int. J. Heat Mass Transfer*, 46:139–148.

Galdi, G. P. and Rionero, S. (1985). *Weighted energy methods in fluid dynamics and elasticity*, volume 1134 of *Lecture notes in mathematics*. Springer, Berlin.

Galenko, P. (2001). Phase-field model with relaxation of the diffusion flux in nonequilibrium solidification of a binary system. *Phys. Letters A*, 287:190–197.

Galenko, P. and Jou, D. (2005). Diffuse-interface model for rapid phase transformations in nonequilibrium systems. *Phys. Rev. E*, 71:046125.

Galenko, P. K. and Danilov, D. A. (1997). Local nonequilibrium effect on rapid dendritic growth in a binary alloy melt. *Phys. Letters A*, 235:271–280.

Garai, M. and Pompoli, F. (2005). A simple empirical model of polyester fibre materials for acoustical applications. *Applied Acoustics*, 66:1383–1398.

Gatti, S., Grasselli, M., Miranville, A., and Pata, V. (2005). On the hyperbolic relaxation of the one-dimensional Cahn-Hilliard equation. *J. Math. Anal. Appl.*, 312:230–247.

Gentili, G. and Giorgi, C. (1993). Thermodynamic properties and stability for the heat flux equation with linear memory. *Quart. Appl. Math.*, 51:343–362.

Ghaleb, A. F. and El-Deen Mohamedein, M. S. (1989). A heat conduction equation with three relaxation times. *Int. J. Engng. Sci.*, 27:1367–1377.

Ghosh, M. K. and Kanoria, M. (2009). Analysis of thermoelastic response in a functionally graded spherically isotropic hollow sphere based on Green - Lindsay theory. *Acta Mech.*, 207:51–67.

Gilbarg, D. and Trudinger, N. S. (1977). *Elliptic partial differential equations of second order.* Springer, Heidelberg.

Glass, D. E., Ozisik, M. N., McRae, D. S., and Vick, B. (1986). Hyperbolic heat conduction with temperature - dependent thermal conductivity. *J. Appl. Phys.*, 59:1861–1865.

Goldstein, J. A. (1985). *Semigroups of linear operators and applications.* Oxford University Press, New York.

Goodman, M. A. and Cowin, S. C. (1972). A continuum theory for granular materials. *Arch. Rational Mech. Anal.*, 44:249–266.

Gouin, H. and Ruggeri, T. (2008). Identification of an average temperature and a dynamical pressure in a multi-temperature mixture of fluids. *Phys. Rev. E*, 78:016303.

Grad, H. (1949). On the kinetic theory of rarefied gases. *Comm. Pure Appl. Math.*, 2:331–407.

Graffi, D. (1928a). Sui problemi della ereditarietà lineare. *Nuovo Cimento*, 5:53–71.

Graffi, D. (1928b). Sulla teoria delle oscillazioni elastiche con ereditarietà. *Nuovo Cimento*, 5:310–317.

Graffi, D. (1936a). Sopra alcuni fenomeni ereditari dell'elettrologia. *Rend. Ist. Lomb. Sc. Lett.*, 19:151–166.

Graffi, D. (1936b). Una teoria ereditaria dell'effetto Lussemburgo. *Rend. Sem. Mat. Univ. Padova*, 14:36–64.

Graffi, D. (1960). Sul teorema di unicità nella dinamica dei fluidi. *Annali di Matematica Pura ed Applicata*, 50:379–388.

Graffi, D. (1984). Personal communication to Prof. F. Franchi. From the Accademia dei Lincei, Roma.

Graffi, D. (1999a). Sui problemi della ereditarietà lineare. In Fabrizio, M., Grioli, G., and Renno, P., editors, *Opere Scelte*, pages 1–19. C.N.R. Gruppo Nazionale per la Fisica Matematica, Bologna.

Graffi, D. (1999b). Sul teorema di unicità nella dinamica dei fluidi. In Fabrizio, M., Grioli, G., and Renno, P., editors, *Opere Scelte*, pages 263–271. C.N.R. Gruppo Nazionale per la Fisica Matematica, Bologna.

Graffi, D. (1999c). Una teoria ereditaria dell'effetto Lussemburgo. In Fabrizio, M., Grioli, G., and Renno, P., editors, *Opere Scelte*, pages 36–54. C.N.R. Gruppo Nazionale per la Fisica Matematica, Bologna.

Grasselli, M., Petzeltová, H., and Schimperna, G. (2007). Asymptotic behaviour of a nonisothermal viscous Cahn-Hilliard equation with inertial term. *J. Differential Equations*, 239:38–60.

Green, A. E. (1972). A note on linear thermoelasticity. *Mathematika*, 19:69–75.

Green, A. E. and Laws, N. (1972). On the entropy production inequality. *Arch. Rational Mech. Anal.*, 45:47–53.

Green, A. E. and Lindsay, K. A. (1972). Thermoelasticity. *J. Elasticity*, 2:1–7.

Green, A. E. and Naghdi, P. M. (1977). On thermodynamics and the nature of the second law. *Proc. Roy. Soc. London A*, 357:253–270.

Green, A. E. and Naghdi, P. M. (1991). A re-examination of the basic postulates of thermomechanics. *Proc. Roy. Soc. London A*, 432:171–194.

Green, A. E. and Naghdi, P. M. (1992). On undamped heat waves in an elastic solid. *J. Thermal Stresses*, 15:253–264.

Green, A. E. and Naghdi, P. M. (1993). Thermoelasticity without energy-dissipation. *J. Elasticity*, 31:189–208.

Green, A. E. and Naghdi, P. M. (1995a). A new thermoviscous theory for fluids. *J. Non-Newtonian Fluid Mech.*, 56:289–306.

Green, A. E. and Naghdi, P. M. (1995b). A unified procedure for construction of theories of deformable media. I. Classical continuum physics. *Proc. Roy. Soc. London A*, 448:335–356.

Green, A. E. and Naghdi, P. M. (1995c). A unified procedure for construction of theories of deformable media. II. Generalized continua. *Proc. Roy. Soc. London A*, 448:357–377.

Green, A. E. and Naghdi, P. M. (1995d). A unified procedure for construction of theories of deformable media. III. Mixtures of interacting continua. *Proc. Roy. Soc. London A*, 448:379–388.

Green, A. E. and Naghdi, P. M. (1996). An extended theory for incompressible viscous fluid flow. *J. Non-Newtonian Fluid Mech.*, 66:233–255.

Green, A. E. and Rivlin, R. S. (1964). Multipolar continuum mechanics. *Arch. Rational Mech. Anal.*, 17:113–147.

Gultop, T. (2006). On the propagation of acceleration waves in incompressible hyperelastic solids. *J. Sound Vibration*, 462:409–418.

Gurtin, M. E. and Pipkin, A. C. (1968). A general theory of heat conduction with finite wavespeeds. *Arch. Rational Mech. Anal.*, 31:113–126.

Guyer, R. and Krumhansl, J. (1964). Dispersion relation for second sound in solids. *Phys. Review*, 133:1411–1417.

Guyer, R. and Krumhansl, J. (1966a). Solution of the linearized Boltzmann phonon equation. *Phys. Review*, 148:766–778.

Guyer, R. and Krumhansl, J. (1966b). Thermal conductivity, second sound, and phonon hydrodynamic phenomena in nonmetallic crystals. *Phys. Review*, 148:778–788.

Han, P., Tang, D., and Zhou, L. (2006). Numerical analysis of two-dimensional lagging thermal behaviour under short-pulse-laser heating on surface. *Int. J. Engng. Sci.*, 44:1510–1519.

Hardy, R. J. and Jaswal, S. S. (1971). Velocity of second sound in NaF. *Phys. Rev. B*, 3:4385–4387.

Hashimoto, T., Morikawa, J., Kurihara, T., and Tsuji, T. (1997). Frequency dependent thermal diffusivity of polymers by temperature wave analysis. *Thermochimica Acta*, 304:151–156.

Herrera, L. and Falcón, N. (1995). Heat waves and thermohaline instability in a fluid. *Physics Letters A*, 201:33–37.

Hetnarski, R. B. and Ignaczak, J. (1996). Soliton - like waves in a low - temperature nonlinear thermoelastic solid. *Int. J. Engng. Sci.*, 34:1767–1787.

Hetnarski, R. B. and Ignaczak, J. (1997). On soliton - like thermoelastic waves. *Applicable Analysis*, 65:183–204.

Hetnarski, R. B. and Ignaczak, J. (1999). Generalized thermoelasticity. *J. Thermal Stresses*, 22:451–461.

Hetnarski, R. B. and Ignaczak, J. (2000). Nonlinear dynamical thermoelasticity. *Int. J. Solids Structures*, 37:215–224.

Hillen, T. and Levine, H. A. (2003). Blow-up and pattern formation in hyperbolic models for chemotaxis in 1-D. *ZAMP*, 54:1–30.

Hochbaum, A. I., Chen, R., Delgado, R. D., Liang, W., Garnett, E. C., Najarian, M., Majumdar, A., and Yang, P. (2008). Enhanced thermoelectric performance of rough silicon nanowires. *Nature*, 451:163–167.

Horgan, C. O. (1989). Recent developments concerning Saint-Venant's principle: An update. *Applied Mechanics Reviews*, 42:295–303.

Horgan, C. O. (1996). Recent developments concerning Saint-Venant's principle: A second update. *Applied Mechanics Reviews*, 49:101–111.

Horgan, C. O. and Knowles, J. K. (1983). Recent developments concerning Saint-Venant's principle. *Advances in Applied Mechanics*, 23:179–269.

Horgan, C. O. and Payne, L. E. (1997). Spatial decay estimates for a coupled system of second-order quasilinear partial differential equations arising in thermoelastic finite anti-plane shear. *J. Elasticity*, 47:3–21.

Horgan, C. O. and Quintanilla, R. (2005). Spatial behaviour of solutions of the dual - phase - lag heat equation. *Math. Meth. Appl. Sci.*, 28:43–57.

Hwang, K. S., Lee, J. H., and Jang, S. P. (2007). Buoyancy driven heat transfer of water based Al_2O_3 nanofluids in a rectangular cavity. *Int. J. Heat Mass Transfer*, 50:4003–4010.

Iesan, D. (1980). Incremental equations in thermoelasticity. *J. Thermal Stresses*, 3:41–56.

Iesan, D. (1988). Thermoelasticity of initially heated bodies. *J. Thermal Stresses*, 11:17–38.

Iesan, D. (2002). On the theory of heat conduction in micromorphic continua. *Int. J. Engng. Sci.*, 40:1859–1878.

Iesan, D. (2004). *Thermoelastic models of continua.* Kluwer, Dordrecht.

Iesan, D. (2005). Second-order effects in the torsion of elastic materials with voids. *ZAMM*, 85:351–365.

Iesan, D. (2006). Nonlinear plane strain of elastic materials with voids. *Math. Mech. Solids*, 11:361–384.

Iesan, D. (2008). Thermopiezoelectricity without energy dissipation. *Proc. Roy. Soc. London A*, 464:631–656.

Iesan, D. and Nappa, L. (2005). On the theory of heat for micromorphic bodies. *Int. J. Engng. Sci.*, 43:17–32.

Iesan, D. and Quintanilla, R. (2009). On thermoelastic bodies with inner structure and microtemperatures. *J. Math. Anal. Appl.*, 354:12–23.

Iesan, D. and Scalia, A. (1996). *Thermoelastic deformations.* Kluwer, Dordrecht.

Iesan, D. and Scalia, A. (2006). Propagation of singular surfaces in thermo-microstretch continua with memory. *Int. J. Engng. Sci.*, 44:845–858.

Ignaczak, J. (1990). Soliton - like solutions in a nonlinear dynamic coupled thermoelasticity. *J. Thermal Stresses*, 13:73–98.

Ignaczak, J. (1998). Saint-Venant type decay estimates for transient heat conduction in a composite rigid semispace. *J. Thermal Stresses*, 21:185–204.

Ignaczak, J. (2000). A spatial decay estimate for a transient thermoelastic process in a composite semispace. *J. Thermal Stresses*, 23:1–14.

Ignaczak, J. (2002). Saint-Venant's principle for a microperiodic composite thermoelastic semispace. *J. Thermal Stresses*, 25:1065–1079.

Ignaczak, J. (2009). Modelling heat transfer in metal films by a third - order derivative - in - time dissipative and dispersive wave equation. *J. Thermal Stresses*, 32:847–861.

Ignaczak, J. and Ostoja-Starzewski, M. (2009). *Thermoelasticity with finite wave speeds*. Oxford Univ. Press.

Iovane, G. and Passarella, F. (2004a). Saint-Venant's principle in dynamic porous thermoelastic media with memory for heat flux. *J. Thermal Stresses*, 27:983–999.

Iovane, G. and Passarella, F. (2004b). Spatial behaviour in dynamical thermoelasticity backward in time for porous media. *J. Thermal Stresses*, 27:97–109.

Irmscher, T. and Racke, R. (2006). Sharp decay rates in parabolic and hyperbolic thermoelasticity. *IMA J. Appl. Math.*, 71:459–478.

Jackson, H. E., Walker, C. T., and McNelly, T. F. (1970). Second sound in NaF. *Phys. Rev. Lett.*, 25:26–29.

Jaisaardsuetrong, J. and Straughan, B. (2007). Thermal waves in a rigid heat conductor. *Physics Letters A*, 366:433–436.

Jiang, J. (2009). Convergence to equilibrium for a fully hyperbolic phase-field model with Cattaneo heat flux law. *Math. Meth. Appl. Sci.*, 32:1156–1182.

Johnson, D. L., Plona, T. J., and Kajima, H. (1994). Probing porous media with first and second sound. II. Acoustic properties of water saturated porous media. *J. Appl. Phys.*, 76:115–125.

Jordan, P. M. (2004). An analytical study of Kuznetsov's equation: diffusive solitons, shock formation, and solution bifurcation. *Phys. Letters A*, 326:77–84.

Jordan, P. M. (2005a). Growth and decay of acoustic acceleration waves in Darcy-type porous media. *Proc. Roy. Soc. London A*, 461:2749–2766.

Jordan, P. M. (2005b). Growth and decay of shock and acceleration waves in a traffic flow model with relaxation. *Physica D*, 207:220–229.

Jordan, P. M. (2006). Finite amplitude acoustic travelling waves in a fluid that saturates a porous medium: Acceleration wave formation. *Phys. Letters A*, 355:216–221.

Jordan, P. M. (2007). Growth, decay and bifurcation of shock amplitudes under the type-II flux law. *Proc. Roy. Soc. London A*, 463:2783–2798.

Jordan, P. M. (2008a). A note on Burgers' equation with time delay: instability via a finite-time blow-up. *Phys. Lett. A*, 372:6363–6367.

Jordan, P. M. (2008b). On the growth and decay of transverse acceleration waves on a nonlinear, externally damped string. *J. Sound Vibration*, 311:597–607.

Jordan, P. M. (2010a). On the application of the Cole-Hopf transformation to hyperbolic equations based on second-sound models. *Mathematics and Computers in Simulation*, 81:18–25.

Jordan, P. M. (2010b). Second - sound propagation in rigid, nonlinear conductors: Instability and temperature - rate wave formation. manuscript, 11 pages.

Jordan, P. M. and Christov, C. I. (2005). A simple finite difference scheme for modelling the finite-time blow-up of acoustic acceleration waves. *J. Sound Vibration*, 281:1207–1216.

Jordan, P. M., Dai, W., and Mickens, R. E. (2008). A note on the delayed heat equation: Instability with respect to initial data. *Mech. Res. Comm.*, 35:414–420.

Jordan, P. M. and Feuillade, C. (2004). On the propagation of harmonic acoustic waves in bubbly liquids. *Int. J. Engng. Sci.*, 42:1119–1128.

Jordan, P. M., Meyer, M. R., and Puri, A. (2000). Causal implications of viscous damping in compressible fluid flows. *Phys. Review E*, 62:7918–7926.

Jordan, P. M. and Puri, A. (1999). Digital signal propagation in dispersive media. *J. Appl. Phys.*, 85:1273–1282.

Jordan, P. M. and Puri, A. (2005). Growth/decay of transverse acceleration waves in nonlinear elastic media. *Phys. Letters A*, 341:427–434.

Jordan, P. M. and Puri, P. (2001). Thermal stresses in a spherical shell under three thermoelastic models. *J. Thermal Stresses*, 24:47–70.

Jordan, P. M. and Straughan, B. (2006). Acoustic acceleration waves in homentropic Green and Naghdi gases. *Proc. Roy. Soc. London A*, 462:3601–3611.

Joseph, D. D. and Preziosi, L. (1989). Heat waves. *Rev. Mod. Phys.*, 61:41–73.

Joseph, D. D. and Preziosi, L. (1990). Heat waves: addendum. *Rev. Mod. Phys.*, 62:375–391.

Jou, D., Casas-Vázquez, J., and Lebon, G. (2010a). *Extended irreversible thermodynamics*. Springer, Berlin, fourth edition.

Jou, D., Cimmelli, V. A., and Sellitto, A. (2009). Nonequilibrium temperatures and second-sound propagation along nanowires and thin layers. *Physics Letters A*, 373:4386–4392.

Jou, D. and Criado-Sancho, M. (1998). Thermodynamic stability and temperature overshooting in dual-phase-lag heat transfer. *Phys. Letters A*, 248:172–178.

Jou, D., Lebon, G., and Criado-Sancho, M. (2010b). Variational principles for thermal transport in nanosystems with heat slip flow. *Phys. Review E*, 82:031128.

Jou, D. and Sellitto, A. (2009). Focussing of heat pusles along nonequilibrium nanowires. *Physics Letters A*, 374:313–318.

Jou, D., Sellitto, A., and Alvarez, F. X. (2011). Heat waves and phonon-wall collisions in nanowires. *Proc. Roy. Soc. London A*, 640:10.1098/rspa.2010.0645.

Kalpakides, V. K. and Maugin, G. A. (2004). Canonical formulation and conservation laws of thermoelasticity without dissipation. *Reports on Mathematical Physics*, 53:371–391.

Kameyama, N. and Sugiyama, M. (1996). Analysis of acceleration waves in crystalline solids based on a continuum model incorporating microscopic thermal vibration. *Continuum Mech. Thermodyn.*, 8:351–359.

Kaminski, W. (1990). Hyperbolic heat conduction equation for materials with a non-homogeneous inner structure. *ASME J. Heat Transfer*, 112:555–560.

Keller, E. F. and Segel, L. A. (1970). Initiation of slime mold aggregation viewed as an instability. *J. Theor. Biol.*, 26:399–415.

Keller, E. F. and Segel, L. A. (1971a). Model for chemotaxis. *J. Theor. Biol.*, 30:225–234.

Keller, E. F. and Segel, L. A. (1971b). Traveling bands of chemotactic bacteria: a theoretical analysis. *J. Theor. Biol.*, 30:235–248.

Kim, S. H., Choi, S. R., and Kim, D. (2007). Thermal conductivity of metal - oxide nanofluids: particle size dependence and effect of laser irradiation. *ASME J. Heat Transfer*, 129:298–307.

Kirchoff, G. (1868). Ueber den Einfluss der Waermeleitung in einem Gas auf die Schallbewegung. *Pogg. Ann. Phys.*, 134:177–193.

Knops, R. J. and Payne, L. E. (1970). On uniqueness and continuous dependence in dynamical problems of linear thermoelasticity. *Int. J. Solids Structures*, 6:1173–1184.

Knops, R. J. and Payne, L. E. (1971a). Growth estimates for solutions of evolutionary equations in Hilbert space with applications in elastodynamics. *Arch. Rational Mech. Anal.*, 41:363–398.

Knops, R. J. and Payne, L. E. (1971b). *Uniqueness theorems in linear elasticity*, volume 19 of *Tracts in Natural Philosophy*. Springer, Berlin.

Knops, R. J. and Payne, L. E. (2005). Alternative spatial growth and decay for constrained motion in an elastic cylinder. *Math. Mech. Solids*, 10:281–310.

Knops, R. J. and Wilkes, E. W. (1973). Theory of elastic stability. In Truesdell, C., editor, *Handbuch der Physik*, pages 125–302. Springer, Berlin. Vol. VIa/3.

Kumar, R. and Mukhopadhyay, S. (2010). Effects of thermal relaxation time on plane wave propagation under two - temperature thermoelasticity. *Int. J. Engng. Sci.*, 48:128–139.

Kuznetsov, A. V. and Nield, D. A. (2010a). Effect of local thermal non-equilibrium on the onset of convection in a porous medium layer saturated by a nanofluid. *Transport in Porous Media*, 83:425–436.

Kuznetsov, A. V. and Nield, D. A. (2010b). Thermal instability in a porous medium layer saturated by a nanofluid: Brinkman model. *Transport in Porous Media*, 81:409–422.

Kwak, K. and Kim, C. (2005). Viscosity and thermal conductivity of copper oxide nanofluid dispersed in ethylene glycol. *Korea-Australia Rheology Journal*, 17:35–40.

Lakes, R. and Wojciechowski, K. W. (2008). Negative compressibility, negative Poisson's ratio, and stability. *Physica Status Solidi b*, 245:545–551.

Lakes, R. S. (2008). Negative Poisson's ratio materials. http://silver.neep.wisc.edu/ lakes/Poisson.html.

Lazzari, B. and Nibbi, R. (2007). On the energy decay of a linear hyperbolic thermoelastic system with dissipative boundary. *J. Thermal Stresses*, 30:1–14.

Lazzari, B. and Nibbi, R. (2008). On the exponential decay in thermoelasticity without energy dissipation and of type III in the presence of an absorbing boundary. *J. Math. Anal. Appl.*, 338:317–329.

Lebeau, G. and Zuazua, E. (1999). Decay rates for the three-dimensional linear systems of thermoelasticity. *Arch. Rational Mech. Anal.*, 148:179–231.

Lebon, G. and Cloot, G. (1984). Bénard - Marangoni instability in a Maxwell - Cattaneo fluid. *Physics Letters A*, 105:361–364.

Lebon, G. and Dauby, P. C. (1990). Heat transport in dielectric crystals at low temperature: A variational formulation based on extended irreversible thermodynamics. *Phys. Review A*, 42:4710–4715.

Lebon, G., Jou, D., and Casas-Vázquez, J. (2008). *Understanding non-equilibrium thermodynamics*. Springer, Berlin.

Leseduarte, M. C., Magana, A., and Quintanilla, R. (2010). On the time decay of solutions in porous - thermo - elasticity of type II. *Discrete Continuous Dynamical Systems B*, 13:375–391.

Levine, H. A. (1970). On a theorem of Knops and Payne in dynamical linear thermoelasticity. *Arch. Rational Mech. Anal.*, 38:290–307.

Lighthill, M. J. (1963). I. Introduction, real and ideal fluids. In Rosenhead, L., editor, *Laminar boundary layers*, pages 1–45. Oxford University Press.

Lighthill, M. J. and Whitham, G. B. (1955). On kinematic waves. II. Theory of traffic flow on long crowded roads. *Proc. Roy. Soc. London A*, 229:317–345.

Lin, C. and Payne, L. E. (2004a). Continuous dependence of heat flux on spatial geometry for the generalized Maxwell-Cattaneo system. *ZAMP*, 55:575–591.

Lin, C. and Payne, L. E. (2004b). Spatial decay bounds in time dependent pipe flow of an incompressible viscous fluid. *SIAM J. Appl. Math.*, 65:458–474.

Lin, C. C. and Segel, L. A. (1974). *Mathematics applied to deterministic problems in the Natural Sciences*. Macmillan, New York.

Lin, H. and Szeri, A. J. (2001). Shock formation in the presence of entropy gradients. *J. Fluid Mech.*, 431:161–188.

Lin, Y. and Lin, C. H. (2008). Phragmen-Lindelof alternative and continuous dependence type results for thermoelasticity of type II. *Applicable Analysis*, 87:431–449.

Lindsay, K. A. and Straughan, B. (1976). Temperature waves in a rigid heat conductor. *ZAMP*, 27:653–662.

Lindsay, K. A. and Straughan, B. (1978). Acceleration waves and second sound in a perfect fluid. *Arch. Rational Mech. Anal.*, 68:53–87.

Lindsay, K. A. and Straughan, B. (1979). Propagation of mechanical and temperature acceleration waves in thermoelastic materials. *ZAMP*, 30:477–490.

Linton-Johnson, D., Plona, T. J., and Kajima, H. (1994). Probing porous media with first and second sound. Acoustic properties of water saturated porous media. *J. Appl. Phys.*, 76:115–125.

Liu, H., Bussmann, M., and Mostaghimi, J. (2009). A comparison of hyperbolic and parabolic models of phase change of a pure metal. *Int. J. Heat Mass Transfer*, 52:1177–1184.

Liu, J. (2000). Preliminary survey on the mechanisms of the wave-like behaviours of heat transfer in living tissues. *Forschung in Ingenieurwesen*, 66:1–10.

Liu, J., Chen, X., and Xu, L. X. (1999). New thermal wave aspects on burn evaluation of skin subjected to instantaneous heating. *IEEE Trans. Biomedical Engng.*, 46:420–428.

Liu, J., Ren, Z., and Wang, C. (1995). Interpretation of living tissue's temperature oscillations by thermal wave theory. *Chinese Sci. Bull.*, 40:1493–1495.

Liu, Z. Y. and Quintanilla, R. (2010a). Analyticity of solutions in type ii thermoelastic plates. *IMA J. Applied Math.*, 75:637–646.

Liu, Z. Y. and Quintanilla, R. (2010b). Energy decay rate of a mixed type II and type III thermoelastic system. *Discrete Cont. Dyn. Sys. B*, 14:1433–1444.

Loh, J. S., Azid, I. A., Seetharamu, K. N., and Quadir, G. A. (2007). Fast transient thermal analysis of Fourier and non-Fourier heat conduction. *Int. J. Heat Mass Trans.*, 50:4400–4408.

López Molina, J. A., Rivera, M. J., Trujillo, M., and Berjano, E. J. (2008). Effect of the thermal wave in radiofrequency ablation modelling: an analytical study. *Phys. Med. Biol.*, 53:1447–1462.

Lor, W. B. and Chu, H. S. (1999). Hyperbolic heat conduction in thin-film high T_c superconductors with interface thermal resistance. *Cryogenics*, 39:739–750.

Lord, H. W. and Shulman, Y. (1967). A generalized dynamical theory of thermoelasticity. *J. Math. Phys. Solids*, 15:299–309.

Magaña, A. and Quintanilla, R. (2006a). On the exponential decay of solutions in one-dimensional generalized porous - thermo - elasticity. *Asymptotic Analysis*, 49:173–187.

Magaña, A. and Quintanilla, R. (2006b). On the time decay of solutions in one-dimensional theories of porous materials. *Int. J. Solids Structures*, 43:3414–3427.

Magaña, A. and Quintanilla, R. (2007). On the time decay of solutions in porous - elasticity with quasi - static microvoids. *J. Math. Anal. Appl.*, 331:617–630.

Maiga, S. E. B., Palm, S. J., Nguyen, C. T., Roy, G., and Galanis, N. (2005). Heat transfer enhancement by using nanofluids in forced convection flows. *Int. J. Heat Fluid Flow*, 26:530–546.

Majumdar, A. and Yang, P. (2008). Silicon nanowire thermoelectrics: electricity from waste heat. Berkeley Lab Research Highlight http://www.lbl.gov/msd/assets/docs/highlights/08_Majumdar_YangSi_nanowires.pdf.

Mallik, S. H. and Kanoria, M. (2007). Generalized thermoelastic functionally graded solid with a periodically varying heat source. *Int. J. Solids Structures*, 44:7633–7645.

Malysiak, K., Borys, P., and Grzywna, Z. J. (2007). On the ball and chain model by simple and hyperbolic diffusion - an analytical approach. *Acta Physica Polonica B*, 38:1865–1879.

Marasco, A. (2009a). On the first-order speeds in any direction of acceleration waves in prestressed second - order isotropic, compressible, and homogeneous materials. *Mathematical and Computer Modelling*, 49:1644–1652.

Marasco, A. (2009b). Second - order effects on the wave propagation in elastic, isotropic, incompressible, and homogeneous media. *Int. J. Engng. Sci.*, 47:499–511.

Marasco, A. and Romano, A. (2009). On the acceleration waves in second - order elastic, isotropic, compressible, and homogeneous materials. *Mathematical and Computer Modelling*, 49:1504–1518.

Mariano, P. M. and Sabatini, L. (2000). Homothermal acceleration waves in multifield theories of continua. *Int. J. Non-Linear Mech.*, 35:963–977.

Marshall, J. S. and Naghdi, P. M. (1989a). A thermodynamical theory of turbulence. I. Basic developments. *Phil. Trans. Roy. Soc. London A*, 327:415–448.

Marshall, J. S. and Naghdi, P. M. (1989b). A thermodynamical theory of turbulence. II. Determination of constitutive coefficients and illustrative examples. *Phil. Trans. Roy. Soc. London A*, 327:449–475.

Masoumi, M., Sohrabi, N., and Behzadmehr, A. (2009). A new model for calculating the effective viscosity of nanofluids. *J. Phys. D: Appl. Phys.*, 42:055501–055501.

Massoudi, M. (2005). An anisotropic constitutive relation for the stress tensor of a rod-like (fibrous-type) granular material. *Mathematical Problems in Engineering*, 6:679–702.

Massoudi, M. (2006a). On the heat flux vector for flowing granular materials. Part I: Effective thermal conductivity and background. *Math. Meth. Appl. Sci.*, 29:1585–1598.

Massoudi, M. (2006b). On the heat flux vector for flowing granular materials. Part II: Derivation and special cases. *Math. Meth. Appl. Sci.*, 29:1599–1613.

Maugin, G. A. (1990). Internal variables and dissipative structures. *J. Non-Equilibrium Thermodynamics*, 15:173–192.

Maugin, G. A. and Muschik, W. (1994a). Thermodynamics with internal variables. Part I. General concepts. *J. Non-Equilibrium Thermodynamics*, 19:217–249.

Maugin, G. A. and Muschik, W. (1994b). Thermodynamics with internal variables. Part II. Applications. *J. Non-Equilibrium Thermodynamics*, 19:250–289.

Maysenhölder, W., Berg, A., and Leistner, P. (2004). Acoustic properties of aluminium foams - measurements and modelling. CFA/DAGA'04, Strasbourg, 22–25/03/2004. See: www.ibp.fhg.de/ba/forschung/aluschaum/aluschaum.pdf.

McCarthy, M. F. (1972). Singular surfaces and waves. In Eringen, A. C., editor, *Continuum Physics*, volume II, pages 449–521. Academic Press.

McCartin, B. J. and Causley, M. F. (2006). Angled derivative approximation of the hyperbolic heat conduction equations. *Appl. Math. Comp.*, 182:1581–1607.

McNelly, T. F., Rogers, S. J., Channin, D. J., Rollefson, R. J., Goubau, W. M., Schmidt, G. E., Krumhansl, J. A., and Pohl, R. O. (1970). Heat pulses in NaF: onset of second sound. *Phys. Rev. Lett.*, 24:100–102.

Meixner, J. (1974). Coldness and temperature. *Arch. Rational Mech. Anal.*, 57:281–290.

Mendez, V. and Camacho, J. (1997). Dynamics and thermodynamics of delayed population growth. *Physical Review E*, 55:6476–6482.

Mentrelli, A., Ruggeri, T., Sugiyama, M., and Zhao, N. (2008). Interaction between a shock and an acceleration wave in a perfect gas for increasing shock strength. *Wave Motion*, 45:498–517.

Messaoudi, S. A. and Said-Houari, B. (2008). Energy decay in a Timoshenko-type system of thermoelasticity of type III. *J. Math. Anal. Appl.*, 348:298–307.

Messaoudi, S. A. and Said-Houari, B. (2009). Energy decay in a Timoshenko-type system with history in thermoelasticity of type II. *Advances in Differential Equations*, 14:375–400.

Metzler, R. and Compte, A. (1999). Stochastic foundation of normal and anomalous Cattaneo - type transport. *Physica A*, 268:454–468.

Meyer, R. J. (2006). Ultrasonic drying of saturated porous solids via second sound. http://www.freepatentsonline.com/6376145.html.

Mian, A. M. and Spencer, A. J. M. (1998). Exact solutions for functionally graded and laminated elastic materials. *J. Mech. Phys. Solids*, 46:2283–2295.

Mielke, A. (1992). On Saint - Venant's problem for an elastic strip. *ZAMP*, 43:449–470.

Mills, J. N. and Childs, J. E. (1998). Ecologic studies of rodent reservoirs: their relevance for human health. *Emerging Infectious Diseases*, 4:529–534.

Mills, J. N., Ksiazek, T. G., Peters, C. J., and Childs, J. E. (1999). Long - term studies of hantavirus reservoir populations in the South Western United States. *Emerging Infectious Diseases*, 5:135–142.

Miranville, A. and Quintanilla, R. (2009). A generalization of the Caginalp phase - field system based on the Cattaneo law. *Nonlinear Analysis; Theory, Methods and Applications*, 71:2278–2290.

Miranville, A. and Quintanilla, R. (2010). A Caginalp phase - field system with nonlinear coupling. *Nonlinear Analysis; Real World Applications*, 11:2849–2861.

Mitra, K., Kumar, S., Vedavarz, A., and Moallemi, M. K. (1995). Experimental evidence of hyperbolic heat conduction in processed meat. *Trans. ASME, J. Heat Transfer*, 117:568–573.

Morikawa, J. and Hashimoto, T. (2005). Thermal diffusivity measurement of low-k dielectric thin film by temperature wave analysis. *Thermochimica Acta*, 432:216–221.

Morro, A. (1978). Interaction of shock waves with acoustic waves. *Rend. Accad. Naz. Lincei*, 64:177–185.

Morro, A. (1980). Acceleration waves in thermo-viscous fluids. *Rend. Sem. Matem. Univ. Padova*, 63:169–184.

Morro, A. (2006). Jump relations and discontinuity waves in conductors with memory. *Math. Computer Modelling*, 43:138–149.

Morro, A. (2010). Evolution equations and thermodynamic restrictions for dissipative solids. *Math. Computer Modelling*, 52:1869–1876.

Morro, A., Payne, L. E., and Straughan, B. (1990). Decay, growth, continuous dependence and uniqueness results in generalized heat conduction theories. *Applicable Analysis*, 38:231–243.

Morro, A. and Ruggeri, T. (1987). Second sound and internal energy in solids. *Int. J. Nonlinear Mech.*, 22:27–36.

Morro, A. and Ruggeri, T. (1988). Non-equilibrium properties of solids obtained from second - sound measurements. *J. Phys. C: Solid State Phys.*, 21:1743–1752.

Muñoz Rivera, J. E. (1992). Energy decay rates in linear thermoelasticity. *Funkcialaj Ekvacioj*, 35:19–30.

Muñoz Rivera, J. E. (1997). Asymptotic behaviour in n-dimensional thermoelasticity. *Appl. Math. Lett.*, 10:47–53.

Muñoz Rivera, J. E. and Qin, Y. (2002). Global existence and exponential stability in one-dimensional nonlinear thermoelasticity with thermal memory. *Nonlinear Anal.*, 51:11–32.

Muñoz Rivera, J. E. and Quintanilla, R. (2008). On the time polynomial decay in elastic solids with voids. *J. Math. Anal. Appl.*, 338:1296–1309.

Müller, I. (1967a). On the entropy inequality. *Arch. Rational Mech. Anal.*, 26:118–141.

Müller, I. (1967b). Zum Paradoxon der Wärmeleitungstheorie. *Zeitschrift für Physik*, 198:329–344.

Müller, I. (1971a). The coldness, a universal function in thermoelastic bodies. *Arch. Rational Mech. Anal.*, 41:319–332.

Müller, I. (1971b). Die Kältefunktion, eine universelle Funktion in der Thermodynamik viskoser wärmeleitender Flüssigkeiten. *Arch. Rational Mech. Anal.*, 40:1–36.

Müller, I. and Ruggeri, T. (1998). *Rational extended thermodynamics*, volume 37 of *Tracts in Natural Philosophy*. Springer, New York, second edition.

Murray, J. D. (2003a). *Mathematical Biology. I: An Introduction*, volume 17 of *Interdisciplinary Appl. Math.* Springer, New York, third edition.

Murray, J. D. (2003b). *Mathematical Biology. II: Spatial models and biomedical applications*, volume 17 of *Interdisciplinary Appl. Math.* Springer, New York, third edition.

Muschik, W. (2007). Why so many "schools" of thermodynamics? *Forsch. Ingenieurwes*, 71:149–161.

Nappa, L. (1998). Spatial decay estimates for the evolution equations of linear thermoelasticity without energy dissipation. *J. Thermal Stresses*, 21:581–592.

Narayanamurti, V. and Dynes, R. C. (1972). Observation of second sound in Bismuth. *Phys. Rev. Lett.*, 28:1401–1405.

Nield, D. A. and Bejan, A. (2006). *Convection in Porous Media.* Springer, New York, third edition.

Niu, T. and Dai, W. (2009). A hyperbolic two-step model based finite difference scheme for studying thermal deformation in a double-layered thin film exposed to ultrashort-pulsed lasers. *Int. J. Thermal Sciences*, 48:34–49.

Niwa, H. (1998). Migration dynamics of fish schools in heterothermal environments. *J. Theor. Biol.*, 193:215–231.

Nunziato, J. W. and Cowin, S. C. (1979). A nonlinear theory of elastic materials with voids. *Arch. Rational Mech. Anal.*, 72:175–201.

Nunziato, J. W. and Walsh, E. K. (1977). On the influence of void compaction and material non-uniformity on the propagation of one-dimensional acceleration waves in granular materials. *Arch. Rational Mech. Anal.*, 64:299–316.

Nunziato, J. W. and Walsh, E. K. (1978). Addendum, On the influence of void compaction and material non-uniformity on the propagation of one-dimensional acceleration waves in granular materials. *Arch. Rational Mech. Anal.*, 67:395–398.

Ogden, R. W. (1997). *Non-linear elastic deformations.* Dover, Mineola, New York.

Ostoja-Starzewski, M. (2009). A derivation of the Maxwell - Cattaneo equation from the free energy and dissipation potentials. *Int. J. Engng. Sci.*, 47:807–810.

Ostoja-Starzewski, M. and Trebicki, J. (1999). On the growth and decay of acceleration waves in random media. *Proc. Roy. Soc. London A*, 455:2577–2614.

Ostoja-Starzewski, M. and Trebicki, J. (2006). Stochastic dynamics of acceleration waves in random media. *Mechanics of Materials*, 38:840–848.

Othman, M. I. A. and Singh, B. (2007). The effect of rotation on generalized micropolar thermoelasticity for a half-space under five theories. *Int. J. Solids Structures*, 44:2748–2762.

Ouellette, J. (2004). Seeing with sound. Acoustic microscopy advances beyond failure analysis. American Institute of Physics Website http://www.aip.org/tip/INPHFA/vol-10/iss-3/p14.html.

Pamplona, P. X., Muñoz Rivera, J. E., and Quintanilla, R. (2009). Stabilization in elastic solids with voids. *J. Math. Anal. Appl.*, 350:37–49.

Papanicolaou, N. C., Christov, C. I., and Jordan, P. M. (2009). Two-gradient convection in a vertical slot with Maxwell - Cattaneo heat conduction. In Todorov, M. D. and Christov, C. I., editors, *Applications of Mathematics in Technical and Natural Sciences*, pages 231–239. American Institute of Physics.

Papanicolaou, N. C., Christov, C. I., and Jordan, P. M. (2011). The influence of thermal relaxation on the oscillatory properties of two-gradient convection in a vertical slot. *European Journal of Mechanics B/Fluids*, 30:68–75.

Passarella, F. and Zampoli, V. (2007). Some exponential decay estimates for thermoelastic mixtures. *J. Thermal Stresses*, 30:25–41.

Passarella, F. and Zampoli, V. (2010). On the spatial behaviour for transversely isotropic plates. In Ciarletta, M., Fabrizio, M., Morro, A., and Rionero, S., editors, *New trends in fluid and solid models*, pages 103–112. World Scientific. Proceedings of the International Conference, Vietri sul Mare, 28 February - 1 March, 2008.

Passarella, F. and Zampoli, V. (2011). Reciprocal and variational principles in micropolar thermoelasticity of type II. *Acta Mechanica*, 216:29–36.

Payne, L. E. (1975). *Improperly Posed Problems in Partial Differential Equations*, volume 22 of *CBMS-NSF Regional Conference Series in Applied Mathematics*. SIAM.

Payne, L. E. and Schaefer, P. W. (2002). Energy bounds for some nonstandard problems in partial differential equations. *J. Math. Anal. Appl.*, 273:75–92.

Payne, L. E., Schaefer, P. W., and Song, J. C. (2004). Improved bounds for some nonstandard problems in generalized heat conduction. *J. Math. Anal. Appl.*, 298:325–340.

Payne, L. E., Schaefer, P. W., and Song, J. C. (2005). Some non-standard problems in generalized heat conduction. *ZAMP*, 56:931–941.

Payne, L. E. and Song, J. C. (1996). Phragmén - Lindelöf and continuous dependence type results in generalized heat conduction. *ZAMP*, 47:527–538.

Payne, L. E. and Song, J. C. (1997a). Continuous dependence on initial-time and spatial geometry in generalized heat conduction. *J. Math. Anal. Appl.*, 214:173–190.

Payne, L. E. and Song, J. C. (1997b). Continuous dependence on initial-time geometry in generalized heat conduction. *Math. Models Meth. Appl. Sci.*, 7:125–138.

Payne, L. E. and Song, J. C. (2001). Convergence results for generalized heat conduction as the relaxation time tends to zero. *J. Math. Anal. Appl.*, 256:175–189.

Payne, L. E. and Song, J. C. (2002). Growth and decay in generalized thermoelasticity. *Int. J. Engng. Sci.*, 40:385–400.

Payne, L. E. and Song, J. C. (2004a). Spatial decay estimates for the Maxwell-Cattaneo equations with mixed boundary conditions. *ZAMP*, 55:962–973.

Payne, L. E. and Song, J. C. (2004b). Temporal decay bounds in generalized heat conduction. *J. Math. Anal. Appl.*, 294:82–95.

Payne, L. E. and Song, J. C. (2005). Improved spatial decay bounds in generalized heat conduction. *ZAMP*, 56:805–820.

Payne, L. E. and Song, J. C. (2006). On a proposed model for heat conduction. *IMA J. Appl. Math.*, 71:590–599.

Payne, L. E. and Song, J. C. (2007a). Spatial decay in a cross-diffusion problem. *IMA J. Appl. Math.*, 72:854–864.

Payne, L. E. and Song, J. C. (2007b). Spatial decay in a double diffusive convection problem in Darcy flow. *J. Math. Anal. Appl.*, 330:864–875.

Payne, L. E. and Song, J. C. (2008). Spatial decay bounds for double diffusive convection in Brinkman flow. *J. Differential Equations*, 244:413–430.

Payne, L. E. and Straughan, B. (2009). Decay for a Keller - Segel chemotaxis model. *Studies in Applied Mathematics*, 123:337–360.

Peshkov, V. (1944). Second sound in liquid helium. *Journal of Physics*, 8:381.

Peshkov, V. (1947). Determination of the velocity of propagation of the second sound in helium II. In *Report of an international conference on fundamental particles and low temperatures: Low temperatures*, volume 2, pages 19–32.

Pilgrim, N. J., Batty, W., Kelsall, R. W., and Snowden, C. M. (2004). Nanoscale electrothermal co-simulation: compact dynamic models of hyperbolic heat transport and self-consistent device Monte Carlo. *Microelectronics Journal*, 35:823–830.

Polikarpov, Y. I. and Slutsker, A. I. (1997). Application of the temperature wave technique for the study of relaxation transitions in polymers. *Thermochimica Acta*, 304:277–282.

Popov, E. B. (1967). Dynamic coupled problem of thermoelasticity for a half-space taking account of the finiteness of the heat propagation velocity. *J. Appl. Math. Mech.*, 31:349–356.

Puri, P. and Jordan, P. M. (1999a). Stokes's first problem for a dipolar fluid with nonclassical heat conduction. *J. Engng. Math.*, 36:219–240.

Puri, P. and Jordan, P. M. (1999b). Wave structure in Stokes' second problem for a dipolar fluid with nonclassical heat conduction. *Acta Mechanica*, 133:145–160.

Puri, P. and Jordan, P. M. (2004). On the propagation of plane waves in type-III thermoelastic media. *Proc. Roy. Soc. London A*, 460:3203–3221.

Puri, P. and Jordan, P. M. (2006). On the propagation of harmonic plane waves under the two - temperature theory. *Int. J. Engng. Sci.*, 44:1113–1126.

Puri, P. and Kythe, P. K. (1997). Discontinuities in velocity gradients and temeprature in the Stokes' first problem with nonclassical heat conduction. *Quart. Appl. Math.*, 55:167–176.

Puri, P. and Kythe, P. K. (1998). Stokes' first and second problems for Rivlin - Ericksen fluids with nonclassical heat conduction. *J. Heat Transfer, ASME*, 120:44–50.

Putra, N., Roetzel, W., and Das, S. (2003). Natural convection of nanofluids. *Heat and Mass Transfer*, 39:775–784.

Qin, Y. and Muñoz Rivera, J. E. (2004). Global existence and exponential stability of solutions of thermoelastic equations of hyperbolic type. *J. Elasticity*, 75:125–145.

Quintanilla, R. (1996). A spatial decay estimate for the hyperbolic heat equation. *SIAM J. Math. Anal.*, 27:78–91.

Quintanilla, R. (1997). Spatial bounds and growth estimates for the heat equation with three relaxation times. *Math. Meth. Appl. Sci.*, 20:1335–1344.

Quintanilla, R. (1999). On the spatial behaviour in thermoelasticity without energy dissipation. *J. Thermal Stresses*, 22:213–224.

Quintanilla, R. (2001a). Damping of end effects in a thermoelastic theory. *Applied Math. Letters*, 14:137–141.

Quintanilla, R. (2001b). Instability and non-existence in the nonlinear theory of thermoelasticity without energy dissipation. *Continuum Mech. Thermodyn.*, 13:121–129.

Quintanilla, R. (2001c). Structural stability and continuous dependence of solutions in thermoelasticity of type III. *Discrete and Continuous Dynamical Systems B*, 1:463–470.

Quintanilla, R. (2002a). Exponential stability in the dual - phase - lag heat conduction theory. *J. Non-Equilib. Thermodyn.*, 27:217–227.

Quintanilla, R. (2002b). On existence in thermoelasticity without energy dissipation. *J. Thermal Stresses*, 25:195–202.

Quintanilla, R. (2002c). Thermoelasticity without energy dissipation of materials with microstructure. *Appl. Math. Modelling*, 26:1125–1137.

Quintanilla, R. (2002d). Uniqueness in exterior domains for the generalized heat conduction. *Appl. Math. Letters*, 15:473–479.

Quintanilla, R. (2003). Slow decay for one-dimensional porous dissipation elasticity. *Appl. Math. Letters*, 16:487–491.

Quintanilla, R. (2004). Exponential stability and uniqueness in thermoelasticity with two temperatures. *Dyn. Cont. Discrete Impulsive Sys. A*, 11:57–68.

Quintanilla, R. (2007a). Impossibility of localization in linear thermoelasticity with voids. *Mech. Research Communications*, 34:522–527.

Quintanilla, R. (2007b). On the impossibility of localization in linear thermoelasticity. *Proc. Roy. Soc. London A*, 463:3311–3322.

Quintanilla, R. (2008a). Polynomial and exponential spatial decay estimates in elasticity. *J. Elasticity*, 93:279–290.

Quintanilla, R. (2008b). A well - posed problem for the dual - phase - lag heat conduction. *J. Thermal Stresses*, 31:260–269.

Quintanilla, R. (2009). A well - posed problem for the three - dual - phase - lag heat conduction. *J. Thermal Stresses*, 32:1270–1278.

Quintanilla, R. (2010a). A note on a non-standard problem for an equation with a delay term. *Applied Mathematics and Computation*, 216:2759–2765.

Quintanilla, R. (2010b). On the spatial behaviour of constrained motion in type III thermoelasticity. *J. Thermal Stresses*, 33:694–705.

Quintanilla, R. and Jordan, P. M. (2009). A note on the two temperature theory with dual-phase-lag delay: Some exact solutions. *Mechanics Research Communications*, 36:796–803.

Quintanilla, R. and Racke, R. (2003). Stability in thermoelasticity of type III. *Discrete and Continuous Dynamical Systems B*, 1:383–400.

Quintanilla, R. and Racke, R. (2006a). A note on stability in dual-phase-lag heat conduction. *Int. J. Heat Mass Transfer*, 49:1209–1213.

Quintanilla, R. and Racke, R. (2006b). Qualitative aspects in dual-phase-lag thermoelastcity. *SIAM J. Appl. Math.*, 66:977–1001.

Quintanilla, R. and Racke, R. (2007). Qualitative aspects in dual-phase-lag heat conduction. *Proc. Roy. Soc. London A*, 463:659–674.

Quintanilla, R. and Racke, R. (2008). A note on stability in three-phase lag heat conduction. *Int. J. Heat Mass Transfer*, 51:24–29.

Quintanilla, R. and Straughan, B. (2000). Growth and uniqueness in thermoelasticity. *Proc. Roy. Soc. London A*, 456:1419–1429.

Quintanilla, R. and Straughan, B. (2002). Explosive instabilities in heat transmission. *Proc. Roy. Soc. London A*, 458:2833–2837.

Quintanilla, R. and Straughan, B. (2004). Discontinuity waves in type III thermoelasticity. *Proc. Roy. Soc. London A*, 460:1169–1175.

Quintanilla, R. and Straughan, B. (2005a). Bounds for some non-standard problems in porous flow and viscous Green-Naghdi fluids. *Proc. Roy. Soc. London A*, 461:3159–3168.

Quintanilla, R. and Straughan, B. (2005b). Energy bounds for some non-standard problems in thermoelasticity. *Proc. Roy. Soc. London A*, 461:1147–1162.

Quintanilla, R. and Straughan, B. (2008). Nonlinear waves in a Green-Naghdi dissipationless fluid. *J. Non-Newtonian Fluid Mech.*, 154:207–210.

Racke, R. (2002). Thermoelasticity with second sound - exponential stability in linear and non-linear 1-D. *Math. Meth. Appl. Sci.*, 25:409–441.

Racke, R. (2003). Asymptotic behaviour of solutions in linear 2 or 3-D thermoelasticity with second sound. *Quart. Appl. Math.*, 61:315–328.

Racke, R. and Shibata, Y. (1991). Global smooth solutions and asymptotic stability in one dimensional nonlinear thermoelasticity. *Arch. Rational Mech. Anal.*, 116:1–34.

Racke, R. and Wang, Y. G. (2005). Asymptotic behaviour of discontinuous solutions to thermoelastic systems with second sound. *Zeitschrift fur Analysis und Ihre Anwendungen*, 24:117–135.

Racke, R. and Wang, Y. G. (2008). Nonlinear well-posedness and rates of decay in thermoelasticity with second sound. *J. Hyperbolic Differential Equations*, 5:25–43.

Rai, A. (2003). Breakdown of acceleration waves in radiative magnetic fluids. *Defence Science Journal*, 53:425–430.

Raiser, G. F., Wise, J. L., Clifton, R. J., Grady, D. E., and Cox, D. E. (1994). Plate impact response of ceramics and glasses. *J. Appl. Phys.*, 75:3862–3869.

Rayleigh, L. (1896). *The theory of sound*, volume 2. Macmillan and Co. Ltd., London, second edition.

Reissig, M. and Wang, Y. G. (2005). Cauchy problems for linear thermoelastic systems of type III in one space variable. *Math. Meth. Appl. Sci.*, 28:1359–1381.

Reverberi, A. P., Bagnerini, P., Maga, L., and Bruzzone, A. G. (2008). On the Maxwell - Cattaneo equation with non - constant diffusivity. *Int. J. Heat Mass Transfer*, 51:5327–5332.

Rionero, S. and Chirita, S. (1987). The Lagrange identity method in linear thermoelasticity. *Int. J. Engng. Sci.*, 25:935–947.

Rionero, S. and Galdi, G. P. (1976). On the uniqueness of viscous fluid motions. *Arch. Rational Mech. Anal.*, 62:295–301.

Roy, S., Murthy, A. S. V., and Kudenatti, R. B. (2009). A numerical method for the hyperbolic - heat conduction equation based on multiple scale technique. *Appl. Numer. Math.*, 59:1419–1430.

Roy Choudhuri, S. K. (2007). On a thermoelastic three-phase-lag model. *J. Thermal Stresses*, 30:231–238.

Ruggeri, T. (2001). The binary mixtures of Euler fluids: a unified theory of second sound phenomena. In Straughan, B., Greve, R., Ehrentraut, H., and Wang, Y., editors, *Continuum Mechanics and Applications in Geophysics and the Environment*. Springer.

Ruggeri, T. (2010). A classical approach of a multi-temperature mixture of fluids. In Ciarletta, M., Fabrizio, M., Morro, A., and Rionero, S., editors, *New trends in fluid and solid models*, pages 123–130. World Scientific. Proceedings of the International Conference, Vietri sul Mare, 28 February - 1 March, 2008.

Ruggeri, T., Muracchini, A., and Seccia, L. (1990). Shock waves and second sound in a rigid heat conductor: a critical temperature for NaF and Bi. *Phys. Review Letters*, 64:2640–2643.

Ruggeri, T., Muracchini, A., and Seccia, L. (1996). Second sound and characteristic temperature in solids. *Phys. Review B*, 54:332–339.

Ruggeri, T. and Simić, S. (2005). On the hyperbolic system of a mixture of Eulerian fluids: A comparison between single and multi - temperature models. *Math. Meth. Appl. Sci.*, 38:827–849.

Ruggeri, T. and Sugiyama, M. (2005). Hyperbolicity, convexity and shock waves in one-dimensional crystalline solids. *J. Phys. A: Math. Gen.*, 38:4337–4347.

Russo, R. (1987). Uniqueness theorems for a partial differential system of parabolic type in unbounded domains. *Commun. Part. Diff. Equations*, 12:883–902.

Sabatini, L. and Augusti, G. (2001). Homothermal acceleration waves in nematic liquid crystals. *Int. J. Solids Struct.*, 38:1227–1242.

Saggio-Woyansky, J., Scott, C. E., and Minnear, W. P. (1992). Processing of porous ceramics. *American Ceramic Society Bulletin*, 71:1674–1682.

Saidane, A., Aliouat, S., Benzohra, M., and Ketata, M. (2005). A transmission line matrix (TLM) study of hyperbolic heat conduction in biological materials. *J. Food Engineering*, 68:491–496.

Saleh, A. and Al-Nimr, M. (2008). Variational formulation of hyperbolic heat conduction problems applying Laplace transform technique. *Int. Communications in Heat and Mass Transfer*, 35:204–214.

Sanderson, T., Ume, C., and Jarzynski, J. (1995). Hyperbolic heat conduction effects caused by temporally modulated laser pulses. *Ultrasonics*, 33:423–427.

Sare, H. D. F., Muñoz Rivera, J. E., and Racke, R. (2008). Stability for a transmission problem in thermoelasticity with second sound. *J. Thermal Stresses*, 31:1170–1189.

Sare, H. D. F. and Racke, R. (2009). On the stability of damped Timoshenko systems: Cattaneo versus Fourier law. *Arch. Rational Mech. Anal.*, 194:221–251.

Sauvage, F., Langlais, M., Yoccoz, N. G., and Pontier, D. (2003). Modelling hantavirus in fluctuating populations of bank voles: the role of indirect transmission on virus persistence. *J. Animal Ecology*, 71:1–13.

Savino, R. and Paterna, D. (2008). Thermodiffusion in nanofluids under different gravity conditions. *Physics of Fluids*, 20:017101–017101.

Schmaljohn, C. and Hjelle, B. (1997). Hantaviruses: a global disease problem. *Emerging Infectious Diseases*, 3:95–104.

Sellitto, A., Alvarez, F. X., and Jou, D. (2011). Phonon-wall interactions and frequency - dependent thermal conductivity in nanowires. *J. Appl. Phys.*, 109:064317.

Serdyukov, S. I. (2001). A new version of extended irreversible thermodynamics and dual-phase-lag model in heat transfer. *Phys. Letters A*, 281:16–20.

Serdyukov, S. I., Voskresenskii, N. M., Bel'nov, V. K., and Karpov, I. I. (2003). Extended irreversible thermodynamics and generation of the dual-phase-lag model in heat transfer. *J. Non-Equilib. Thermodyn.*, 28:207–219.

Shablovskii, O. N. (1984). Strong temperature - field discontinuities in a nonlinear medium. *J. Engng. Physics and Thermophysics*, 46:607–611.

Shablovskii, O. N. (1985). Propagation of a planar shock thermal wave in a nonlinear medium. *J. Engng. Physics and Thermophysics*, 49:1070–1076.

Shablovskii, O. N. (1987). The study of nonlinear problems of high - intensity nonstationary heat transfer. *J. Engng. Physics and Thermophysics*, 52:237–243.

Sharma, J. N. and Thakar, M. D. (2006). Effect of rotation on Rayleigh - Lamb waves in magneto-thermoelastic media. *J. Sound Vibration*, 296:871–887.

Shen, B. and Zhang, P. (2003). Notable physical anomalies manifested in non - Fourier heat conduction under the dual - phase - lag model. *Int. J. Heat Mass Transfer*, 46:1713–1727.

Shnaid, I. (2003). Thermodynamically consistent description of heat conduction with finite speed of heat propagation. *Int. J. Heat Mass Transfer*, 46:3853–3863.

Slemrod, M. (1981). Global existence, uniqueness and asymptotic stability of classical smooth solutions in one-dimensional nonlinear thermoelasticity. *Arch. Rational Mech. Anal.*, 76:97–133.

Sneddon, I. N. (1980). *Special functions of mathematical physics and chemistry.* Longman, London, third edition.

Sneddon, I. N. (1995). *Fourier Transforms.* Dover.

Sobolev, S. L. (1997). Rapid solidification under local nonequilibrium conditions. *Phys. Review E*, 55:6845–6854.

Spencer, A. J. M. (1980). *Continuum Mechanics.* Longman, London.

Stokes, G. G. (1851). An examination of the possible effect of the radiation of heat on the propagation of sound. *Phil. Mag. (4th Ser.)*, 1:305–317.

Straughan, B. (1974). *Qualitative analysis of some equations in contemporary continuum mechanics.* PhD thesis, Heriot-Watt University.

Straughan, B. (1986). Nonlinear waves in a general magnetic fluid. *ZAMP*, 37:274–279.

Straughan, B. (2004). *The energy method, stability, and nonlinear convection*, volume 91 of *Appl. Math. Sci.* Springer, New York, second edition.

Straughan, B. (2008). *Stability, and wave motion in porous media*, volume 165 of *Appl. Math. Sci.* Springer, New York.

Straughan, B. (2009a). Nonlinear acceleration waves in porous media. *Mathematics and Computers in Simulation*, 80:763–769.

Straughan, B. (2009b). Oscillatory convection and the Cattaneo law of heat conduction. *Ricerche di Matematica*, 58:157–162.

Straughan, B. (2010a). Acoustic waves in a Cattaneo-Christov gas. *Physics Letters A*, 374:2667–2669.

Straughan, B. (2010b). Green-Naghdi fluid with non-thermal equilibrium effects. *Proc. Roy. Soc. London A*, 466:2021–2032.

Straughan, B. (2010c). Porous convection with Cattaneo heat flux. *Int. J. Heat Mass Transfer*, 53:2808–2812.

Straughan, B. (2010d). Thermal convection with the Cattaneo - Christov model. *Int. J. Heat Mass Transfer*, 53:95–98.

Straughan, B. (2011). Tipping points in Cattaneo-Christov thermohaline convection. *Proc. Roy. Soc. London A*, 467:7–18.

Straughan, B. and Franchi, F. (1984). Bénard convection and the Cattaneo law of heat conduction. *Proc. Roy. Soc. Edinburgh A*, 96:175–178.

Su, S. and Dai, W. (2006). Comparison of the solutions of a phase-lagging heat transport equation and damped wave equation with a heat source. *Int. J. Heat Mass Transfer*, 49:2793–2801.

Su, S., Dai, W., Jordan, P. M., and Mickens, R. E. (2005). Comparison of the solutions of a phase-lagging heat transport equation and damped wave equation. *Int. J. Heat Mass Transfer*, 48:2233–2241.

Sugiyama, M. (1994). Statistical mechanical study of wave propagation in crystalline solids at finite temperatures. *Current Topics in Acoustics Research*, 1:139–158.

Sun, Y., Fang, D., and Soh, A. K. (2006). Thermoelastic damping in micro-beam resonators. *Int. J. Solids Structures*, 43:3213–3229.

Tarkenton, G. M. and Cramer, M. S. (1994). Nonlinear second sound in solids. *Phys. Review B*, 49:11794–11798.

Tibullo, V. and Vaccaro, M. (2008). Spatial behaviour for constrained motion of a cylinder made of strongly elliptic anisotropic material. *J. Mechanics of Materials and Structures*, 3:983–993.

Tibullo, V. and Zampoli, V. (2011). A uniqueness result for the cattaneo-christov heat conduction model applied to incompressible fluids. *Mechanics Research Communications*, 38:77–79.

Torvi, D. A. and Dale, J. D. (1994). A finite element model of skin subjected to a flash fire. *ASME J. Biomechanical Engineering*, 116:250–255.

Triani, V., Sellitto, A., and Cimmelli, V. A. (2010). Exploitation of the second law for second grade fluids: Generalized Coleman - Noll and Liu procedures in comparison. In Ciarletta, M., Fabrizio, M., Morro, A., and Rionero, S., editors, *New trends in fluid and solid models*, pages 169–176. World Scientific. Proceedings of the International Conference, Vietri sul Mare, 28 February - 1 March, 2008.

Truesdell, C. (1953). Precise theory of the absorption and dispersion of forced plane infinitesimal waves according to the Navier-Stokes equations. *J. Rational Mech. Anal.*, 2:643–741.

Truesdell, C. (1976). Correction of two errors in the kinetic theory of gases which have been used to cast unfounded doubt upon the principle of material frame - indifference. *Meccanica*, 11:196–199.

Truesdell, C. and Noll, W. (1992). *The non-linear field theories of mechanics*. Springer, Berlin, second edition.

Truesdell, C. and Rajagopal, K. R. (1999). *An introduction to the mechanics of fluids*. Birkhäuser, Basel.

Truesdell, C. and Toupin, R. A. (1960). The classical field theories. In Flügge, S., editor, *Handbuch der Physik*, volume III/1, pages 226–793. Springer.

Tung, M. M., Trujillo, M., López Molina, J. A., Rivera, M. J., and Berjano, E. J. (2009). Modelling the heating of biological tissue based on the hyperbolic heat transfer equation. *Mathematical and Computer Modelling*, 50:665–672.

Tzou, D. Y. (1995a). The generalized lagging response in small-scale and high-rate heating. *Int. J. Heat Mass Transfer*, 38:3231–3234.

Tzou, D. Y. (1995b). A unified approach for heat conduction from macro to micro-scales. *J. Heat Transfer*, 117:8–16.

Tzou, D. Y. (2008). Thermal instabilty of nanofluids in natural convection. *Int. J. Heat Mass Transfer*, 51:2967–2979.

Tzou, D. Y., Chen, J. K., and Beraun, J. E. (2002). Hot - electron blast induced by ultrashort - pulsed lasers in layered media. *Int. J. Heat Mass Transfer*, 45:3369–3382.

Vadasz, J. J., Govender, S., and Vadasz, P. (2005). Heat transfer enhancement in nano-fluids suspensions: possible mechanisms and explanations. *Int. J. Heat Mass Transfer*, 48:2673–2683.

Vadasz, P. (2005). Lack of oscillations in dual - phase - lagging heat conduction for a porous slab subject to heat flux and temperature. *Int. J. Heat Mass Transfer*, 48:2822–2828.

Vadasz, P. (2006). Heat conduction in nanofluid suspensions. *ASME J. Heat Transfer*, 128:465–477.

Valenti, G., Curro, C., and Sugiyama, M. (2004). Acceleration waves analysed by a new continuum model of solids incorporating microscopic thermal vibrations. *Continuum Mech. Thermodyn.*, 16:185–198.

Vedavarz, A., Mitra, K., Kumar, S., and Moallemi, M. K. (1992). Effect of hyperbolic heat conduction on temperature distribution in laser irradiated tissue with blood perfusion. *Adv. Biological Heat Mass Transfer, ASME HTD*, 231:7–16.

Vila Bravo, J. C. and Muñoz Rivera, J. E. (2009). The transmission problem to thermoelastic plates of hyperbolic type. *IMA J. Appl. Math.*, 74:750–762.

Walia, V., Sharma, J. N., and Sharma, P. K. (2009). Propagation characteristics of thermoelastic waves in piezoelectric (6 mm class) rotating plate. *European J. Mech. A/Solids*, 28:569–581.

Wang, Y. G. and Yang, L. (2006). $L^p - L^q$ decay estimates for Cauchy problems of linear thermoelastic systems with second sound in three dimensions. *Proc. Roy. Soc. Edinburgh A*, 136:189–207.

Wang, Z. and Hillen, T. (2008). Shock formation in a chemotaxis model. *Math. Meth. Appl. Sci.*, 31:45–70.

Warren, W. E. and Chen, P. J. (1973). Wave propagation in the two temperature theory of thermoelasticity. *Acta Mechanica*, 16:21–33.

Webber, M. (2007). *Instability of fluid flows, including boundary slip.* PhD thesis, Durham University.

Webber, M. (2008). Instability of thread - annular flow with small characteristic length to three - dimensional disturbances. *Proc. Roy. Soc. London A*, 464:673–690.

Weingartner, B., Osinov, V. A., and Wu, W. (2006). Acceleration wave speeds in a hypoplastic constitutive model. *Int. J. Non-linear Mech.*, 41:991–999.

Weingartner, B., Osinov, V. A., and Wu, W. (2008). Effect of inherent anisotropy on acceleration wave speeds in hypoplasticity. *Int. J. Engng. Sci.*, 46:286–292.

Weinmann, O. (2009). The equations of thermoelasticity with time-dependent coefficients. *J. Math. Anal. Appl.*, 350:81–99.

Whitham, G. B. (1974). *Linear and nonlinear waves.* John Wiley and Sons, New York.

Wilson, D. K. (1997). Simple, relaxational models for the acoustic properties of porous media. *Applied Acoustics*, 50:171–188.

Wong, K. F. V. and Kurma, T. (2008). Transport properties of alumina nanofluids. *Nanotechnology*, 19:345702–345702.

Wu, H., Grasselli, M., and Zheng, S. (2007). Convergence to equilibrium for a parabolic-hyperbolic phase-field system with dynamical boundary condition. *J. Math. Anal. Appl.*, 329:948–976.

Xinchun, S. and Lakes, R. S. (2007). Stability of elastic material with negative stiffness and negative Poisson's ratio. *Physica Status Solidi b*, 244:1008–1026.

Xuan, Y., Li, Q., and Hu, W. (2003). Aggregation structure and thermal conductivity of nanofluids. *AIChE Journal*, 49:1038–1043.

Xuan, Y. and Roetzel, W. (2000). Conceptions for heat transfer correlation of nanofluids. *Int. J. Heat Mass Transfer*, 43:3701–3707.

Yang, L. and Wang, Y. G. (2006). Well-posedness and decay estimates for Cauchy problems of linear thermoelastic systems of type III in 3-D. *Indiana Univ. Math. J.*, 55:1333–1362.

Yang, W. H. (1993). Thermal (heat) shock biothermomechanical viewpoint. *J. Biomechanical Engineering*, 115:617–621.

Zelati, M. C., Pata, V., and Quintanilla, R. (2010). Regular global attractors of type III thermoelastic extensible beams. *Chinese Annals of Mathematics, series B*, 31:619–630.

Zhang, X. and Zuazua, E. (2003). Decay of solutions of the system of thermoelasticity of type III. *Comm. Contemporary Math.*, 5:25–84.

Zhang, Y. (2009). Generalized dual-phase lag bioheat equations based on nonequilibrium heat transfer in living biological tissues. *Int. J. Heat Mass Transfer*, 52:4829–4834.

Index